This volume provides a thought-provoking perspective on the empirical and analytic study of body form and composition. The techniques used for measuring body components such as fat, water, lean tissue, bone mass and bone density are evaluated against potential 'gold standards'. The nature of regional differences, developmental changes, pathological abnormalities, and the impact of heredity and environment in shaping body composition are discussed in the context of human evolution. All those concerned with biological anthropology, both clinicians and researchers, will find this book of great interest.

T0276193

Cambridge Studies in Biological Anthropology   6

Body composition in biological anthropology

Cambridge Studies in Biological Anthropology

Series Editors

G. W. Lasker
Department of Anatomy, Wayne State University, Detroit, Michigan, USA

C. G. N. Mascie-Taylor
Department of Biological Anthropology, University of Cambridge

D. F. Roberts
Department of Human Genetics, University of Newcastle upon Tyne

Also in the series

G. W. Lasker *Surnames and Genetic Structure*

C. G. N. Mascie-Taylor and G. W. Lasker (editors) *Biological Aspects of Human Migration*

Barry Bogin *Patterns of Human Growth*

J. E. Lindsay Carter and Barbara Honeyman Heath *Somatotyping – development and applications*

J. A. Kieser *Human adult odontometrics – The study of variation in adult tooth size*

Body composition in biological anthropology

ROY J. SHEPHARD

*School of Physical and Health Education
and
Department of Preventive Medicine and
Biostatistics, Faculty of Medicine
University of Toronto*

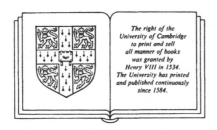

The right of the
University of Cambridge
to print and sell
all manner of books
was granted by
Henry VIII in 1534.
The University has printed
and published continuously
since 1584.

CAMBRIDGE UNIVERSITY PRESS

*Cambridge
New York   Port Chester
Melbourne   Sydney*

CAMBRIDGE UNIVERSITY PRESS
Cambridge, New York, Melbourne, Madrid, Cape Town, Singapore, São Paulo

Cambridge University Press
The Edinburgh Building, Cambridge CB2 2RU, UK

Published in the United States of America by Cambridge University Press, New York

www.cambridge.org
Information on this title: www.cambridge.org/9780521362672

First published 1991
This digitally printed first paperback version 2005

A catalogue record for this publication is available from the British Library

Library of Congress Cataloguing in Publication data

Shephard, Roy J.
Body composition in biological anthropology/Roy J. Shephard.
   p.   cm. – – (Cambridge studies in biological anthropology; 6)
Includes bibliographical references and index.
ISBN 0 521 36267 9
1. Body composition.   2. Physical anthropology. I. Title.
II. Series.
GN66.S46   1990
573– –dc20   90-36079   CIP

ISBN-13 978-0-521-36267-2 hardback
ISBN-10 0-521-36267-9 hardback

ISBN-13 978-0-521-01903-3 paperback
ISBN-10 0-521-01903-6 paperback

Contents

Preface

A previous contribution to the present series of monographs (Carter & Heath, 1988) has thoroughly explored the issue of somatotyping. The present volume provides a counterpoint to such a 'Gestalt' of body form by focussing upon an empirical and analytic approach to body composition.

A brief introduction compares empirical and intuitive analyses of body form and composition. It considers also the difficult question of proposing 'reference' and 'normal' standards at various stages of life, touching upon such issues as the reasons for data collection, an optimal statistical treatment of the results, concepts of proportionality and the 'unisex phantom'. The next four chapters look at methods for the determination of body fat, body water, lean tissue, bone mass and bone density, available techniques being evaluated against the potential 'gold standards' provided by detailed anthropometry and cadaver dissection.

The book then moves on to examine definitive issues, including regional differences in the relative amounts of fat, muscle and bone, developmental changes in the proportions of these three body components, and the respective influences of heredity and environment in shaping body composition. The implications of inter-individual differences of body composition are considered in the specific contexts of acute and chronic adaptations to a variety of hostile habitats, including situations characterized by extremes of heat and cold, malnutrition, and an over-abundance of food. A subsequent chapter explores pathological abnormalities of body composition, including gross obesity, muscle wasting, and osteoporosis. Finally, information is summarized in the broader context of human evolution; in particular, the text explores how far hostile features of the world (extremes of environmental temperature, starvation, an excessive food supply and disease) have encouraged the emergence of genetic isolates with unusual features of body composition.

In writing the monograph, I have drawn quite extensively upon the research experience of the University of Toronto laboratories. I would like to acknowledge my debt to studies of body composition shared with Drs Joan Harrison and Ken McNeill, field investigations of the Inuit

Preface

conducted in collaboration with Andris Rode and members of the Canadian International Biological Programme team, measurements of fat loss during acute cold exposure shared with Dr W. O'Hara, Stan Murray, and the Defence and Civil Institute of Environmental Medicine, research on anorexia nervosa shared with Drs J. Garfinkel, J. Garner and S. Shinder, and studies of body fat distribution completed with Drs Hugues Lavallée, John Brown, R. Forsyth and P. Kofsky-Singer.

Much still remains for the physical anthropologist to clarify. But if the present book provokes further empirical research upon body composition, allowing more definitive conclusions to be reached, I shall be well-pleased with the outcome of this endeavour.

Toronto   Roy J. Shephard

1   *Introduction*

Empirical versus intuitive analysis

Origins of positivism

The study of human phenomena has provided the occasion for a sustained – and at times heated – battle between empiricists, eager to quantify both body form and function, and other investigators who have preferred a more intuitive approach. The positivist concept that body structures conform to natural, mathematically describable laws had a strong grounding in the empirical philosophy of such early writers as John Locke (1632–1704), George Berkeley (1685–1753), David Hume (1711–1776) and Auguste Comte (1798–1857). Much patient investigation finally established beyond question that fundamental physical laws such as the conservation of matter and energy applied to human body constituents.

Application of empiricism to studies of body composition

An important first step towards proving energy conservation *in vivo* was taken by Santorio Santorio (1561–1636). He apparently spent much of his life eating and sleeping in a specially constructed weighing chair (Fig. 1.1), accumulating valuable data on the mass of his ingested food and excreta over a period of some 30 years. A century later, Audry (1658–1742) gave tacit acceptance to the principle of energy conservation by prescribing increased exercise for patients who needed to 'lose weight', but still in the early nineteenth century people as well educated as Charles Dickens were wont to discuss the possibility that a person could die of spontaneous combustion. Disproof of such ideas required the development of closed-circuit metabolic chambers, in which animals (Von Regnault, 1810–1878) and humans (Voit, 1831–1909) could live, eat and work, while observers kept careful records of energy balance for the system.

Fig. 1.1. The weighing chair used by Santorio in many of his long-term studies of body mass. Source: O. L. Bettman, 1956.

Information theory, empiricism and body composition

Arguments favouring an empirical approach to the study of body composition have now moved from philosophy and theology to the firmer ground of information theory. Information reduces uncertainty about an object or a phenomenon, and is usually conveyed most economically through the use of specific, numerical data rather than through a semantic or intuitive approach (Shephard, 1974).

In the context of body composition, a young woman might be described as 'pleasantly plump', but the uncertainty of a reader would be lessened if the author stated that the woman carried 20 kg of fat, and that this figure was two standard deviations above the actuarial ideal for her height and age.

Parnell (1965) has distinguished 'compositionists' who are interested largely in the chemical constituents of the body, 'nutritionists' who measure the proportions of the various tissue constituents, and 'behaviourists' who study the selective advantages of various types of body form and composition. The nutritionist or the behaviourist might note that a young man was 'very fat' or 'an extreme endomorph'. However, for most purposes the description would become more meaningful if reported as an average subcutaneous fat thickness of 30 mm relative to the norm for a young adult of 11 ± 6 mm.

Patterns of body composition

Pattern recognition is the main weakness of the empirical approach in general and computer analysis of body composition data in particular. There thus remains some justification for a descriptive approach when considering patterns of body composition. For example, Kissebah *et al.* (1982) and Krotkiewski *et al.* (1983) each noted the association between a 'masculine' distribution of accumulated body fat and vulnerability to heart attacks. It is easy to form a 'Gestalt' of the masculinity or femininity of body form, but the argument for a descriptive approach is not very strong; it remains statistically more satisfying to document any departure from the anticipated sexual dimorphism by comparing the thickness of specific skinfolds such as the hips (a typical location for a female accumulation of fat) and the front of the abdomen (a typical location for male fat accumulation).

Semantic descriptions of body form

More than 200 years ago, Hippocratic physicians distinguished the short, thick-set person with a 'habitus apoplecticus', typically a red-faced, jovial and forceful individual, liable to die of apoplexy, from the long and thin patient with a 'habitus phthysicus', commonly a more introspective subject who was liable to die of phthisis (tuberculosis). Halle (1797) noted four different body types: the fat 'abdominal' person, the strong 'muscular' individual, the long, slender-chested 'thoracic' form, and the large-headed 'cephalic' type. An appropriate body build is important to success in athletic competition (Tanner, 1964), and already in 1936 Kretschmer & Enke (1936) were distinguishing the round and compact form of the 'pyknic' contestant from the muscular 'athletic' build and the long, thin, asthenic 'leptosome'.   More recently, Škerlj (1959) proposed developing vectors to describe four aspects of body form (sex, frame-size, amount and distribution of soft tissues). Likewise, Sheldon (1963) used standard photographs to classify his population in terms of fatness ('endomorphy'), muscularity ('mesomorphy') and linearity ('ectomorphy'). Each

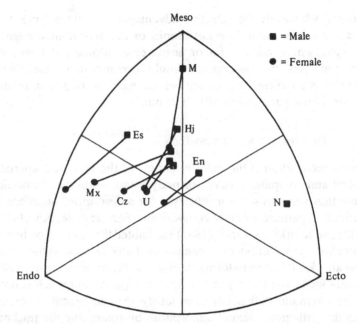

Fig. 1.2. Three dimensional plot to show ectomorphy, mesomorphy and endomorphy of selected national samples. Es = Eskimo, Mx = Mexican,
Cz = Czechoslovak, U = United States, M = Manus, Hj = Hawaian Japanese,
En = English and N = Nilote. Source: J. E. L. Carter (1980). Reproduced
with permission of the publishers, University Park Press.

variable was rated on an arbitrary seven point scale, yielding a three digit
'somatotype' (Fig. 1.2). Unfortunately, a large proportion of subjects
occupy an intermediate position with respect to each of the rated
characteristics. Sheldon estimated that in a well-nourished adult population there would be 7% endomorphs, 12% mesomorphs, 9% ecto-
morphs, and 72% of individuals who shared characteristics of two if not
three of the proposed body types.

Subsequent development of somatotyping has seen a useful amalga-
mation of traditional subjective impressions with quantitative data
(Carter, 1980, 1984). Technical issues of this hybrid methodology are
reviewed in a companion volume (Carter & Heath, 1988). There is a close
correspondence between 'endomorphy' and estimates of body fat, and an
equally close relationship between 'mesomorphy' and empirical
measurements of lean tissue mass per unit of stature. There thus seems
much to commend a replacement of the semantic 'Gestalt' by specific

numerical data describing body composition in precise mathematical terms.

'Reference' and 'normal' standards of body composition

'Reference' versus 'normal' standards

A 'reference' standard is an arbitrary value, usually based on data for healthy and well-nourished subjects (Waterlow, 1986); it offers a convenient basis for the comparison of populations, without making any judgements about the desirability of the chosen reference. For example, when assessing the body mass of young children, the World Health Organization (1983) has proposed adoption of data currently available from the US National Center for Health Statistics (Table 1.1), while the somewhat dated figures of Baldwin (1925, Table 1.2) have been commended for assessing the mass of older adolescents (World Health Organization, 1985).

Other authors have called for population-specific reference standards. These may be quite helpful when assessing selected groups such as athletes who are committed to a particular discipline. On the other hand, population-specific figures have more doubtful application when applied to average citizens, particularly in the less-developed parts of the world.

Table 1.1. *Desirable body mass of children.*

Age (years)	Girls (kg)			Boys (kg)		
	−2SD	Median	+2SD	−2SD	Median	+2SD
1	2.2	3.2	4.0	2.4	3.3	4.3
1/2	5.5	7.2	9.0	5.9	7.9	9.8
1	7.4	9.5	11.6	8.1	10.2	12.4
2	9.4	11.9	14.5	9.9	12.6	15.2
3	11.2	14.1	18.0	11.4	14.6	18.3
4	12.6	16.0	20.7	12.9	16.7	20.8
5	13.8	17.7	23.2	14.4	18.7	23.5
6	15.0	19.5	26.2	16.0	20.7	26.6
7	16.3	21.8	30.2	17.6	22.9	30.2
8	17.9	24.8	35.6	19.1	25.3	34.6
9	19.7	28.5	42.1	20.5	28.1	39.9
10	21.9	32.5	49.2	22.1	31.4	46.0

(Based on data of US Public Health Service HIHS growth charts, HRA 76–1120, 25, 3.)

Table 1.2. *Median body mass of adolescent boys and girls (kg) in relation to standing height.*

Height (cm)	Age (years)																	
	10		11		12		13		14		15		16		17		18	
	M	F	M	F	M	F	M	F	M	F	M	F	M	F	M	F	M	F
120	—	22.3																
125	24.2	24.6	—	24.7														
130	26.8	27.1	27.0	27.9	—	27.3												
135	29.3	30.1	29.4	30.1	29.6	30.7	—	31.5										
140	32.2	32.9	32.2	33.1	32.4	33.2	32.4	34.1	—	34.8								
145	34.9	36.6	35.7	36.4	35.4	36.6	35.8	37.2	36.3	39.3	—	41.4						
150	38.1	38.8	38.5	40.2	39.0	39.9	39.1	41.1	39.3	43.0	39.2	44.6	—	45.9				
155			41.5	44.0	42.1	44.8	42.7	45.0	43.4	47.0	43.5	48.1	44.8	50.2	—	46.4	—	51.4
160					46.2	48.9	46.7	49.2	47.4	49.8	48.0	51.5	49.8	51.9	51.5	50.4	53.9	53.1
165						52.6	50.9	53.1	51.4	54.0	52.3	54.2	53.1	54.8	55.1	52.8	57.1	55.9
170								56.8	55.6	57.6	56.5	58.0	58.1	58.9	59.1	55.4	60.5	60.1
175									59.7	60.0	60.4	60.8	61.9	61.2	63.5	58.9	64.7	62.9
180										61.3	65.1	62.2	65.7	63.0	66.1	62.1	67.7	64.4
185													69.5		70.3	63.9	71.3	

(Based on data of Baldwin, 1925.)

In such circumstances, it is difficult to undertake representative sampling and standards are quickly out-dated by secular changes in both growth patterns and adult size.

Some investigators have argued that if a world-wide 'reference' value were established for well-nourished subjects, this could be equated with a 'normal' standard, since the impact of any inherited differences of body composition is small relative to the departures from normality caused by an adverse environment (Graitcer & Gentry, 1981). In contrast, pragmatists have argued that if a world-wide 'normal' value were to be adopted, it would imply the existence of a standard that was both attainable and should be attained; in poor parts of the world, the setting of what might be perceived as unrealistically high figures could hamper practical planning.

'Ideal' body composition

Can one assume that the status quo, as seen in developed nations provides either a 'normal' or an 'ideal' standard of body composition? The concept of an 'average man' dates back to Quetelet (1796–1874), who applied concepts of error distribution when analysing the height of Belgian army recruits. However, the substantial social-class-related differences of stature in the nineteenth century raised the difficult issue as to whether a professional person or a labourer should be chosen as the reference standard.

There are good reasons to infer that the average body mass currently observed in North American adults exceeds an acceptable 'ideal' statistic. The dilemma of selecting an appropriate body mass for height has been faced by several generations of physicians, as they have compared tables of 'average' and 'ideal' weights. The average figures reported for insured North Americans show a steady increase over the span of adult life; on the other hand, 'ideal weights' such as those proposed by the Society of Actuaries (1959) and the Metropolitan Life Insurance Company (1983) generally assume a body mass that is independent of age.

There are many possible definitions of an 'ideal' body build and composition. One study of Canadian women university students from the early 1960s (Yuhasz, unpublished data, 1965) rated the ideal percentage of body fat in terms of personal appearance as judged by a male observer (Table 1.3). Current investigators would probably condemn this approach as sexist, although interestingly, the 'ideal' carried more lean tissue as well as less fat. Moreover, it is arguable that in the culture of the period, failure to conform to a masculine ideal of appearance may have

Table 1.3. *Comparison of average and 'ideal' body composition of female university students, judged on the criterion of 'optimal' physical appearance.*

	'Ideal' (17 women)	Average (75 women)
Body mass (kg)	54.8	59.4
Self-judged mass (kg)	56.5	56.5
Average skinfold thickness (6 sites, mm)	16.1	18.7
Fat mass (kg)	7.9	10.2
Per cent fat	14.3	17.3
Lean mass (kg)	50.0	49.0

(Based on unpublished data of M. Yuhasz, 1965.)

Table 1.4. *Ideal body mass, based on an average of the 'ideal' values proposed by (a) Society of Actuaries (1959) and (b) Kemsley et al. (1962), both adjusted for shoe height.*

Height (cm, no shoes)	Ideal body mass (kg, indoor clothing)			
	Male		Female	
	(a)	(b)	(a)	(b)
147.3	—	—	48.5	49.0
149.9	—	—	49.9	50.5
152.4	—	53.6	51.2	51.8
155.0	—	55.5	52.6	53.2
157.5	57.6	57.3	54.2	54.5
160.0	58.9	59.1	55.8	56.4
162.6	60.3	60.5	57.8	57.7
165.1	61.9	62.3	60.0	59.1
167.6	63.7	63.6	61.7	60.5
170.2	65.7	65.5	63.5	61.8
172.7	67.6	67.3	65.3	63.2
175.3	69.4	68.6	66.8	64.5
177.8	71.4	70.5	68.5	65.9
180.3	73.5	72.3	—	—
182.9	75.5	73.6	—	—
185.4	77.5	75.5	—	—
188.0	79.8	77.3	—	—
190.5	82.1	—	—	—
193.0	84.3	—	—	—

had a significant impact upon the marriage prospects of female students and thus the perpetuation of particular anthropometric characteristics.

'Ideal' body mass

Life assurance companies and epidemiologists have chosen as their 'ideal' a body mass associated with minimal mortality over a specified interval, or with maximal life expectancy (Fig. 1.3; Table 1.4). Actual values have been biassed by the technicalities of data collection (Jarrett, 1986), including:

(i)   use of a self-reported mass, which in some populations deviates systematically from the measured body mass (Table 1.5; Perry & Leonard, 1963; Wing *et al.*, 1979; Stunkard & Albaum, 1981);

(ii)   day-to-day variations of body mass (Adams *et al.*, 1961; Durnin, 1961; Khosla & Billewicz, 1964; Robinson & Watson, 1965; Edholm *et al.*, 1974);

(iii)   uncertain allowances for the height of shoes and the mass of clothing;

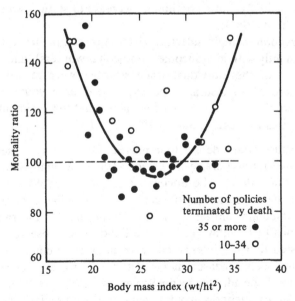

Fig. 1.3. Relationship between mortality rate and body mass index. Source: R. Andres (1985). Reproduced with permission of publishers, McGraw-Hill.

Table 1.5. *Relationship of observed to reported body mass (kg).*

Measurement location	Body mass		Coefficient of correlation
	Measured (kg)	Reported (kg)	
US gynaecologist's office	60.2	59.7	0.99
US internist's office	66.3	66.5	0.99
US obesity treatment			
– medical school	97.7	95.7	0.98
– medical school	95.1	93.3	0.99
– union	85.5	84.6	0.99
– psychiatrist	91.5	90.0	0.98
US pre-employment	78.6	77.4	0.99
Danish men <40 yrs	77.0	75.9	0.95
>40 yrs	79.4	78.1	0.96
Danish women <40 yrs	61.1	59.5	0.97
>40 yrs	62.7	60.3	0.91

(Based in part on data of Stunkard & Albaum, 1981.)

(iv)   a secular trend to an increase of stature in young adults (Waller & Brooks, 1972), but a continuing decrease of stature with aging (Shephard, 1987);

(v)   non-random sample selection (insurance companies deal predominantly with high-income white-collar workers), and

(vi)   the age of the individual when statistics were recorded (on purchase of an insurance policy, usually as a young adult), whereas there is now growing evidence that the optimum body mass rises with age, (Andres, 1985; Fig. 1.4).

Bones are substantially denser than muscle. One important determinant of body mass is thus the size of a person's frame (Welham & Behnke, 1942; Clark *et al.*, 1977). The Society of Actuaries (1959) arbitrarily classified their data into ranges of ideal mass for individuals rated as having small, medium and large frames. Unfortunately, the rating of frame size tends to be influenced by the build of the observer. Pryor (1940) suggested that the interpretation of body mass could be made more precise if objective measurements of frame size were taken, and this proposal has now been adopted by the Society of Actuaries (1980). While a person with large bones now avoids penalisation, a muscular person can still be classed as 'overweight', despite a relatively low percentage of body fat (Clark *et al.*, 1977).

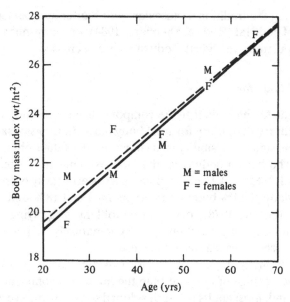

Fig. 1.4. Body mass index in relation to age. Source: R. Andres (1985). Reproduced with permission of publishers, McGraw-Hill.

Despite such problems, the 1912 and the 1959 standards of the Society of Actuaries (Society of Actuaries, 1959) have been widely adopted. Recent modifications include an allowance for shoes and clothing, and omission of the arbitrary adjustment for frame size (Bray, 1975; Royal College of Physicians, 1983).

Secular trends have modified the actuarial 'ideal' – upwards in developed societies (Society of Actuaries, 1980) and downwards in populations undergoing acculturation to urban living (Rode & Shephard, 1984). Lightness was once associated with diseases such as tuberculosis, and more recently it has been linked to a heavy consumption of cigarettes (Khosla & Lowe, 1971; Howell, 1971; but not Pincherle, 1971). On the other hand, the current interest in physical fitness has increased muscle development among active individuals, thus increasing the optimum body mass for developed western societies, even after allowance for changing patterns of cigarette consumption (Garrison *et al.*, 1983). Nevertheless, it remains arguable that the onset of chronic disease symptoms was the signal that encouraged some lighter members of the insured sample to stop smoking; the chronic effects of former smoking might thus persist, leaving an association between a light body mass and an unfavourable actuarial experience. In contrast, acculturation to urban living is replacing the traditional muscular Inuit hunter by a person who has an excess of body fat (Rode & Shephard, 1984); one may thus suspect that in this population, the 'ideal' body mass has decreased.

'Ideal' body fat

'Ideal' standards for individual tissue components such as body fat are even more arbitrary than those for total body mass. In terms of skinfold thicknesses, one simple possibility is to commend the values observed in young adults who have a body mass that approximates to the actuarial ideal (Shephard, 1982a). A British military survey arbitrarily proposed a desirable standard of 14% body fat in young men and 18% body fat in young women (Amor, 1978); one major problem in applying such a criterion is that body fat predictions vary systematically, depending on the measuring techniques that have been used.

Irrespective of methodology, many North American samples have a much higher percentage of body fat than the military recommendation. On the other hand, most adults in less developed societies carry very little body fat, apparently without any adverse influence upon either their health or their ability to withstand extremely low environmental temperatures (Shephard, 1978a), Likewise, certain classes of athletes have percentages of body fat as low as 5% in males and 10% in females; the one potential health complication seems a temporary interruption of sperm production in the male and of menstruation in the female. Thus, from some points of view, the 'ideal' body might contain an extremely small percentage of fat.

Evidence from distribution curves

A further possible method of defining normality is to test whether any of the anthropometric variables are bimodally distributed. If so, the overall population would comprise a group of 'normal' individuals and a second group who are in some respect 'abnormal' – the 'stunted', who are abnormally short, the 'obese', who are excessively fat, the 'wasted', with a lack of muscular development, and so on. A frequency diagram of population data for height, body fat, or muscle mass may show two distinct peaks, a cumulative percentage plot (Fig. 1.5) may fail to reach a Z-score of zero when at the 50th percentile, and a probit plot may show a sudden break at the separation between 'normal' and 'abnormal' members of the sample (Shephard, 1982b; Fig. 1.6).

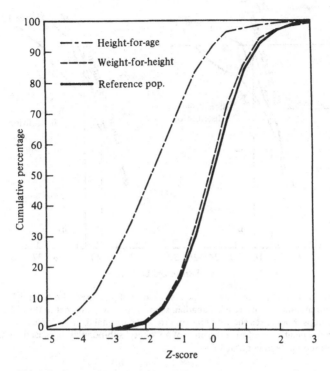

Fig. 1.5. Cumulative distribution of Z-scores for a population that includes 'stunted' individuals, but is free of wasting malnutrition. Note that the weight for height curve conforms to a normal distribution curve, but that the height for age curve is displaced to the left. Source: J. C. Waterlow (1986).

Influence of growth and aging on 'ideal' values

The problem of specifying an 'ideal' body composition is made yet more complex by growth and aging.

In childhood, nutritionists might call for a body composition associated with such 'indicators' of population health as an acceptable mass of the newborn infant in relation to the duration of gestation, a normal rate of post-natal growth, or a normal attained height for a given age (Waterlow, 1986).

In elderly subjects, a hunching of the back (kyphosis), a narrowing of the inter-vertebral discs and actual collapse of the vertebrae lead to a gradual decrease of stature. Such changes rob the physical anthropologist of the simplest method of standardizing data for overall body size. Moreover, normal body composition seems to change with aging,

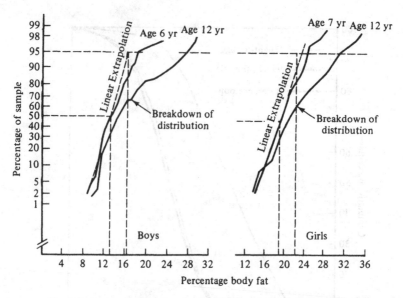

Fig. 1.6. Probit plot showing the distribution of percentage body fat values within a sample of French Canadian schoolchildren. Note that in the older age groups, both sexes show a departure from the normal distribution curve which could be attributed to emergence of an 'obese' sub-population (based on data presented by Shephard, 1979).

although it remains unclear how far fat accumulation, protein loss and mineral loss reflect the lesser habitual activity of the elderly, and how far they are inevitable, if not an 'ideal' accompaniment of aging.

Is prolonged survival an evolutionary 'ideal'?

From both a social and an anthropological perspective, one could finally question the value of survival into extreme old age. An ergonomist would wish to consider also functional capacity (which is sometimes very low in those weakened by senescence or malnourishment), while a social analyst would look at the quality of the resultant life, calculating an 'ideal' in terms of quality-adjusted life expectancy. The quality of life is inevitably subjective, for instance, a gourmet cook might be prepared to accept the body composition associated with a 20% worsening of calendar lifespan if the alternative was to abandon the perceived pleasures of preparing and eating certain categories of food.

In communities where food supplies remain very limited, the immediate need is to define not an ideal ceiling for long-term survival, but rather a minimum standard of body composition which has both sensitivity and specificity as a means of distinguishing those who are at risk of early death from malnutrition (Habicht, 1980). The minimum values appropriate for this purpose depend upon the age of the individual and the prevalence of debilitating diseases in the community (Kielmann & McCord, 1978). In populations faced by starvation, a body build that increases fertility or prolongs senescence could be viewed as a disadvantageous adaptation to the local habitat. From an evolutionary standpoint, the 'ideal' for such a community would be a pattern of body composition that maximized the individual's ability to acquire food and assured escape from prey during the reproductive phase of life, but which led to death shortly thereafter.

Reasons for data collection

The type of body composition data that is required depends upon the purpose for which information is being collected. Garn (1963) argued that the main stimuli to research on body composition were linkages between obesity and the probability of death, an interest in the minimum protein supply needed for muscle growth, and a concern about osteoporosis in the elderly. Such issues remain important to current students of body composition. However, the primary focus of this volume is upon the anthropological and evolutionary implications of body composition. Cross-sectional studies compare findings from differing habitats, while longitudinal research examines the impact of changing nutrition and other secular trends upon body composition. Such research requires consistent methodology, both between observers and also within the same laboratory over time (Weiner & Lourie, 1981). Thus, several chapters are devoted to reviewing methods for the examination of body composition.

Waterlow (1986) noted potential applications of body composition data in developing societies. Figures for young children are a sensitive indicator of overall nutritional status within a given population. Surveys of body composition allow comparisons between habitats, with identification of target groups or regions where food supplementation is desirable. Periodic repetition of surveys allows the planning of nutritional and agricultural strategies, particularly if changes in nutritional status are correlated with changes in the prevalence of health and disease. Finally, the periodic repetition of observations helps to assess the effectiveness of programmes seeking to improve nutritional status.

In developed nations, community surveys may identify regions where over-nutrition is particularly prevalent. Any optimization of body composition seen in follow-up studies can further be correlated with governmental initiatives designed to improve lifestyle and any changes in community health (Shephard, 1986).

Methods of data analysis

Centiles

Centiles provide a simple, and reasonably valid assessment of body composition if the individual concerned lies relatively close to the median reference value. However, such statistics are inappropriate when dealing with individuals at the extremes of the distribution curve (for instance, international-class athletes, or those suffering from severe malnutrition). Centiles are normally defined only from the 3rd to the 97th; attempts to extrapolate beyond these limits are complicated by issues of sample size and the skewing of data.

Z-scores

If the average figure for a population is to be described, the Z-score (the deviation from the anticipated mean, expressed as a ratio to the standard deviation for the reference population) is reported, together with the standard deviation of the Z-score. Alternatively, a cumulative distribution of Z-scores can be plotted. The latter approach may bring out interesting differences between anthropometric data – for instance, in a population that is stunted, but without wasting, the weight for height curve will conform to a normal probit plot, but the height for age curve will be displaced to the left (Fig. 1.5). The main disadvantage of basing the presentation upon Z-scores is that a part of the apparent variation within a population is really due to measurement variability.

Percentages of reference value

Percentages of a reference median yield a similar impression to Z-scores if deviations from the norm are small. However, when calculating the mean percentage score for a sample that includes people at both extremes of the distribution curve, there is the problem that a negative deviation carries less weight than a positive deviation of equal absolute size.

The concept of proportionality

Importance to data interpretation

When considering the relative growth of various body constituents, sex differences in body composition, secular changes or differences between populations, it is immediately necessary to make some adjustment of data sets for differences of body size.

The question of data standardization has more than academic interest. Thus Garn (1965) commented on an earlier report that Guamese children were relatively slender. Since this particular population was much shorter than its US counterpart, it was certainly 'slender' in terms of a simple mass to height ratio, but on the other hand, if the data had been expressed as a mass to height^3 ratio, the Guamese children would have been regarded as having a 'normal' body build. Recent comparisons of Canadian Indian, Inuit, Anglophone and Francophone children have encountered similar problems (Shephard, in preparation 1989).

Beginning with Quetelet (1835) and Rohrer (1908), a variety of mass to height ratios have been adopted in quantifying human physical characteristics and athletic performance (Hirata, 1979). The simplest approach has been to use a mass to height ratio (see, for example, Maas, 1974). Although arguments abound in favour of more complex adjustments, the simple form of standardization may well suffice if the difference of height between two samples is relatively small.

Theories of proportionality

Theoretical justification for the use of complex mass to height ratios derives from Archimedes' laws for geometrically similar bodies (Ross *et al.*, 1980). Galileo Galilei suggested that if animals had a similar shape and general body composition but differed in body size, then their mass would be proportional to the cube of their linear dimensions, while the cross-sectional area of their muscles and supporting structures would be proportional to the square of their height. Borelli noted further that geometrically similar animals could jump to the same height, while the maximum velocity of running was independent of a creature's body size. These last observations have implications for interactions between body build and survival of the species.

Sheldon (1963) scaled ectomorphy in terms of the reciprocal of the 'ponderal index', the ratio of stature to the cube root of body mass, while

Hebbelinck *et al.* (1973) adopted an equivalent metric scaling for a Belgian study of child growth and development. Scandinavian investigators went further, suggesting that the various body parts grew at a rather similar pace, so that volumetric variables such as total body mass or blood volume should increase in proportion to the third power of standing height (Von Döbeln, 1966; Asmussen & Christensen, 1967; Åstrand & Rodahl, 1985).The assumptions underlying such propositions include not only a regional consistency of growth patterns, but also a consistent cubic relationship between the length and the volume of a body part (that is to say, a regular spherical shape), and a consistency of density within that volume (Gunther, 1975). As we shall see, none of these conditions is fully satisfied.

Following the views of Borelli, it has been argued that any evolutionary selective effects are directed primarily at locomotor factors essential to survival (McMahon, 1975; Grand, 1977; Ross & Ward, 1984). Nevertheless, mechanical, hydrodynamic and thermal grounds can be advanced supporting a cubic relationship (Gunther, 1975). Furthermore, on the basis that the organism must adapt to elastic and static stressors, a case can be made for a relationship of body mass to height^4 (H^4) or even H^5 (McMahon, 1975).

Objections to H^3 *standardization*

The cubic relationship is now quite widely accepted as a method of comparing body mass in the face of inter-individual differences of body size. However, there remain practical and theoretical objections to this type of data standardization (Tanner, 1964; Ross *et al.*, 1982).

The calculation of a ratio presupposes that the relationship passes through the origin of both variables, and that the ratio can be distorted by an abnormality in either the numerator or the denominator. Moreover, the total variance is an inextricable combination of the variance contributed by its two constituents. The data thus defy any simple statistical analysis.

Inherent mathematical problems are usually worsened by a skewing in the distribution of body mass and other anthropometric variables (Jéquier *et al.*, 1977). Furthermore, much of body form is arranged as a sequence of truncated cones rather than as the simple spheres implied by a cubic relationship (Van der Walt *et al.*, 1986).

In children, additional problems arise from a differential growth of the various body segments. The import of standing height changes progressively over the first few years of life, as the relative size of the head

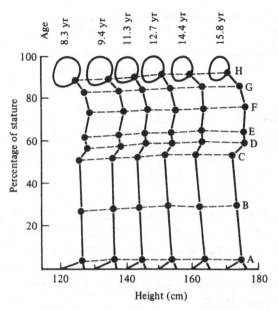

Fig. 1.7. To illustrate the overall growth in body size versus the growth of individual body parts (based on a concept of Erling Asmussen, 1973). Landmarks: A, lateral malleolus; B, lateral femoral epicondyle; C, greater trochanter of femur; D, posteriosuperior iliac spine; E, deepest point of lumbar lordosis; F, highest point of dorsal kyphosis; G, vertebra prominens; H, external auditory meatus.

diminishes (Behnke & Wilmore, 1974, Fig. 1.7). Moreover, at puberty the height spurt antecedes the rapid increase of body mass by an interval of 1–2 years (Shephard, 1982*b*).

Finally, the density of any given body part varies according to the relative proportions of fat, muscle and bone that it contains.

Empirical evidence

Since theoretical arguments can be advanced for and against each of the proposed power functions of height, some authors have attempted to resolve the issue by recourse to empirical data. If a hydrostatic estimate of body fat is compared to various mass to height ratios, the third power of height yields no closer a correlation than the second power or even a simple mass to height ratio (Table 1.6). Again, if body mass is used to compare function in differing species, variation is minimized if data are expressed in terms of mass$^{0.67}$ or mass$^{0.75}$ rather than mass$^{0.33}$ (Dreyer *et*

Table 1.6. *Coefficients of correlation between percentages of body fat as determined hydrostatically and simple ratios of body mass to height.*

Ratio	Coefficient of correlation	
	Boys	Girls
Mass to height	0.41	0.38
Mass to height^2	0.37	0.51
Mass to height^3	0.24	0.47

(Based on data of Shephard *et al.*, 1971.)

al., 1912; Krogh, 1916; Lambert & Tessier, 1927; Brody, 1945; Kleiber, 1947, 1975). However, except when the difference of mass between groups is threefold or more, mass$^{0.75}$ offers no advantage over a linear function of mass. Forbes (1974) commented further that the fitting of a polynomial curve did not help in describing the relationship between body mass and standing height.

Circumferential scaling

There are a variety of alternatives to height standardization. Behnke *et al.* (1959) proposed using the product of height and an averaged cross-sectional area. The latter could be derived from a total of 11 circumference measurements (Behnke, 1961). In his subjects, the proposed formula showed a correlation of 0.98 with body mass, and after application of a suitable scaling factor, body mass (M) could be estimated to ±2–3%:

$$M \text{ (kg)} = 1.11 \ H \ (C/100)^2$$

where H was the stature of the subject, in metres, and C was the sum of the individual circumferences, measured in centimetres.

Proportional body mass

Ross & Ward (1984) proposed the calculation of a proportional body mass, using the formula $M(180.18/H)^3$, where M is the observed mass and H the observed height.

Critics to this approach have noted that problems arise if the heights of the population under investigation deviate substantially from the reference value of 170.18 cm. When examining contestants in a South African festival games, Van der Walt *et al.* (1986) found it prudent to substitute for the previously proposed figure of 170.18 cm, a standard height of 174.14 cm (the latter value being more applicable to the elite white minority participants in their competition).

Multivariate techniques

Investigators such as van Gerven (1972) and Kowalski (1972) have endeavoured to demonstrate a communality among anthropometric measurements, and thus to isolate a general 'size factor', but in practice, the principal components isolated by factor analysis have proven complex and correspondingly difficult to interpret. Tanner (1964) suggested the alternative approach of co-variance analysis, in essence using a linear or a power function of height as a co-variate when making comparisons of variables such as body mass between populations.

Body mass as a basis of standardization

The dilemma of an appropriate method of size standardization has been partially resolved for variables such as aerobic power by relating results to body mass rather than stature. Gross body weight is also widely used in calculating nutritional requirements. Nevertheless, body mass is not an entirely satisfactory standard, since the denominator can be influenced by such differing factors as an accumulation of metabolically inert fat or an unusual development of skeletal muscle.

For the present, there seems no internationally accepted technique for the standardizing of body composition measurements. Investigators should thus insist upon the publication of base data, as well as the results of any standardizations that are attempted.

The unisex phantom

Principle of calculation

The 'unisex phantom' (Ross *et al.*, 1982; Ross & Ward, 1984) is one further attempt at data standardization which requires a specific critique. The concept apparently evolved from the ideas of proportionality and biological similarity discussed above. All anthropometric data, whether for male and female subjects, are scaled to a standard stature of 170.18 cm, and deviations from scaled values are expressed as ratios to an empirically determined standard deviation for the variable in question:

$$Z = 1/S \ (v[170.18/H]^d - P)$$

where Z is the 'proportionality ratio' (Ross *et al.*, 1982). Choice of the letter Z to denote this ratio is unfortunate, since it is not, strictly speaking, a Z statistic. S is the standard deviation of the variable in the reference sample, v is the magnitude of the variable in the subject under consideration, H is the individual's height, and P is the reference value for the same variable. Positive Z values indicate that a particular variable is larger than expected, and negative values indicate that it is smaller than expected. In the case of body mass, the actual numerical values used by Ross *et al.* (1982) are:

$$Z = 1/8.60 \ (M[170.18/H]^3 - 64.58)$$

Application to body composition

Ross & Marfell-Jones (1982) have suggested that their unisex phantom technique can be used when describing the main body compartments. The mean and the standard deviation of the proposed reference values are for fat mass 12.13 ± 3.25 kg, for lean body mass 52.45 ± 6.14 kg, for skeletal mass 10.49 ± 1.57 kg, for muscle mass 25.55 ± 2.99 kg, and for residual mass 16.41 ± 1.90 kg. The authors of the technique maintain that because data have been scaled to a common standard, and expressed as ratios to the corresponding standard deviation, different anthropometric variables can be compared both between and within samples.

Ross and his associates (1982) further argue that departures from their unisex model provide insights into individual differences, whether these be related to sex, growth, maturation, training, nutritional factors or secular trends. They consider the phantom as a metaphorical reference that provides mathematically consistent rules, thereby allowing for differences in body size between samples.

Critique of phantom approach

One may question the wisdom of applying a univariate basis of standardization (a power function of the height observed in a convenience sample of North American adults) to the multivariate problems of changes in body shape and composition throughout the entire periods of growth and aging (Shephard *et al.*, 1985a). In essence, the regional development of a child is compared to adult norms of body form, using mean values derived from a somewhat arbitrary heterogeneous and non-random mixture of men and women which the developing individual has no biological reason to emulate, together with a 'population variance' calculated for this adult unisex artifice.

It seems highly illogical to suppress one clearly defined biological variable (sex) in order to create a single reference phantom. The immediate practical consequence is that for many anthropometric variables, 'normal' adults of a given sex have a Z-score of 1.0 or more, rather than showing what one might anticipate, a normal distribution of values about a Z-score of 0.

Even if the idea of a unisex adult phantom could be accepted as a reasonable means of data standardization, the necessary data should have been drawn from a large, well-defined and appropriately stratified population with 'ideal' characteristics of body composition. However, Ross and his associates used a relatively small heterogeneous convenience sample to create their reference phantom.

There are also important mathematical and statistical objections to the unisex approach. The use of Z statistics presupposes a normal distribution of the measurements which are tested by the model, both at maturity and over the presumed growth curve, whereas in practice, there is an often substantial skewing of anthropometric data. The interpretation of the supposed Z-scores is further complicated when these are scaled by an arbitrary power function of stature. Even if such scaling were permissible and there was agreement on an appropriate theoretical exponent, it would remain necessary to face mounting evidence that exponents vary about the theoretical average, depending on age, sex and patterns of physical activity (Shephard, 1982b).

While the phantom approach may find some application in looking at the regional growth and development of body constituents, objections to it seem sufficient that it has little place in the standardizing of data on overall body composition.

2 *Epidemiological indices, anthropometric and cadaver estimates of body composition*

Simple epidemiological indices of malnutrition

Clinicians and epidemiologists have for a long time sought simple anthropometric indicators of body composition and nutritional status (Gray & Gray, 1980; Neumann *et al.*, 1982). Thus, the upper arm circumference has been used to assess protein/energy malnutrition (Jelliffe, 1966; Wolanski, 1969*a*; Fomon, 1978; Bertrand *et al.*, 1984). Unfortunately, a simple circumference measurement necessarily confounds the contributions of fat, muscle and bone. In the context of protein/energy malnutrition, muscle is the important component. Gurney & Jelliffe (1973) thus proposed calculating the muscle cross-section of the upper arm as:

$$\text{Muscle cross-section} = (C - \pi T)^2/4\pi$$

where C was the maximum circumference over the triceps, and T was the thickness of the triceps skinfold, both measured in cm. The limitations of this formula are that it assumes a circular cross-section for the upper arm, with a uniform layer of subcutaneous fat and a negligible bone content (Lerner *et al.*, 1985). The error arising from ignoring the contribution of the humerus to the total cross-section has been examined by comparing the Gurney & Jelliffe formula against computed tomography (Heymsfield *et al.*, 1982*b*). At the maximum circumference of the triceps, bone accounted for some 10 cm^2 in the men (18% of the total cross-sectional area), while in the women it accounted for 6.5 cm^2 (17% of the total cross-sectional area).

Unfortunately, the proportional contribution of the humerus varies with nutritional status. This limits the possibility of allowing for the presence of bone by making an arbitrary scaling of the cross-section computed from such formulae. Jelliffe & Jelliffe (1982) suggested that the muscle volume of the upper arm could be estimated as 20% of the upper arm length, multiplied by the muscle cross-section, calculated from the original formula as described above.

Plainly, a similar equation can be used to estimate body fat (Trowbridge *et al.*, 1982):

Fat cross-section $= (C^2/4\pi) - (C - \pi T)^2/4\pi$

As suggested by Gray & Gray (1980), the simple circumference formula $(CT/2)$ is corrected by a skinfold thickness factor of $(\pi T^2/4)$.

Himes *et al.* (1979) commented that estimates of the cross-sectional area of fat made in this fashion did not provide any better estimate of body density than the original skinfold. At a first inspection, this is a little surprising, since the circumference does scale the fat thickness for body size. Presumably, this advantage is outweighed by the fact that an increase of limb circumference can reflect an increase of muscle or bone as well as fat. On the other hand, Himes *et al.* (1979) found that relative to skinfolds alone, the use of circumference data did not improve their estimate of total fat mass.

Anthropometric measurements of body form and composition

In 1921, the Czech anthropologist Matiegka used the available cadaver data of the period (Vierordt, 1906) to develop a series of equations. These equations allowed him to make an empirical estimation of the four main body components (muscle, skin plus subcutaneous fat, deep fat plus viscera, and bone) from external measurements of body form:

Muscle mass (g) $= 6.5\, r^2\, H$

(where *r* was the mean of the radii calculated from the maximal circumferences of the arm, forearm, thigh and calf, and *H* was the stature, all values being measured in cm).

Skin and subcutaneous fat mass (g) $= 0.065\, \Sigma\, S/6\, A$

(where $\Sigma\, S$ was the sum of six skinfolds, and *A* was the body surface area (cm^2), as calculated by the Dubois nomogram).

Bone mass (g) $= 1.2\, o^2\, H$

(where *o* is the average of diameters for the femoral and humeral condyles, the wrist and the ankle, all measured in cm).

Deep fat and visceral mass (g) $= 0.206\, M$

(where *M* is the body mass in g).

Katch *et al.* (1979) suggested further that total fat mass could be estimated from surface area (*A*), skinfold thicknesses ($\Sigma\, S$) and a population specific constant (k) which varied with the sum of 11 girth measurements:

$$\text{Fat mass (kg)} = A \times \Sigma S \times k$$

Drinkwater & Ross (1980) expanded Matiegka's original analysis, developing an alternative formula for the direct estimation of residual tissue mass from biacromial breadth (a), transverse chest diameter (b), bi-iliocristal breadth (c), and antero-posterior chest diameter (d):

$$\text{Residual mass} = 0.35 \, [\{(a + b + c)/3 + d\}/2]^2 H$$

Some minor changes of anthropometric site were necessary in order to apply Matiegka's method to data which had already been collected on athletes participating in the Montreal Olympic Games (Drinkwater & Ross, 1980). The constants for muscle (6.41) and bone (1.25) were essentially unchanged from the original figures of 6.5 and 1.2, but a smaller coefficient of 0.036 rather than 0.065 was proposed for the estimation of skin and subcutaneous fat mass. The original Matiegka (1921) formulae showed a systematic error of 8.0% in predicting the total body mass of this somewhat atypical and highly athletic population sample, but when revised constants were substituted, the sum of the Matiegka (1921)measurements predicted total body mass with an accuracy of 0.8%; using the Drinkwater & Ross (1980) modifications, the prediction was accurate to an average of 1.3%.

The final step in the validation of Matiegka's concepts – the correlation of the anthropometric estimates with the results of some recent Brussels cadaver dissections – was reported by Drinkwater et al. (1986). The original Matiegka equations underestimated the mass of muscle in the cadavers by an average of 8.5%; skin and fat were also underestimated, by 21.9% on average, while the mass of visceral tissue was underestimated by 11.6%. In contrast, bone mass was overestimated by a substantial 24.8%. Attempts were made to develop a further set of coefficients for the equations, using the Brussels cadaver data, but it was quickly recognized that any results thus obtained would be largely specific to the new sample, which was relatively elderly. Even within the sample group of cadavers dissected in Brussels, a need for sex-specific equations was demonstrated. The new coefficients overestimated the fat content of women by 15.9%, while underestimating that of the men by 12.3%.

Several authors (Standard et al., 1959; Jones & Pearson, 1969; Katch & Katch, 1973; Katch & Michael, 1973; Katch et al., 1973; Wartenweiler et al., 1974; Anderson & Saltin, 1985; Shephard et al., 1988) have applied similar anthropometric principles to the calculation of regional fat, lean tissue, bone mass, and overall limb volume. In general, calculations have been restricted to the limbs; these have been considered as formed from four or more truncated cones, each of which contains concentric cylinders of fat, muscle and bone.

Jones & Pearson (1969) calculated the volume of limb segments as $\frac{1}{3}L(a + \sqrt{(ab)} + b)$ where a and b were the areas of two parallel cross-sections as estimated from circumference readings, and L was the distance separating the two cross-sections. The foot volume was estimated as $\frac{1}{2}L(h.b)$, where L was the length of the foot, h was the height from the sole of the foot to the first cross-section at the ankle, and b was the average breadth of the foot. The total limb volume was compared with a hydrostatic criterion, the correlation coefficients being 0.98 and 0.99 in men and women, respectively. Muscle plus bone volume was calculated by taking skinfold readings for the thigh and calf, and correcting cone dimensions for overlying fat. Fat was calculated as total limb volume, minus 'muscle plus bone'. Comparison of fat was with a radiographic criterion (Chapter 3); the coefficients of correlation were for the thigh 0.95 and 0.85, and for the calf 0.83 and 0.86 in men and women, respectively.

Shephard *et al*. (1988) proposed the following formulae:

$$\text{Limb volume (ml)} = (\Sigma\, c^2)L/62.8$$

where $\Sigma\, c^2$ was the sum of the square of five individual circumference readings, measured in cm, and L was the length of the limb (cm).

$$\text{Limb muscle (ml)} = \text{Total limb volume} - (\text{fat volume} + \text{bone volume}).$$

$$\text{Limb fat (ml)} = (\Sigma\, c/5)(\Sigma\, s/2n)L$$

where $\Sigma\, c$ was the sum of the five limb circumferences (cm), $\Sigma\, s$ was the sum of the skinfolds measured over the limb (cm), and n was the number of skinfold readings taken over the limb.

$$\text{Limb bone (ml)} = 3.14\, R^2 L$$

where R was the average bone radius for the limb, 21% of the humeral intercondylar diameter corrected for overlying fat, and 23.5% of the corrected femoral intercondylar diameter.

Typical volumes for the three limb compartments are summarized in Table 2.1. In validation of the suggested equations, a good correlation was demonstrated between maximum oxygen intake and muscle volume in the lower limbs, while the percentage of fat that was calculated for the

Table 2.1. *Estimates of regional body composition derived from anthropometric measurements. Data of Shephard* et al. *(1988) for upper and lower limbs of young adult men and women (excluding the volumes of the hands and feet, respectively).*

Variable	Arm		Leg	
	Men	Women	Men	Women
Muscle (l)	1.81	1.08	7.68	5.66
	±0.20	±0.12	±1.29	±0.63
(%)	71.50	62.10	77.20	64.30
Fat (l)	0.31	0.41	1.17	2.38
	±0.08	±0.05	±0.52	±0.31
(%)	12.30	23.70	11.80	27.10
Bone (l)	0.41	0.25	1.10	0.75
	±0.09	±0.03	±0.24	±0.91
(%)	16.30	14.20	11.10	8.60
Total (l)	2.53	1.74	9.96	8.79
	±0.28	±0.16	±1.83	±0.91

limbs coincided fairly closely with the percentage for the whole body, as estimated from skinfold formulae (Shephard *et al.*, 1988).

One possible refinement of these various anthropometric formulae, suggested by Young (1965), would be to apply a correction factor (he suggested 10/7) to allow for compression of the skinfolds by the measuring calipers (see Chapter 3).

Cadaver data

The available cadaver data can be divided into two categories, chemical and anatomical. Chemical analyses have determined water content by desiccation, fat content by ether extraction, and mineral residue by ashing. In contrast, anatomists make a painstaking dissection of the body into what superficially appears to be muscle, fat, bone and other tissues. The distinction between ether extractable and anatomical fat is both important and substantial; this issue is discussed further in the next chapter.

Chemical analyses

Chemical analyses are of value to the molecular biologist, but are also of interest to the gross anatomist who wishes to translate data on mineral content of the body into a corresponding estimate of tissue mass. Thus,

Table 2.2. *Chemical constituents of the body as estimated by Widdowson (1965) in cadaver analyses of fat-free tissue; mean and range for five adults and typical neonatal figures.*

Variable	Adult Mean	Adult Range	Neonatal	Adult whole body if 18% fat
Nitrogen (g/kg)	34	31.0–37.5	23	27.9
Sodium (mE/kg)	(80)*	78.2–97.0	82	65.6
Potassium (mE/kg)	69	66.5–73.0	53	56.6
Chloride (mE/kg)	50	43.9–55.5	55	41.0
Calcium (g/kg)	22.4	20.7–24.0	9.6	19.7
Magnesium (g/kg)	0.47	0.43–0.49	0.26	0.39

*Two subjects only.

determinations of body potassium count can yield an estimate of lean tissue mass (Chapter 4), body nitrogen measurements provide an estimate of muscle mass (Chapter 4) and body calcium values give an indication of bone mass (Chapter 6). An understanding of changes in the chemical composition of the body (due, for instance, to physical inactivity or aging), also emphasizes to the investigator the errors that can arise when the body content of various chemical elements is used to predict the corresponding tissue mass and thus body composition.

Widdowson (1965) presented typical values for the concentrations of various chemical constituents in unit tissue mass, as obtained from an analysis of fat-free tissue in five adult cadavers and a series of newborn infants (Table 2.2). Nitrogen, potassium, calcium and magnesium concentrations all increase with maturation. Cross-sectional comparisons of averaged results for younger and older adults are complicated by the usual decline of habitual activity and general health as a person becomes older. However, available data suggest that there is a progressive loss of tissue potassium, nitrogen and calcium over the span of adult life.

One figure of particular importance in studies of gross body composition is the potassium content of lean tissue. The normative figure proposed for this variable (69 mE/kg) is unfortunately based on the analysis of only four cadavers; in addition to small sample size, figures may well have been biassed by ante-mortem change. Similar figures have been reported from chemical analyses of rats, rabbits and pigs (Forbes & Hursh, 1963), but lower concentrations are found in humans when *in vivo* measurements of body potassium are related to total body water content. Using the latter approach, some authors have found potassium concentrations as low as 54 mE/kg of lean tissue (Boddy *et al.*, 1972), although in a group of relatively active older women we calculated a figure of 59 mE/kg from total body potassium and hydrostatic estimates of body fat (Shephard *et al.*, 1985*b*).

Aging normally leads to a decrease of both intracellular potassium concentrations and the ratio of total potassium to total body water (Sagild, 1956; Sievert, 1956; Anderson & Langham, 1959; Allen *et al.*, 1960; Novak, 1972; Noppa *et al.*, 1979; Bruce *et al.*, 1980; Cohn *et al.*, 1980; Pierson *et al.*, 1982). These changes reflect in part a lesser efficiency of the sodium/potassium pump at the muscle cell membrane, and in part a replacement of muscle by connective and visceral tissue with a lower potassium content (Von Kriegel & Airscherl, 1964; Tzankoff & Norris, 1978). While muscle, liver and brain all have a potassium content of 95–101 mE/l, bone contains only 6–21 mE/l, tendon 13 mE/l and skin 21 mE/l (Forbes & Hursh, 1963). Cohn *et al.* (1980) reported an average 23% loss of potassium from age 25 to 95 years (0.33%/yr). Other estimates are of a similar order (0.25%/yr, Allen *et al.*, 1960; 0.43%/yr, Forbes & Reina, 1970; and 0.50%/yr, Shephard *et al.*, 1985*b*). However, in some older adults a combination of well-conserved muscles and demineralization of bone can lead to a seemingly anomalous increase in the potassium/lean mass ratio (Shephard *et al.*, 1985*b*).

The relationship between total body nitrogen and lean tissue mass depends on muscularity, body hydration and the protein and mineral content of the skeleton. Aging effects are uncertain. Our data show an apparent increase from the usually accepted figure of about 3% in young women, to about 3.5% of lean mass in active older female subjects (Shephard *et al.*, 1985*b*). In contrast, Cohn *et al.* (1980) reported a loss of nitrogen with aging, amounting over the age span 25 to 95 years to 14% in men and 21% in women.

An age-related calcium loss (Cohn *et al.*, 1976*b*; Harrison *et al.*, 1979) is associated with a demineralization of bone. We observed a correlation of −0.50 between body calcium and age in years (Shephard *et al.*, 1985*b*). The process is considered further in Chapter 6.

Anatomical dissections

Cadaver dissection following sudden or accidental death might seem the ideal method of setting norms of human body composition. However, there are certain practical and legal obstacles to this approach. While there is likely to be the least ante-mortem change of body composition in those who die suddenly, such an occurrence requires an inquest and often a forensic post-mortem examination. For these reasons, the anatomist and the student of body composition have often been obliged to accept bodies where the tissues were affected by protracted disease processes. Because of the physical bulk of the tissue involved, an enormous amount of labour is also required to dissect an entire human body.

A substantial amount of the available information on body composition has thus been collected through the dissection of smaller mammalian species (Pearson, 1963; Reid *et al.*, 1963; Weil & Wallace, 1963). Unfortunately, there are substantial inter-species differences in the water, protein, fat and mineral content of the body, making it difficult to extrapolate animal results to human subjects (Pearson, 1963). Moreover, prolonged feeding with a rather unappetizing chow and very restricted movement patterns plus the trauma associated with slaughter can lead to considerable discrepancies between the body composition of a laboratory animal and a free-living member of the same species (Reid *et al.*, 1963). Despite the technical problems involved in obtaining suitable material and in carrying out the necessary dissections, the physical anthropologist thus prefers to base conclusions about human body composition on cadaver studies rather than on animal specimens.

A review of the literature in this area of research was undertaken by Clarys *et al.* (1984, 1987). They were able to supplement their personal series of 25 whole body and 7 regional dissections of human cadavers, by reports describing no more than 16 other subjects. These observations spanned a period of 140 years. It is possible that an exhaustive search of the literature might have unearthed a little more information. For instance, it is unclear whether the early work of Moleschott (1859) or the extensive data on limb musculature published by Frohse & Frankel (1908) were included in the Vierordt (1906) handbook which was cited by Clarys *et al.* (1984). The Belgium team also did not refer to the interesting work of Alexander (1964), which was based on the partial dissection of 86 male and 69 female bodies. However, there can be little argument with their main thesis. Because accurate dissection and subsequent chemical analysis of the body constituents are very time-consuming, and suitable cadavers are difficult to obtain, there is still a dearth of data on directly measured tissue masses (Table 2.3) and their chemical composition (Widdowson *et al.*, 1951).

The overall body size of the average adult has increased substantially over the past 140 years. Unfortunately, some early investigators neglected to indicate the body mass of their subjects, thus limiting the possibility of comparisons between their data and present-day results. There are other difficulties in interpreting available figures. Because of the limited quantity of information that is available, some authors such as

Table 2.3. *Body constituents of cadavers.*

Type of subject	Total (kg)	Skin (%)	Adipose (%)	Muscle (%)	Bone (%)	Residue (%)
Alexander (1964)	—	—	20.0	33.0	17.2	29.8
Other early work (7M, 1F)	64.7	6.8	15.5	39.4	16.8	21.5
Clarys et al. (1984)						
Men: embalmed	61.7	5.7	27.8	36.1	15.0	15.3
unembalmed	70.7	5.4	28.4	38.8	13.5	14.0
Women: embalmed	60.1	6.0	37.0	29.7	13.4	14.0
unembalmed	64.6	5.0	43.6	27.7	12.0	11.7

Note: Values for muscle include ligaments, values for bone include cartilage and marrow; Alexander's data for adipose also include skin.

Clarys *et al.* (1984) have pooled results for men and women, including in their averages data for both embalmed and unembalmed specimens. Given the well-recognized sexual dimorphism of body composition and the substantial impact of embalming upon the fluid content of a corpse, such data treatment seems statistically unwise and potentially misleading. Moreover, the age of the subjects dissected and their causes of death have varied widely (Table 2.4). Indeed, in a number of early reports, the cause of death was unspecified. The ideal arrangement would be to obtain for dissection the fresh bodies of young adults who had been recent victims of accidental death or execution, but in fact the known diagnoses of dissected cadavers have included heart and renal disease (where the body fluid content might have been increased by oedema) and carcinoma (where there might have been substantial ante-mortem wasting). Wasting has a differential effect upon body composition, fat being lost more rapidly than muscle or bone; on the other hand, prolonged ante-mortem bed-rest can also lead to a loss of muscle and bone (Chapter 10). In selecting cadavers for dissection, Alexander (1964) claimed to have avoided the grossly oedematous, the grossly emaciated, and a few subjects where height was difficult to determine. His refined formulae were based on 17 cases of violent death. Clarys *et al.* (1984) likewise stated that they avoided cadavers that were 'severely' emaciated, had physical deformities or had 'obviously deteriorated' subsequent to death. However, such judgements on the condition of a corpse are inevitably highly subjective.

Clarys *et al.* (1984) had some initial hesitation in dissecting bodies that had not been embalmed, on the grounds of odour and an increased risk of infection for the anatomists concerned. Thus the first half of the corpses

Table 2.4. *Conditions leading to death in cadaver series of Clarys et al.* (1984).

Type of disorder	Number of cases
Heart disease	12
Natural causes	4
Renal disease	1
Respiratory disease	1
Neoplasia	4
Accident/suicide	2
Unknown	1

that they dissected were embalmed. They found that the embalming fluid (a substantial 6 l) was distributed unevenly in the corpse. Far from restoring normal hydration (as suggested by Todd & Lindala, 1928), embalming appeared to increase the variance in their results. When selecting bodies for the second half of their sample, Clarys *et al.* (1984) thus rejected embalmed specimens. By recruiting a substantial team of assistants, they were able to complete the dissections within 24 hours; problems of odour and infection were thus avoided.

Another significant problem for the anatomist is dehydration, which develops progressively over the course of a protracted dissection. Clarys *et al.* (1984) minimized water loss by covering the body parts with a sheet of plastic when they were not actively being dissected. Nevertheless, some 2 kg of weight was lost over the day that their 10–15 assistants required in order to complete the dissection of a single cadaver. Their final data were thus scaled upwards in a proportional manner (although it seems inherently unlikely that the rate of evaporative loss was the same for all tissues).

A further problem when looking at regional body composition is to make an appropriate decision on the line to be adopted when severing body parts (Clarys & Marfell-Jones, 1986). In particular, a decision must be made between an incision at the joint line (Clauser *et al.*, 1969) and an arbitrary assignment of whole muscles to the proximal or the distal body part (Grand, 1977). There remains no agreement on an appropriate approach, and this leads to substantial inter-laboratory differences in results.

Alexander (1964) adopted a simple statistical ruse in partitioning the body constituents. He argued that total body mass should be predicted by appropriately weighted measures of body fat (F), muscle (M) and bone (B):

$$\text{Body mass (kg)} = a(F) + b(M) + c(B) + d$$

where a, b, c and d are the constants to a multiple regression equation. We may note immediately that the term d includes both the error of the estimate and a constant allowance for the mass of the viscera and other body tissues.

In his first experiments, Alexander (1964) estimated body fat from the depth of subcutaneous fat observed in three incisions (F, mm), multiplied by the surface area of the body (SA, m^2). Muscle was represented by the summed mass (M, g) of the two psoas muscles and bone by the mass (B, g) of the sternum, corrected for attached intercostal muscle:

Fat (kg) = 0.293 $F(SA)$
Muscle (kg) = 0.061 M
Bone (kg) = 0.033 B
Residue (kg) = 16.0 men, 11.5 (women)

In a refinement of the analysis, the number of subcutaneous fat measurements was increased to five, the mass of the psoas was corrected by an empirical factor based on the subcutaneous fat readings, and bone mass was estimated from stature (H) and the square of four bone diameters (B). The equations then became:

Fat (kg) = 0.173 $F(SA)$
Muscle (kg) = 0.057 $M(1 - F/1000)$
Bone (kg) = 0.107 $HB2$
Residue (kg) = 18.7 (men), 14.5 (women)

Despite the various problems associated with whole cadaver analysis, there is a good general agreement of body composition data between the classical nineteenth-century anatomists and more recent investigators such as Alexander (1964) and Clarys *et al.* (1984), particularly if results are expressed relative to adipose-tissue-free body mass. On the other hand, the earlier samples carried substantially less body fat than the cadavers examined by Clarys *et al.* (1984). What is disconcerting in the recent series is the wide range of inter-individual variation. Thus, the bones have accounted for 16.3–25.7% of adipose-tissue-free body mass in individual subjects, while the muscle component has ranged from 41.9–59.4% of adipose-tissue-free mass. The wide scatter of values has important practical implications for the two component models commonly used when estimating body fat by hydrostatic weighing (see Martin, 1984, described in Chapter 3). It clearly indicates a continuing need to provide a more solid 'gold standard' of body composition against which the various alternative methodologies can be judged.

3   Determination of body fat

General considerations

Problems of semantics

Unfortunately, different investigators have interpreted 'body fat' in differing ways. In his early investigations of body composition, Behnke included some 10% of 'essential fat' in the figure for lean body mass; the remaining tissue was regarded as 'excess fat' (Lesser *et al.*, 1960). The 'essential fat' included the lipid content of the central nervous system, bone marrow and (in females) the mammary glands. More recent estimates suggest it amounts to about 3% of body mass or 20% of body fat in the male, and 9% of body mass or almost 30% of body fat in the female (Lohman, 1981). Other sites of fat accumulation include subcutaneous tissue (normally accounting for about a third of total body fat in both sexes, although varying from 20–70% with age and the degree of obesity), intramuscular depots (about 10% of total body fat in the male and 4% in the female), and the thoracic and abdominal fat depots (about 12% of total body fat in the male and 8% in the female), with a substantial residue of intermuscular fat, 30% in the male and 20% in the female.

Given that it is not possible to reduce total body fat below 4% of body mass without damage to essential structures such as cell membranes, there is a certain logic to the above definitions of 'essential fat' and 'excess fat'. However, many more recent authors have estimated total body fat as the value coinciding with the mass of ether extractable material, whether this is essential to survival or not. The difference between total body mass and fat mass is then the fat-free mass. As Keys & Brozek (1953) and Lohman (1981) have pointed out, the fat-free mass is not identical with lean body mass (since the latter contains some 'essential fat'). Likewise, Brozek *et al.* (1963) have stressed that 'obese tissue' contains material other than chemically extractable fat – particularly the protein and water content of the fat cells. Therefore, whether the density of fat tissue is estimated from the changes of overall body density that occur during an increase or a decrease of body fat stores, or whether the calculation is based upon a cross-sectional comparison between fat and thin subjects, the figure obtained is greater than the density of pure ether-extracted fat. Failure to understand this point has sometimes led to errors in the calculation of fat mass (see, for example, Dauncey *et al.*, 1977).

Choice of reference criterion

Given the problems associated with cadaver dissection (see Chapter 2), the reference criterion adopted to validate alternative methods for the estimation of body fat has often been based upon hydrostatic weighing. However, as we shall see, there are many limitations to the hydrostatic approach. Thus Lohman (1984) and Coombs *et al.* (1986) have prudently suggested using the average of estimates made by three independent techniques (hydrostatic weighing, determination of total body water and determination of total body potassium). Ideally, there should be close agreement between the results obtained by the three procedures (Krzy-wicki *et al.*, 1974; Lukaski *et al.*, 1981; Presta *et al.*, 1983a), and in the future, the reference criterion may well become some mathematical combination of the three data sets (Lohman, 1981), as in the equation of Selinger (1977):

$$\text{Fat } (\%) = 100(2.747/D - 0.714W + 1.146B - 2.050)$$

where D is the hydrostatic estimate of density, W is the body water content, and B the body mineral content.

Whole body estimates of fat

Underwater weighing

General principles
Legend traces the concepts of water displacement and buoyancy back to Archimedes, as he lay in his bath puzzling over a suitable method to detect the adulteration of gold. Practical application to human subjects dates from 1757, when a London physician named J. Robertson wrote of 'bribing ten middling-sized men' to 'duck under water' in order to determine their specific gravity (Falkner, 1963). Many subsequent authors have regarded underwater weighing as the most reliable of available techniques for the estimation of body fat, and alternative procedures such as the measurement of skinfold thicknesses have been judged severely against this supposed 'gold standard' (Wilmore, 1983),

sometimes without considering the possibility of error in the hydrostatic value.

The prime requirements of underwater weighing are a tank that is easy to enter and leave, and is large enough to submerge the subject, a well-stirred system that provides water of a known and uniform temperature, and a chair or cradle of known submerged weight in which the subject can be weighed accurately while underwater. Approximate values for a subject's submerged weight can be obtained using a corner of a swimming pool, particularly if baffles are used to restrict water movement (Katch *et al.*, 1967; Clark & Mayher, 1980), but it is difficult to obtain a representative figure for water temperature in such a setting. Most authors find the prone position gives more consistent results than sitting, presumably because body movement is minimized (Katch *et al.*, 1967; Williams *et al.*, 1984; Warner *et al.*, 1986). Subjects who are unfamiliar with the process may require many submersions to yield consistent data. Mayhew *et al.* (1981) reported the largest of ten readings, but the average of the final three readings usually provides a more stable basis for calculations (Shephard *et al.*, 1985*b*).

Body density D is estimated from the weight in air (W_a, kg) and the weight while submersed (W_s, kg). The latter is best determined by a load-cell with a digital read-out (Akers & Buskirk, 1969); the observer can then read off the average of any oscillations induced by the forced emptying of the lungs underwater. Appropriate allowances must be made for the additional buoyancy contributed by the residual gas volume remaining in the lungs (V_R, l) and gas in the intestines (Greenwald *et al.*, 1969). The latter is often assumed to be a standard figure of 0.1 l (Buskirk, 1961), although it may be a smaller volume in children; a height-scaled value of $0.115 [170.18/h]^3$ l may thus be preferred. Density is then calculated according to the formula:

$$D = W_a/\{[(W_a - W_s)/D_w] - (V_R + 0.1)\}$$

where D_w is the density of water at the temperature of submersion, as obtained from standard tables.

Estimation of residual volume
The residual gas volume accounts for about 2% of the submerged volume, and the accuracy of this value becomes critical to the calculation, where a very small change of density produces a much larger change in the predicted percentage of body fat.

There is an appreciable day-to-day biological variation of residual volume in any given subject; in itself, this can change the estimated body fat content by as much as 1% (Mendez & Lukaski, 1981; Marks & Katch, 1986). Superimposed upon the biological variation are certain technical errors in the measurement of both residual volume and submerged mass. Let us suppose that a person has a body mass of 70 kg when weighed in air. If there is a technical error of 200 ml when measuring a true residual volume of 1200 ml, this will affect the estimate of body density by about 0.0025 units, equivalent to a further 1.5–1.8% error in the estimated percentage of body fat.

Most of the commonly proposed underwater weighing procedures require considerable cooperation from the subjects. As Robertson soon discovered, some volunteers are 'more interested in the bribe than in the experiment'. Some investigators have suggested that because it is difficult to make a complete expiration while submerged, determinations should be made at functional residual volume, 50% of vital capacity (Welch & Crisp, 1958), or at total lung capacity (Thomas & Etheridge, 1980; Weltman & Katch, 1981; Donnelly *et al.*, 1984; Timson & Coffman, 1984; Latin & Ruhling, 1986). Arguments continue on the relative merits of these tactics, but the consensus seems that they all have a rather similar accuracy.

Another way of making the experiment easier for the subject is to avoid submersion of the face during the residual volume measurement (Robertson *et al.*, 1978). A head-out technique has the advantage of allowing the use of tubing with a smaller dead-space (Brandon *et al.*, 1981), although purists have argued that it is difficult to make an accurate estimate of head volume from a limited number of anthropometric measurements.

Possible options for measuring the residual gas volume in the lungs include the dilution of a single breath of a helium gas mixture (Forster *et al.*, 1955), equilibration of the residual gas during the steady-state rebreathing of a helium mixture (Comroe *et al.*, 1962), the rapid rebreathing of oxygen (Wilmore *et al.*, 1980), a calculation based upon the progressive wash-out of nitrogen from the lungs (Wilmore, 1969), and planimetry of chest radiographs (Shephard & Seliger, 1969). Alternatively, residual volume can be predicted from height and age, anthropometric data (Polgar & Promadht, 1971) or vital capacity (Wilmore, 1969; Shephard & Seliger, 1969; Forsyth *et al.*, 1988*b*). Unfortunately, these various procedures yield systematically different estimates of residual volume and thus body fat (Table 3.1), due to such technicalities as the trapping of gas in poorly-ventilated parts of the lungs and solution of helium in lung tissues during the course of equilibration (Forster *et al.*, 1955).

Table 3.1. *Residual volume and resultant estimate of body fat as obtained by various procedures.*

Method	Residual volume (1)		Body fat (%)	
	Men	Women	Men	Women
Helium (sb)	1.76	1.44	9.4	19.5
Oxygen rebreathe	1.24	1.24	12.9	21.2
Nitrogen washout	1.52	1.13	11.11	22.2
*Bass V_R	1.23	0.89	13.2	24.2
*Goldman V_R	1.67	1.63	10.2	18.1
*Wilmore V_R	1.18	1.00	13.5	23.3

Prediction formulae:
Bass $V_R = 0.025\ VC$
Goldman (men) $V_R = 0.027\ (H,\text{cm}) + 0.017\ (A,\text{yr}) - 3.447$
        (women) $V_R = 0.032\ (H,\text{cm}) + 0.009\ (A,\text{yr}) - 3.900$
Wilmore (men) $V_R = 0.24\ VC$
        (women) $V_R = 0.28\ VC$
(Based on data of Forsyth *et al.*, 1988*b*.)

One of the simplest approaches is the single-breath helium method (Forster *et al.*, 1955; Shephard *et al.*, 1958). A single, known volume of inspirate is held in the lungs for 10 seconds, and the volume of lung gas plus inspirate is calculated from the dilution of helium. This approach has the important advantage that the required period of subject cooperation is quite brief, although some people find difficulty in holding a steady inspiration for 10 seconds, particularly while underwater. In young adults, single-breath results agree quite well with predictions based upon steady-state helium equilibration. On the other hand, the findings could be misleading in an older person with obstructive lung disease, as there is no warning that gas in some parts of the lung is failing to mix with the inspirate (Shephard *et al.*, 1958; Forsyth *et al.*, 1988*b*). The method most commonly used at present is probably that of rapid oxygen rebreathing (Wilmore *et al.*, 1980).

In young children, there is considerable difficulty in obtaining adequate cooperation during any method of measuring residual volume, and empirical data show that it is better not to attempt such measurements; the correlation between hydrostatic weighing and other estimates of percentage body fat is increased if residual volume is assumed to be a fixed 24 or 25% of vital capacity (Wilmore, 1969; Rajic, Shephard *et al.*, unpublished data, 1974), or is predicted from anthropometric measurements (Polgar & Promadht, 1971). In contrast, Sinning (1974) found that in adults an additional error of 0.003 density units was introduced if the residual volume was assumed or was predicted from the vital capacity rather than being measured directly. Plainly, significant errors become likely if standard formulae for the prediction of body density are applied to older and more obese individuals or to subjects with chronic obstructive lung disease (Hackney & Deutsch, 1985; Latin & Ruhling, 1986).

A further unresolved issue is the extent to which lung volumes are compressed by the extra-thoracic pressure of the water during submersion (Craig & Ware, 1967; Girandola *et al.*, 1977a; Sawka *et al.*, 1978; Brandon *et al.*, 1981; Ostrove & Vaccaro, 1982; Hsieh *et al.*, 1985). In practice, the chest is commonly submerged to an average depth of 50 cm during hydrostatic weighing; this creates an external pressure of about 5% of an atmosphere, enough to reduce the residual volume by a barely measurable 50–70 ml (Table 3.2, Jarrett, 1965; Agostini *et al.*, 1966; Bondi *et al.*, 1976; Sawka *et al.*, 1978; Ostrove & Vaccaro, 1982). In some instances, congestion of the pulmonary vessels may further minimize this small effect (Shephard *et al.*, 1958; Craig & Ware, 1967; Arborelius *et al.*, 1972; Dahlback & Lundgren, 1972; Sloan & Bredell, 1973; Prefaut *et al.*, 1976; Robertson *et al.*, 1978), and in some instances residual volumes have even increased while the subject is submersed (Girandola *et al.*, 1977a; Brandon *et al.*, 1981). However, compression errors become larger if subjects are short and stout, or if measurements are made at total lung capacity (Weltman *et al.*, 1987a).

The theoretical gain of precision associated with a submerged measurement of residual volume (Timson & Coffman, 1984) is often more than offset by the loss of accuracy associated with undertaking respiratory manoeuvres while underwater. Ross *et al.* (1986) stressed the difficulty in obtaining a complete exhalation while submerged. In consequence, all subjects show a prolonged learning curve; the recorded underwater weight and the resultant estimate of body density can increase over as many as ten trials (Katch *et al.*, 1967; Katch, 1969). It is not surprising that the residual volume measured on dry land provides a fair approximation to the required residual gas volume (Girandola *et al.*, 1977a). Indeed, if a systematic discrepancy is observed, the dry-land value sometimes provides a more accurate figure than that obtained while the subject is submerged. A third option (Fryer, 1960; Young *et al.*, 1963; Brandon *et al.*, 1981; Donnelly & Sintek, 1986), is to make measurements of lung volumes with the head out of water, subsequently applying a correction for the volume of the head. For example, Donnelly & Sintek (1986) proposed a correction factor of:

$$4.61 - 0.2476 \, (H_W) - 0.1829 \, (H_L)$$

Table 3.2. *Differences of residual volume measured in air and while submerged.*

Vol in air (l)	Vol in water (l)	Diff (l)	Authors
1.67	1.39	−0.28	Agostini *et al.* (1966)
1.58	1.45	−0.13	Brozek *et al.* (1949)
1.92	2.02	+0.10	Carey *et al.* (1956)
1.44	1.38	−0.06	Craig & Ware (1967)
1.25	1.15	−0.10	Jarrett (1965)
1.43	1.52	+0.09	Girandola *et al.* (1977)
1.69	1.22	−0.47	Iltis *et al.* (1985)
1.58	1.37	−0.21	Weighted mean

(Based in part on data accumulated by Craig & Ware, 1967; see original paper for details of references.)

where H_W was the width of the head, and H_L its length. This formula had a standard error of the estimate of 0.27 kg.

It is quite important to minimize the volume of unmeasured gas elsewhere in and around the body. For 24 hours prior to submersion, items which encourage the production of intestinal gas (for example, beans) should be excluded from the diet. Measurements are most reproducible if taken in the fasting state (Durnin & Satwanti, 1982). The prior consumption of food can change the estimated body fat content by up to 1% (Durnin & Satwanti, 1982), and figures can also be modified temporarily by hyperhydration or hypohydration (Girandola *et al.*, 1977*b*). As a minimum, testing should be undertaken several hours after a meal, with the subject urinating and defaecating immediately before measurements are made (Noble, 1986). During measurements, the subject should avoid wearing a bathing hat or a rigid brassière (which could trap gas during submersion).

Estimation of percentage of body fat

The body density as calculated by the equation presented at the beginning of this section is converted to a corresponding percentage of body fat on the assumption that the body comprises two homogenous components, each having a consistent density – fat and fat-free tissue (Siri, 1961).

In reality, body fat comprises a complex mixture of glycerides, sterols, glycolipids and phospholipids; as some of these constituents are more labile than others, it is likely that the average density of 'fat' varies, both from one person to another, and within a given person at different times. However, this complication is usually ignored in densitometric calculations.

A fat density of 0.9000 g/cm³ is commonly assumed (Fidanza *et al.*, 1953; Schemmel, 1980); this figure is an appropriate average figure for the density of ether-extractable material at a temperature of 37 °C. The local water temperature may deviate from 37 °C even if it is well-stirred, and it should be noted that a temperature decrease of only 1 °C increases the average density of adipose tissue fat from 0.9000 to 0.9007 g/cm³. Moreover, the density of the ether-extracted cell and interstitial fat obtained from muscles is about 0.93 g/cm³, while that of brain material is as high as 1.03 g/cm³ (Mendez *et al.*, 1960). *In vivo,* the average fat density thus depends on the relative contributions of the several body sources of fat; a typical mixture may have a density as high as 0.915 g/cm³ (Brozek *et al.*, 1963). Taking the mean of values reported in four earlier studies, Leonard *et al.* (1983) adopted a figure of 0.9168 g/cm³.

With regard to the non-fatty tissues, Nadeshdin (1932) reported an average density of 1.049 g/cm³ for muscle at 11.5 °C (equivalent to 1.043 g/cm³ at 37 °C); the corresponding figure for liver was 1.059 g/cm³, while that for nervous tissue was 1.035 g/cm³. These figures were based on the results of some 200 autopsies. Bone, of course, has a much higher density. Lohman (1986) thus estimated a mean value for the density of the lean compartment of 1.080 g/cm³ at an age of 10 years, rising to 1.100 g/cm³ in the adult. The last figure is simple and easy to remember, and perhaps for this reason it is widely used in calculations (Schemmel, 1980). It rests largely upon the data of Brozek *et al.* (1963) for a 'reference man' having an overall body density (D) of 1.064 g/cm³ and a body fat content of 15.3% (Table 3.1). Leonard *et al.* (1983) have recently adopted a figure of 1.0997, based on the average of earlier studies.

The cadaver studies of Clarys *et al.* (1984) suggested that there is also a wide variation in the relative proportions of constituent tissues; muscle accounted for 41.9–59.4% of the total adipose-tissue-free mass and bone accounted for a further 16.3–25.7%, leaving a residue of 24.0–32.4% attributable to other tissues.

The density of muscle at any given age is relatively consistent (for example, in the young adult it has been suggested that there is a variation of only about 1% around a mean value of 1.05 g/cm³). However, the density of the lean-tissue compartment shows short-term changes with variations in the hydration of the body (Robinson & Watson, 1965; Edholm *et al.*, 1974; Girandola *et al.*, 1977b). Over the course of a menstrual cycle, the total body mass of the average woman changes by 1–2 kg (Taggart, 1962) and hydration-related fluctuations in lean tissue density can give an apparent change in body fat content of 0.4–4.0% (Svoboda & Query, 1986). Other possible factors temporarily influencing lean tissue density and thus estimates of body fat content are a temporary energy imbalance and irregular bowel habits (Edholm *et al.*, 1974).

The density of bone varies very widely, from about 1.15–1.60 g/cm^3 in different parts of the skeleton; Ross *et al.* (1986) reported an average figure of 1.236 g/cm^3. The density of the bone mineral component is as high as 3.00 g/cm^3, and the progressive loss of bone mineral with aging leads to a decrease in body density of some 0.020 g/cm^3 per decade of life. If 25% of lean tissue mass is bone, the density of the lean compartment is changed by 0.005 g/cm^3 per decade of life, and if no allowance is made for this, the estimated percentage of body fat changes by 2.5% per decade.

Commonly accepted formulae

Despite the various theoretical problems noted above, the usually accepted densitometric formulae are based on the assumptions of an average density of 0.9000 g/cm^3 for 'fat' and 1.100 g/cm^3 for the fat-free mass. The percentage of fat (F, %) is then given by a hyperbolic equation of the type (Siri, 1961):

$$F = (4.95/D - 4.50)100$$

for a fat density of 0.900 g/cm^3 or making the alternative assumption of a fat density of 0.9007 g/cm^3, the equation would become:

$$F = (4.971/D - 4.519)100$$

One immediate source of difficulty with both of these formulae is that when adipose tissue is laid down by over-eating, or is removed by dieting, the densities of the fat and lean tissue which are gained or lost may not conform to this simple two compartment model.

If the body is considered as having an average density of 1.064 g/cm^3, modified by the addition or subtraction of excess fat (density 0.9007 g/cm^3), a third option for the calculation (Brozek *et al.*, 1963) becomes the equation:

$$F = (4.57/D - 4.142)100$$

Predictions of body fat using the three alternative formulae agree with each other to within 1% over the density range of 1.03–1.09 g/cm^3, but in very fat subjects the Siri equations give substantially higher estimates than the Brozek formula.

Based upon this concept of 'excess fat', Behnke & Wilmore (1974) have suggested a fourth equation:

$$F = (5.053/D - 4.614)100$$

For a woman with an overall body density of 1.045 g/cm^3, the four formulae discussed to this point yield respective body fat estimates of 23.7, 23.8, 23.1 and 24.5%. Other proposed formulae yield more disparate results:

$$F = (5.548/D - 5.044)100 \text{ (Rathbun \& Pace, 1945; } F = 26.5\%)$$

$$F = (4.0439/D - 3.6266)100 \text{ (Grande, 1961; } F = 23.3\%)$$

$$F = (5.50 - 5D)100 \text{ (MacMillan } et\ al., 1965; F = 27.5\%)$$

Population-specific equations

One obvious difficulty in deciding the respective merits of the seven rival equations is that the density of individual body constituents changes with growth, aging and environmental factors. Most of the current formulae were developed on young adult males. The bones are less dense in children, women and older subjects of both sexes (Lohman *et al.*, 1984*a,b*), leading to an overestimation of body fat by all of the standard formulae (Weredin & Kyle, 1960). Thus, Lohman *et al.* (1984*a,b*) proposed a revised formulae for pre-adolescents:

$$F = (5.30/D - 4.89)100$$

Conversely, athletes have dense muscle and bone, leading to an underestimation of body fat (MacDougall *et al.*, 1982). Adams *et al.* (1982) noted that in 8 of 22 professional football players, body densities in excess of 1.1 g/cm^3 led to an apparently negative body fat content! The relative contributions of muscle and bone to fat-free mass also change with age, and the average density of 'lean tissue' further depends on the hydration of the body (as modified by exercise and exposure to heat or cold). Siri (1961) suggested that the fat content of a reference subject would show an overall standard deviation (SD) of 3.8%, reflecting variations in water content (±2.7%), protein/mineral ratio (±2.1%), mean fat content of obese tissue (±1.8%), obese tissue density (±0.5%) and the mean fat content of the reference person (±0.5%). Bakker & Struidenkamp (1977) set the cumulative effect of biological variations in tissue water, skeletal density and composition of essential fat at a similar overall SD of 3.4% fat; they noted that the effect of inter- and intra-individual variations in skeletal density was exacerbated because the skeleton also accounted for a variable fraction of the total lean tissue. Martin (1984) supported this concern, finding twice the previously reported range of variation in bone densities. Pooling results for men and women, Martin (1984) found that the skeletal mass was quite closely correlated with wrist diameter ($r = 0.875$). The best multiple regression equations for the prediction of skeletal mass from bone dimensions were:

$$\text{Skeletal mass (men)} = 28.0A + 0.482B + 1.38C + 4265$$
$$(r^2 = 0.98)$$

$$\text{Skeletal mass (women)} = 0.182D - 6.42E + 1.15F + 787$$
$$(r^2 = 0.79)$$

where

$A = (\text{wrist width})^2 \times \text{ankle width}$
$B = \text{head girth} \times \text{humerus width} \times \text{biacromial width}$
$C = \text{head girth} \times \text{humerus width} \times \text{femur width}$
$D = \text{head girth} \times \text{stature} \times \text{wrist width}$
$E = (\text{femur width})^2 \times \text{wrist width}$
$F = (\text{humerus width})^2 \times \text{ankle width}$

Technical errors

The several sources of biological variation can be compounded by technical errors in the body density measurements. In what we must regard as the easiest group to study (young men), Durnin & Taylor (1960) claimed that body density determinations were 'reproducible' to within 0.0023 units, equivalent to a difference of 1.1% fat if body density is 1.065 g/cm^3.

Such data do not assure the absolute validity of hydrostatic estimates. There have been attempts to check underwater weighing against figures for eviscerated carcasses of animals, although the interpretation of such comparisons is clouded by problems related to both ante-mortem and post-mortem changes in the tissues (see Chapter 2). The supposed 'gold standard' of hydrostatic weighing thus remains a somewhat fallible criterion. One sign of its fallibility is that simple indices of obesity such as total body mass and circumferences sometimes show a closer correlation with skinfold data than with the percentages of body fat as estimated by underwater weighing (Murray & Shephard, 1988).

Water displacement

The principle of the water-displacement procedure is similar to that of underwater weighing. However, the volume of displaced water is estimated not from the decrease of body weight while submerged, but from the alteration of water level in an angulated manometer scale attached to the submersion tank (Garn & Nolan, 1963; MacDougall *et al.*, 1982). Data collection is again facilitated by construction of a rigid-walled and not too large tank with easy entry and egress. If the tank is large, the change of water level may be quite modest and difficult to read accurately. Keys & Brozek (1953) and Durnin & Taylor (1960) both set the 90% limits of a single assay at ±0.005 density units. In a woman with a true density of 1.045, this could decrease the apparent percentage of body fat from 23.8 to 21.5%. The other issues concerning the interpretation of body density are, of course, as for hydrostatic weighing.

Air displacement

An alternative approach to the estimation of body volumes and thus body density is to measure the volume of air displaced by the subject's body (Lim, 1963; Gnaedinger *et al.*, 1963; Gundlach *et al.*, 1980). Noyons & Jongbloed (1935) weighed the body at two different ambient gas pressures. Unfortunately, the decrease in body weight was only 30 g on passing from an ambient pressure of 760 to 1100 Torr; as in the early experiments of Santorio (Chapter 1), the change was within the combined errors arising from respiratory and cutaneous water loss and inaccuracies of the weigh-scale. Gnaedinger *et al.* (1963) measured the increase of pressure in a reference chamber after an animal had been introduced, but they also found a very poor correlation between air-displacement and underwater weighing data; readings of air temperature, pressure and humidity were not sufficiently accurate to estimate the displaced gas volume precisely.

The more usual air-displacement approach has been to enclose a human subject in an air-tight box, determining the change in air content of the box by dilution of an injected marker gas, such as helium (Siri, 1956*a*; Fomon *et al.*, 1963), or the increase of box pressure in response to injection of a known volume of air (Falkner, 1963; Schmid & Schlick, 1976; Gundlach *et al.*, 1980; Gundlach & Visscher, 1986). The problems associated with this type of methodology are well-recognized by the body-box plethysmographists of respiratory physiology. Beeston (1965) found air leaks in 4 of 14 experiments. In addition to the over-riding need for an air-tight system, difficulties arise from a rapid rise of box temperature and humidity. Once the subject is shut in the box, there is a progressive absorption of oxygen, with excretion of a differing volume of carbon dioxide; such changes can affect both helium and pressure estimates of body volume. The helium method assumes further that there is no solution of the marker gas in body tissues (although the lungs usually absorb about 100 ml of helium). All of the air displacement methods also assume a fixed volume of intestinal gas (Bedell *et al.*, 1956).

Falkner (1963) suggested three tactics that would improve the accuracy of pressure measurements – the subject was allowed a preliminary period of thermal equilibration, the box was fitted as closely to the subjects as possible, and any continuing pressure change was extrapolated back to zero time. Nevertheless, he admits to 'considerable trouble with air leaks (and) manometers . . .' More recently, Gundlach *et al.* (1980) suggested that technical problems from heating and humidification could be largely overcome if the residual space in the box was filled with polyurethane foam. In their view, the enclosed volume could then be considered as obeying Boyle's law. However, their published report indicated a surprisingly large standard error in the estimated body volume (1.4 l). In a woman with a body mass of 65 kg and a true density of 1.045, such figures would yield density readings ranging from 1.022–1.069, corresponding to a fat percentage of 13.1–34.5%. Further technical improvements of the body-box approach (Gundlach & Visscher, 1986) have now reduced the standard error of the body volume estimate to a more acceptable 0.33 l; nevertheless, the estimated body fat would still range from 21.3–26.3% in the example cited.

Another option (Garrow *et al.*, 1979) is to combine the water displacement method (for the body) with air displacement (for the head). An SD of only ±0.3 kg fat has been claimed when replicate estimates have been made by this technique. A final gasometric possibility currently undergoing investigation is to test the shift in acoustic resonance frequency of the body-box when it is occupied by the subject (Deskins *et al.*, 1985).

Anthropometric estimates of volume

Sady *et al.* (1978) and Freedson *et al.* (1979) have suggested that total body volume could be approximated from anthropometric measurements made on ten body segments, using three basic geometric shapes (a right-angled parallelepiped, a frustum of a pyramid and a truncated right circular cone). The coefficient of correlation with hydrostatic measurements was high (0.96 in females, 0.98 in males), but there were substantial errors, both systematic and random, relative to the criterion (−5.46 ± 1.86 l in women, 1.74 ± 1.88 l in men). Plainly, from the calculations made in the previous section, the accuracy of this method is too low to provide useful information about body composition.

Raja *et al.* (1978) suggested the further possibility that whole body volume could be approximated from limb volume:

Total body volume (l) = 8.48 + 4.62 leg vol.

$$(r = 0.94, \text{ SE } \pm 3.26 \text{ l})$$

$$= 4.38 + 19.17 \text{ arm vol.}$$

$$(r = 0.93, \text{ SE } \pm 3.49 \text{ l})$$

Again, the error is so large that this approach must be dismissed by those interested in body composition.

Inert gas absorption

A number of inert and/or anaesthetic gases dissolve selectively in body fat. Thus, by determining the rate of gas uptake or the ultimate volume absorbed, an estimate of body fat content can be made (Lesser & Zak, 1963; Lesser *et al.*, 1971). Early studies on rats used propane as the marker gas (Lesser *et al.*, 1952), while Behnke *et al.* (1942) applied a similar approach to the solution of nitrogen in human fat. More recent investigators have also used cyclopropane, krypton, xenon and 85Kr (Lesser & Zak, 1963; Hytten *et al.*, 1966a; Mettau *et al.*, 1977).

In animals, the mean of the propane results agrees with the amount of ether-extractable fat to within 5% (Lesser *et al.*, 1952), but when the gas absorption method is applied to humans, problems arise from differences in the rate of uptake by different types of fatty tissues. In consequence, poorly-perfused body compartments do not attain equilibrium within an acceptable period of rebreathing; indeed, uptake of the foreign gas is still continuing at the end of eight hours. Lesser & Zak (1963) proposed using a mixture of cyclopropane and krypton. They hoped that the differing solubilities and diffusion properties of the two gases would allow the calculation of an accurate averaged value for fat mass. However, their data are again sufficiently imprecise to dismiss the method; there was an average difference of almost 2 kg in estimates derived from solution of the two gases, and in one of eight subjects the discrepancy was as large as 15.4 kg!

Photogrammetry

Sterner & Burke (1986) suggested that expert raters could assess the obesity of human subjects almost as easily from photographs as from skinfold measurements.

Interest in more formal photogrammetry can be traced to the photo mapping of military targets during World War II; more recently, the method has been applied to satellite photography and robotics. The technique has occasionally been applied to the determination of body volume (Pierson, 1963; Baumrind, 1986), although the procedure is complicated, expensive, and difficult to carry out on a monochromatic surface such as human skin.

One simplification of the method has projected contours upon the body, using side-illumination which is passed through coloured but transparent strips of equal width. The volume of the subject is then given by: $V = (a + 2b, 2c \ldots + 2n)/2$(contour interval), where a is the area encompassed by the zero contour line. The main technical difficulty with the simplified photogrammetric method is that the intensity of light must be bright enough to allow a photograph to be taken with a brief exposure, without being so bright as to obscure the differences of colour between individual contours.

Pierson (1963) claimed that his method had an error of less than 2% relative to water displacement. More direct validation of the technique by the photographing and subsequent dissection of a cadaver is difficult, because the skin stretches when it is removed, and refrigeration also changes the volume of the body fluids.

Electrical impedance

The electrical impedance technique was first developed to examine regional blood flow, and at various times has also been used to look at cardiac output and lung volumes; recently it has attracted attention as a possible simple method of estimating body fat.

Impedance comprises both resistance (R) and reactance (X), and in theory is calculated as the square root of $R^2 + X^2$; however, under the usual conditions of body composition determination, the reactance term is small, and resistance is used as the surrogate of impedance. The physical principle then becomes quite simple. At any given electrical frequency, the impedance between the ankle and the wrist is directly correlated with the distance separating the measuring electrodes, and is inversely correlated with the cross-section of interposed lean tissue. Fat, with a water content of only 14–22%, has a much higher impedance than does the lean tissue compartment, so that for any given body dimensions a fat person conducts the electrical signal much less readily than a thin individual. Estimates of body fat can thus be made by applying between ankle and wrist electrodes a weak (100–800 μA) signal in the frequency range 25–100 kHz (Nyboer, 1970; Pethig, 1979; Presta *et al.*, 1983*a,b*; Segal *et al.*, 1985; Kushner & Schoeller, 1986; Lukaski *et al.*, 1986; Guo *et al.*, 1987).

An obvious difficulty arises from the complexity of body shape. Because impedance increases as the reciprocal of the cross-section of a part, the impact upon the recorded impedance is much greater for the upper arm and calf than for the trunk and abdomen (Baumgartner *et al.*, 1987). Moreover, commercially available apparatus automatically adjusts impedance data for the square of stature (Hoffer *et al.*, 1969), apparently ignoring potential differences in the distance between ankle and wrist electrodes, and raising issues of the proportionality between height and effective cross-section of the conducting tissues. Possibly for this reason, systematic differences of body fat estimates have been observed between lean and obese subjects, with over-prediction occurring in lean individuals (Van Loan & Mayclin, 1987; Hodgdon & Fitzgerald, 1987).

Nevertheless, since the impedance method essentially relies upon the conductance of *lean tissue*, it works better in lean than in over-weight subjects (Helenius *et al.*, 1987). Carr *et al.* (1985) noted that the impedance technique provided a low estimate of fat loss relative to hydrostatic weighing in subjects who were dieting.

The test/retest correlation for whole-body impedance measurements is quite high (0.99 over five days, Lukaski *et al.*, 1985). Lawlor *et al.* (1985) reported a correlation with hydrostatic estimates of body fat amounting to 0.88 in men and 0.82 in women. Segal *et al.* (1985) found a disappointingly large standard error of the estimate (SEE; 6.1% body fat), but Lukaski *et al.* (1986) reported a more acceptable SEE of 2.7% fat.

Guo *et al.* (1987) tested the unique value of the impedance measurements in a series of multiple regression equations that also included body mass, skinfold readings and circumferences. They proposed the equations:

$$F = -18.00 - 0.279\,(H^2/R) + 0.346\,M + 0.632\,T\ \text{(men)}$$

and

$$F = -8.48 - 0.845\,(H^2/R) + 0.383\,M$$
$$+ 0.430\,B + 1.341\,C\ \text{(women)}$$

where F is the percentage of body fat, H is the standing height (cm), R is the impedance (ohms), M is the body mass (kg), T is the triceps skinfold thickness (mm), B is the biceps skinfold thickness (mm) and C is the calf circumference (cm). The impedance term made a significant independent contribution to the description of variance in the women, but not in the men. While some authors favour the use of such relatively complex equations, others have suggested that these equations do not improve impedance predictions of body fat (Hodgdon & Fitzgerald, 1987), and that if multiple regression equations include body dimensions, the impedance data add little to simple anthropometric measurements (Helenius *et al.*, 1987). Certainly, the impedance equipment is quite expensive relative to its accuracy, and the fundamental problem of the complexity of body shape can hardly be overcome by improvements in design.

Electrical conductivity

The electrical conductivity technique has certain similarities with the impedance method of estimating body fat, and suffers from some of the same limitations. Like several of the other available methods of examining tissue composition, it was originally developed for the meat industry.

In essence, the observer notes changes in a 2.5–5.0 MHz electromagnetic field induced by the conductivity of interposed body electrolytes (Presta *et al.*, 1983*a,b*; Harrison, 1987). The subject, lying on a stretcher, is placed inside a large electromagnetic coil. Energy is absorbed and is subsequently emitted as heat. The extent of such absorption depends on the quantity of conducting material present in the tissue (Klish *et al.*, 1984), although it should be stressed that the amount of energy involved reaches only a small fraction of current US safety standards for microwave exposure. Van Itallie *et al.* (1985) suggested that results were influenced by the orientation of the body within the coil. However, Harrison (1987) found no effect of body geometry.

At low frequencies, most of the current passes by extracellular pathways. At higher frequencies, the current crosses cell membranes. Thus future designs of electromagnetic conductivity measuring equipment may have the ability to distinguish intra- and extra-cellular water, simply by varying the frequency of the applied oscillation in the electromagnetic coil.

Domeruth *et al.* (1976) reported a good correlation between conductivity data, ^{40}K estimates and carcass analyses of pigs. Over a ten minute period, the test/retest correlation coefficient for the conductivity measurements was as high as 0.98, while correlations of 0.92 and 0.87 have been reported relative to underwater weighing and total body water determinations respectively (Presta *et al.*, 1983*a,b*). Segal *et al.* (1985) found that relative to densitometric estimates of total body fat, the presently available electromagnetic method yielded a closer correlation ($r = 0.96$) than did the impedance method ($r = 0.91$). Harrison (1987) also claimed a close correlation ($r = 0.95$) between electromagnetic and densitometric results; the standard error of the electromagnetic determination was about 3.5%, a figure comparable with many of the other indirect methods of determining body fat.

Regional estimates of body fat

Skinfolds

Nature of skinfolds and contribution of skin

The skinfold, as normally measured by the physical anthropologist, comprises a double fold formed from the skin and (at least in theory) all of the underlying subcutaneous fat. The thickness contributed by the double layer of skin can coveniently be examined on the dorsum of the hand, where there is almost no subcutaneous fat (Roberts *et al.*, 1975). A typical value for the back of the hand in a young adult is 1–3 mm (Edwards *et al.*, 1955; Lee, 1957; Lee & Ng, 1965; Shephard & Memma, 1967). It is less clear that this same reading can be extrapolated to indicate skin thicknesses in other parts of the body. Indeed, recent cadaver studies (Clarys *et al.*, 1987) suggest substantial regional differences; the skin is thinnest over the biceps (0.8 mm in men and 0.5 mm in women) and is thicker over the trunk (2.1 mm in men, 1.7 mm in women), reaching maximum dimensions in the soles of the feet. In general, values are greater in men than in women, and decrease with age (Clarys *et al.*, 1987); in regions such as the palms, the skin may also be thickened by the mechanical stresses of hard physical labour.

The fat content of the skinfold can range all the way from 5–94% of total thickness, although it is typically in the range 60–85%. In the very thin people of developing countries, the skin itself may account for a substantial fraction of the total skinfold reading; as adiposity increases, so does the fat content of the adipose tissue (Pawan & Clode, 1960; Thomas, 1962).

Differences in the composition of the skinfold reflect not only the relative proportions of skin and subcutaneous fat, but also variations in the water content of the adipose compartment (Bliznak & Staple, 1975;

O'Hara *et al.*, 1979; Martin *et al.*, 1985). Increases in water content are seen with the oedema of malnutrition (Grant *et al.*, 1981), although somewhat surprisingly, Womersley & Durnin (1973) observed no variations of skinfold readings over the menstrual cycle.

Instrumentation

The observed skinfold readings depend on both the design of caliper that is used, and the period for which pressure is applied (Sloan & Shapiro, 1972). Edwards *et al.* (1955) recommended that the caliper jaws should exert a standard pressure of 10 ± 2 g/mm2, compressing the skin for no more than two seconds (Le Bideau & Rivolier, 1958); more recently, Becque *et al.* (1986) have allowed four seconds of compression. Some early studies used the Best caliper (30 g/mm2, Parizkova & Goldstein, 1970; Parizkova & Roth, 1972), and other non-standard measuring devices.

In practice, even the widely accepted Harpenden calipers can depart from recommended pressures; thus Leger *et al.* (1982) observed values ranging from 7.6–8.2 g/mm^2. Fortunately, the measured thickness of the skinfold does not seem to vary greatly between pressures of 9 and 20 g/mm^2 (Keys & Brozek, 1953; Sloan & Shapiro, 1972; Leger *et al.*, 1982), although a practical upper limit of 15 g/mm^2 is set by pain and increasing inconsistency of the readings (Keys & Brozek, 1953; Behnke & Wilmore, 1984). A face area of 35 mm^2 is recommended, although again, minor variations from this figure do not seem to change skinfold readings systematically (Edwards *et al.*, 1955). Sloan & Shapiro (1972) and Womersley *et al.* (1973) found good agreement of results between the Harpenden and the Lange calipers, but Lohman *et al.* (1984c) reported that skinfold thicknesses were 1–2 mm greater with the Lange than with the Harpenden instrument; in the series of Lohman *et al.* (1984), the SE of results was greater for the Lange than for either the Harpenden or the Holtain calipers.

In recent years, inexpensive plastic calipers such as the Ross and the McGaw instruments have appeared (Sloan & Koeslag, 1973; Burgert & Anderson, 1979; Leger *et al.*, 1982). These have some attractions for large-scale field surveys; they yield fairly satisfactory results when used by experienced investigators, but are less reliable than the precision Harpenden, Lange and MNL instruments (Sloan & Shapiro, 1972), particularly if the skinfold mesurements are made by those with limited experience (Lohman & Pollock, 1981; Lohman *et al.*, 1983); for example, the Ross caliper gives a 16% overestimate relative to standard devices.

Measurement site

The site of measurement must be defined closely, as both the compressibility of the skinfold (Brozek & Kinsey, 1960; Clegg & Kent, 1967; Lee & Ng, 1965; Himes *et al.*, 1979) and its ultimate thickness (Edwards *et al.*, 1955; Hammond, 1955) can change dramatically if the measuring calipers are displaced by a few centimetres in a proximal, distal, medial or lateral direction (Ruiz *et al.*, 1971).

Site locations must be based on an accurate description of the appropriate anatomical landmarks and/or clear photographs, with careful surface measurements and marking of appropriate sites (Katch & Katch, 1984; Weiner & Lourie, 1981; McNair *et al.*, 1984; Jackson & Pollock, 1985). A given fold must be picked up along the line of the natural skin creases, taking care not to include any underlying muscle, and throughout the measurement the fold must be supported by the fingers at a standard distance of 1 cm from the measuring calipers.

Compressibility

One practical source of difficulty, soon recognized by early observers, was the compressibility of all skinfolds (Fletcher, 1962; Booth *et al.*, 1966; Orpin & Scott, 1984). Most authors have reported this compressibility in terms of the ratio:

(Uncompressed − Compressed) : Uncompressed

but Martin *et al.* (1985) related the compressed caliper readings to direct cadaver measurements:

(Incised depth − caliper reading) : incised depth

Because compression develops progressively, it is important to take skinfold readings rapidly (Himes *et al.*, 1979; Becque *et al.*, 1986). In neonates, the skinfold readings continue to decline exponentially over a period of at least 60 seconds (Brans *et al.*, 1974), while in adults a two component exponential model gives a better description of the process (Becque *et al.*, 1986).

There are regional, sex and age-related changes in the compressibility of the skinfolds (Brozek, 1965; Clegg & Kent, 1967; Lee & Ng, 1965; Clarys *et al.*, 1987). In males, but not in females, the iliac site is more compressible than most others (Becque *et al.*, 1986). The supraspinal and biceps folds show a compressibility of about 60%, while the thigh and medial calf folds have a compressibility of only 30%. At any given site, the extent of compressibility is affected by tissue hydration (Laven, 1983) and nutritional status; after dieting, the subcutaneous tissues seem particularly prone to deformation by the measuring calipers.

Reliability

Weiner & Lourie (1981) commented 'skinfolds are notoriously difficult to measure; a class exercise of test–retest reliability will demonstrate this'. Problems of poor reproducibility are particularly acute when measurements are attempted in the grossly obese (Ashwell *et al.*, 1985); in such subjects, an alternative method of assessment such as the measurement of circumferences may be desirable.

Depending in part on the extent of true intra-population differences, intra-tester correlations for experienced observers can reach 0.92–0.98 (Tanner & Weiner, 1949; Keys & Brozek, 1953). Reliable skinfold measurements are possible after as few as two practice and orientation sessions (Table 3.3). Depending on the sex of the subject, the degree of obesity and the site of measurement, observers who have been trained in the same laboratory can generally duplicate the readings of their colleagues to within 1–2 mm. An electronic fat analyser (Syntex) has recently been developed that automatically disregards all paired readings that do not replicate to 1 mm.

Table 3.3. *Intra-observer reliability of skinfold and circumference readings. Mean values for test and retest measurements in women and men, with probability of difference between the two sets of observations (independent of sex) by paired t-test.*

Variable	Females		Males		Probability of significant test/retest difference
	Test	Retest	Test	Retest	
Skinfolds[1]					
(1) Triceps	17.1	17.2	10.0	10.1	0.97
(2) Subscap.	11.8	11.9	11.3	11.4	0.94
(3) Suprail.	13.5	13.8	21.3	21.9	0.92
(4) Abdominal	11.7	11.8	17.5	16.6	0.91
(5) Ant. thigh	25.3	25.5	15.0	15.5	0.93
Sum of 1–3	42.3	43.0	42.6	43.3	0.92
Sum of 1–5	79.4	80.3	75.0	75.4	0.96
Circumferences[1]					
Knee	36.3	36.0	—	—	0.82
Gluteal	95.9	96.0	—	—	0.97
Abdomen	73.3	72.7	89.4	88.9	0.90
Wrist	14.9	14.8	17.6	17.6	0.96
Flexed up. arm	—	—	34.7	34.4	0.83
Upper arm	—	—	32.0	32.4	0.96
Neck	—	—	38.8	38.8	0.96

[1] Skinfold measurements are in mm.
[2] Circumference measurements are in cm.
(Based on unpublished data of Forsyth, Plyley & Shephard.)

Nevertheless, difficulties can arise from systematic errors between investigators, particularly if they have been trained in different laboratories and still have only limited personal experience of measuring skinfolds (Edwards *et al.*, 1955; Crook *et al.*, 1966; Munro *et al.*, 1966; Ruiz *et al.*, 1971; Womersley & Durnin, 1973; Burkinshaw *et al.*, 1973; Jackson *et al.*, 1978; Prahl-Andersen *et al.*, 1979; Oppliger *et al.*, 1987). Systematic inter-laboratory differences can amount to as much as 7 mm for the triceps fold (Lohman, 1981). The most reproducible sites of measurement are the subscapular fold in men, and the subscapular and triceps in women, while the least reproducible are the suprailiac, the abdomen and thigh, and in women the biceps (Womersley & Durnin, 1973; Lohman *et al.*, 1984c).

Interpretation

Skinfolds may be interpreted in their own right, as an indicator of the amount and distribution of subcutaneous fat (Table 3.4; Mueller & Stallones, 1981). They have, however, also been used to predict body density and thus the overall percentage of body fat (Lohman, 1981; Johnston, 1982), lean mass (Jackson & Pollock, 1976) or total body volume (Weltman & Katch, 1975). Déspres *et al.* (1983) pointed out that one important limitation to the usual interpretation of skinfold readings is a unidimensional approach. For example, if the biceps and abdominal readings are summed to yield an average subcutaneous fat thickness, no account is taken of the relative surface area represented by limb and trunk fat. Another limitation is that the significance of a given fold thickness presumably differs between a child with a height of 1.00 m and an adult with a height of 1.9 m.

Shephard *et al.* (1969) noted that in the simplest two-compartment model, body density (D) was given by:

$$D = 1.100 - 0.2F/M$$

where F was the mass of body fat, and M was the total body mass. They further argued that (ignoring the problem of compressibility), the fat mass corresponding to this equation could be approximated by the relationship:

$$F = [(S - k)/2]\, Ar$$

where S was the skinfold thickness, k was the skin thickness, A was the body surface area, and r was the ratio of body fat to subcutaneous fat. A could be approximated in turn by the formula:

$$A = 50 \times H^{0.75} \times M^{0.5} \times 10^{-4}\,\text{m}^2$$

Table 3.4. *Average thickness of selected skinfolds, as observed in adult Canadians approximating the Actuarial 'Ideal' body mass.*

Skinfold site	Skinfold thickness (mm)	
	Men	Women
Chin	5.8 ± 8.7	7.1 ± 2.8
Triceps	7.8 ± 4.1	15.6 ± 6.2
Chest	12.0 ± 7.9	8.6 ± 3.7
Subscapular	11.9 ± 5.1	11.3 ± 4.2
Suprailiac	12.7 ± 7.0	14.6 ± 8.0
Waist	14.3 ± 8.2	15.3 ± 7.5
Suprapubic	11.0 ± 6.4	20.5 ± 8.2
Knee (medial aspect)	8.6 ± 4.1	11.8 ± 4.2
Average of 8 folds	10.4 ± 4.9	13.9 ± 5.1

(Based on data of Shephard, 1977.)

where H was the standing height and M was the body mass. We may further note that the percentage of body fat (F) would be given by:

$$F = [(S - k)/2]50 \times H^{0.75} \times M^{0.5} \times 10^{-4} \ (r/M)$$

or (putting $M = h^3$)

$$F = [(S - k)/2]50 \times 10^{-4} \ r/H^{0.75}$$

This substantiates the point made above about the need for some height-scaling of skinfold readings.

Lohman (1981) argued that it was preferable to predict body density rather than percentage fat from the skinfold readings, as equations other than those of Siri could then be used to convert density to percentage fat, if this was thought appropriate. Well over 100 skinfold formulae have now been proposed. Most have led to an estimate of body density, although occasional authors have attempted a direct prediction of percentage body fat. Problems in choosing an appropriate equation have arisen from (i) an excessive reliance on stepwise regression analysis (Mukherjee & Roche, 1984), which (if pursued to its ultimate conclusion) can yield a highly specific formula that accounts for much of the variance in a particular population, (ii) a close inter-correlation of the individual skinfold readings (which can invalidate multivariate analyses unless the correlated variables are summed prior to their introduction into the calculation), and (iii) failure to consider possible curvilinearity in the relationships between skinfold data and total body fat. Moreover, the sample sizes which have been used to develop and to validate individual equations have often been too small to comment on their merits relative to previous alternatives, or indeed to offer any possibility of a broad application to populations varying widely in age, sex and nutritional status, and genetic background (Jackson, 1984; Slaughter *et al.*, 1984; Norgan & Ferro-Luzzi, 1985). Perhaps the most widely accepted formulae for healthy adults are those of Durnin & Womersley (1974), developed for the International Biological Programme. After testing various combinations of folds from two to eight, they recommended the equations:

$$\text{Density} = 1.1765 - 0.0744 \, (\log_{10} \Sigma \, _4S) \text{ (males 20–69 years)}$$

$$\text{Density} = 1.1567 - 0.0717 \, (\log_{10} \Sigma \, _4S) \text{ (females 20–69 years)}$$

The base data were collected on a large sample of the Scottish population. The recommended formulae assume a logarithmic relationship between obesity and the sum of four skinfold readings (biceps, triceps, subscapular and suprailiac), thereby meeting several of the criticisms of earlier skinfold technology – measurement of an excessive number of variables relative to the number of subjects, inter-correlation of the individual skinfold readings, skewing of the data, an increase of measurement error as folds become thicker, and a curvilinearity of the generalized relationship between density and subcutaneous fat (Chen, 1953; Allen *et al.*, 1956; Chien *et al.*, 1975a; Parizkova, 1977; Mayhew *et al.*, 1983; Roche, 1984; Jackson, 1984; Cureton, 1984). Notice that as in many previous equations, the variable predicted is body density rather than the amount or the percentage of body fat. Errors can thus arise in both the density prediction itself and its subsequent interpretation as a fat percentage. The error of estimation relative to a hydrostatic criterion is typically ±3.5% in women and ±5.0% in men. The slope of the relationship, but not its intercept, decreases with age; for greater precision, it is thus desirable to use age-specific formulae. Notice also that women have lower intercepts than men. Durnin & Womersley (1974) and Spurr *et al.* (1981) pointed out that their equations tended to overestimate the fat content of very thin or malnourished people; further research has shown that errors in the prediction are also distributed asymmetrically (Forbes, 1987).

Specificity and generality of equations

A major source of difficulty with most skinfold prediction equations is that they are population specific (Davies, 1982). Roche (1984) argued that the stability of regression coefficients should be tested by adding small amounts of bias. Pollock *et al.* (1984) compared densitometry and skinfold data in 249 female and 308 male North Americans; their ages ranged from 10–61 years, and their body fat content from 4–44%. Generalized prediction equations developed from their data had a SE of about ±3.5% in women and ±4.0% in men, similar to the figures cited by Durnin & Womersley (1974). More importantly, the equations appeared to be satisfactory when they were cross-validated in average men (Jackson & Pollock, 1978), average women (Jackson *et al.*, 1980), male athletes (Sinning *et al.*, 1985) and female athletes (Sinning & Wilson, 1984).

As in the earlier studies of Shephard *et al.* (1969) and Parizkova (1977), Pollock *et al.* (1984) found no improvement of precision when the number of skinfolds measured was increased beyond three. This is perhaps not surprising, since principal component analysis suggests that much of the total variance in skinfold data is attributable to a common 'fatness factor' (Shephard *et al.*, 1969; Jackson & Pollock, 1976; Mueller & Reid, 1979; Mueller & Wholleb, 1981; Table 3.5). Roche (1984) went further, pointing out that the decrease in the portion of the variance described was not alarming if predictions of body fat were based upon a single (triceps) skinfold rather than three or four folds. Lohman (1981) likewise commented that use of three or more folds offered no great advantage over two; he combined the data of Sloan (1967), Boileau *et al.* (1971), Sinning (1974) and Lohman *et al.* (1978), developing a quadratic equation, based on the sum of chest, abdominal and thigh folds:

$$\text{Density} = 1.0982 - 0.000815 \ \Sigma \ S + 0.0000084 \ \Sigma \ S^2$$

Comparison with cadaver data suggested to Martin *et al.* (1985) the importance of including observations from the lower limb (for example, the front of the thigh) in prediction equations. Mueller & Stallones (1981) reached a similar conclusion on the basis of principal component analyses. This is an important argument in favour of the quadratic equation developed by Jackson *et al.* (1978); their formula is based on readings for the chest, abdomen and thigh, plus axilla, triceps, subscapular and suprailiac folds;

$$\text{Fat (\%)} = 0.197 \ \Sigma \ 7S - 0.00024(\Sigma \ 7S)2 - 2.2 \text{ (seven folds)}$$

$$\text{Fat (\%)} = 0.376 \ \Sigma \ 3S - 0.00070(\Sigma \ 3S)2 - 1.7 \text{ (three folds)}$$

The quadratic approach, like the logarithmic transformation, avoids the problem of an increase in variance at the extremes of the distribution, and it is thus an effective method of estimating body fat content in specialized groups such as athletes, where the average member of the sample has a very low percentage of body fat.

Table 3.5. Principal component analysis of skinfold measurements (Eigen vectors are shown).

Skinfold	Men (1)	(2)	(3)	(4)	Women (1)	(2)	(3)	(4)	Boys (1)	(2)	(3)	(4)
Chin	0.25	0.91	0.20	0.11	0.33	-0.30	0.77	0.26	0.28	0.58	0.20	0.71
Subscapular	0.37	-0.14	-0.08	-0.49	0.37	-0.18	-0.17	-0.44	0.37	0.27	0.33	-0.28
Chest	0.38	-0.05	-0.33	0.44	0.37	-0.32	-0.26	-0.27	0.31	-0.57	-0.25	0.55
Waist	0.39	-0.03	-0.26	0.21	0.38	0.27	0.01	-0.22	0.39	0.03	0.25	-0.23
Abdomen	0.38	0.07	-0.25	-0.51	0.38	0.02	-0.17	-0.17	0.40	-0.19	-0.01	-0.11
Triceps	0.35	0.01	0.34	0.57	0.33	-0.47	-0.23	0.51	0.39	-0.26	0.14	-0.07
Knee	0.33	-0.32	0.76	-0.23	0.33	0.54	-0.28	0.54	0.26	0.39	-0.84	-0.17
Suprailiac	0.37	-0.20	-0.19	0.66	0.34	0.44	0.38	-0.29	0.40	-0.08	-0.08	-0.14
Cumulative % of variance	73.5	82.5	86.5	89.2	70.9	83.8	76.3	89.0	71.0	81.3	89.6	95.6

Note: Component 1 is 'general fatness', component 2 reflects largely chin fatness, and component 3 knee fatness. Note: In women, Component 3 corresponds with Component 2 in men and boys. (Based on data of Shephard et al., 1969.)

Specific population groups

Body fat is of particular importance to athletes. It is therefore not surprising that specific skinfold prediction equations have been described both for overall populations of athletes (men – Forsyth & Sinning, 1973; Sinning, 1974; Pollock *et al.*, 1977; women – Sinning, 1978; Meleski *et al.*, 1982; Mayhew *et al.*, 1985) and for participants in specific sports including swimming (Hall, 1977; Meleski *et al.*, 1982), distance running (Pollock *et al.*, 1977), gymnastics (Sinning, 1978), wrestling (Sinning, 1974) and team sports (Forsyth & Sinning, 1973). A review of possible options led Thorland *et al.* (1984) to the conclusion that the equations of Jackson & Pollock (1978) remained the most generalizable prediction equations for athletes.

Another situation where some standard formulae have apparently failed to provide answers corresponding to hydrostatic weighing is during dieting. Bray *et al.* (1978) and Bradfield *et al.* (1978) found that changes in the sum of triceps and subscapular readings correlated most closely with the decrease in total body mass.

No adult skinfold prediction equation is likely to be particularly appropriate for very young children. Oakley *et al.* (1977) offered normative data for the thickness of the triceps and subscapular skinfolds of neonates, while Dauncey *et al.* (1977) compared skinfold readings with cadaver estimates in young children. In older children, Lohman *et al.* (1975) found that upper arm and back readings described some 91% of the variance in body fat. Nelson & Nelson (1986) noted that the triceps and subscapular folds were the best predictors of body fatness. However, perhaps because of difficulties in carrying out underwater weighing in children, Corbin & Zuti (1982) found only a moderate correlation between these readings and hydrostatic data.

Size scaling

A few investigators, such as Behnke & Wilmore (1974), Mazess *et al.* (1984) and Katch & Katch (1984) have argued that the usefulness of skinfold readings is increased at all ages if data are scaled to allow for the sex and the body fluid of the individual. Methods based on surface area have been noted above. Behnke & Wilmore (1974) calculated body fat (F) as:

$$F(\%) = \Sigma \, S/\{3(M/H)^{1/2} \, (K)\}$$

where $\Sigma \, S$ was the sum of triceps, subscapular, suprailiac, abdominal and thigh folds, and K was a constant determined as

$$K = \Sigma \, S_r\{3(M_r/H_r)^{1/2} \, (F_r)\}$$

where ΣS_r was the sum of skinfolds for the reference person, M_r was the reference mass, H_r was the reference height and F_r was the reference percentage of body fat. We may note in passing that if M is assumed dimensionally equivalent to H^3, solution of the equation becomes:

$$F(\%) = \Sigma S/3H \ (K)$$

Forsyth *et al.* (1988*a*) evaluated the approach of Behnke & Wilmore (1974) on a large sample of Canadians, finding a correlation of 0.81 relative to hydrostatic estimates of body fat; the precision of Behnke's method compared favourably with their own optimum multiple regression equations. However, the Behnke formula tended to a positive error when making predictions on thin subjects, while the error became negative on fat subjects (Tables 3.6 and 3.7). Scaling procedures have yet to be widely adopted, and some authors still regard them as an unnecessary complication.

Perhaps because ectomorphy is usually linked with tallness, Brozek & Mori (1958) and Benn (1971) found almost no correlation between stature and skinfold thickness in adult subjects; however, this observation does not prove that individual readings should not be scaled for the size of the subject. Hampton *et al.* (1966) commented that obese children were often tall; here, the explanation seems that the endocrine factors favouring fat distribution have speeded maturation.

Distribution of body fat
There is inevitably some lack of generality in estimates of total body fat which are based upon measurements made at a few subcutaneous sites. Regional deposition of subcutaneous fat seems encouraged by local physical inactivity of a body part (Table 3.8), whether this is due to paresis of a limb (Lee, 1959) or pursuit of a particular type of sport such as kayaking (Sidney & Shephard, 1973); however, in the former case, muscle wasting may contribute to the relative fatness of the part. The relationship between waist/hip distribution of subcutaneous fat and cardiovascular disease (Larsson *et al.*, 1984; Jarrett, 1986; DiGiraloma, 1986) is discussed further in Chapter 7, although we may note here the interesting statistic that the lowest risk is yielded by a muscular individual, where the desirable low waist/hip ratio is coupled with a seemingly unfavourable high Quetelet index (Jarrett, 1986).

Any skinfold prediction formula has inherent assumptions regarding the constancy of the relationship between superficial and deep body fat. Unfortunately, the available data suggest that these assumptions are not altogether warranted. Cadaver studies indicate that while skinfold read-

Table 3.6. *A comparison of selected procedures for the prediction of percentage body fat (absolute values and coefficients of correlation with hydrostatic criterion, based on data of Forsyth et al., 1988a, for men with average body fat of 19.1% and women with average body fat of 25.1%).*

Equation	Males Mean fat	r	Females Mean fat	r
Behnke scaling	19.0	0.78	24.9	0.76
Durnin & Womersley	18.7	0.78	27.1	0.73
Jackson & Pollock	14.5	0.79	—	—
Jackson, Pollock & Ward	—	—	22.8	0.79
Lohman	16.3	0.78	—	—
Wright, Dotson & Davis (1)	18.5	0.78	24.6	0.65
(2)	20.4	0.81	—	—

Table 3.7. *Influence of age, sex and fatness (high or low) on average discrepancies (% body fat) of estimates made by the Behnke scaling method from the hydrostatic criterion.*

Age group (years)	Males high	low	Females high	low
20–30	−1.8	+0.8	−1.6	+3.9
30–40	−1.2	+1.0	−1.6	+3.7
40–50	−1.3	+2.3	0.0	+1.7

(Based on data of Forsyth *et al.*, 1988a.)

Table 3.8. *Skinfold thickness in whitewater paddlers (mm). Note thickness of knee fold in women paddlers.*

Skinfold	Men Ideal	Paddlers Young	Older	Women Ideal	Paddlers
Triceps	7.8	5.9	7.7	15.6	10.5
Subscapular	11.9	7.4	12.2	11.3	9.0
Suprailiac	12.7	6.4	9.7	14.6	8.5
Chin	5.8	3.8	9.0	7.1	6.5
Chest	12.0	5.9	10.8	8.6	6.5
Waist	14.3	4.4	7.5	15.3	7.0
Suprapubic	11.0	6.1	10.8	20.5	9.8
Knee	8.6	6.9	7.3	11.8	16.5
Average, 8 folds	10.5	5.8	9.4	13.1	9.3

(Based on data of Sidney & Shephard, 1973.)

ings are well correlated with superficial fat, they are unrelated to the amount of deep body fat (Clarys *et al.*, 1987). On average, subjects accumulate about 200 g of internal adipose tissue for every kg of subcutaneous fat. In cadavers, the coefficient of correlation between deep and superficial fat is statistically significant, but relatively low, about 0.75 for men, and 0.89 for women (Martin, 1984). The instability of the relationship is emphasized by the divergent views that have been expressed. Edwards *et al.* (1955) suggested that thin subjects had a larger proportion of deep fat than obese individuals, but in apparent disagreement with this view, Škerlj (1958), emphasized that with aging internal fat increased more rapidly than subcutaneous fat. Likewise, Garn (1957) maintained that women carried a greater proportion of subcutaneous fat than men, while Durnin & Womersley (1974) came to the opposite conclusion.

Final evaluation of skinfold data

Some authors have become very pessimistic about the possibility of making an accurate prediction of the percentage body fat by any manipulation of skinfold data. Parizková (1977) wrote that in older adults the 'percentage of body fat cannot be derived so far from skinfolds'. At least one journal in the exercise sciences still has an editorial policy to refuse articles where such predictions have been used. In the present author's view, this is an over-reaction to the limitations of current equations.

Admittedly, inter-observer errors increase in older subjects (Chumlea *et al.*, 1984). Age-specific predictions are also needed, since the proportion of subcutaneous fat decreases from birth to adult life, particularly in women. However, skinfold data have considerable value in clinical and epidemiological research. Moreover, there are many circumstances where skinfold data are quite closely correlated with independent measures of fat-free weight (Beddoe *et al.*, 1984; Mazess *et al.*, 1984) and indeed provide a better picture of total body fat content than can be obtained by underwater weighing (Table 3.9).

Lohman (1981) has summarized the main sources of error in the skinfold prediction of body fat as: biological variations in the proportion of subcutaneous fat ($\pm 2.5\%$), biological variations in the distribution of subcutaneous fat ($\pm 1.8\%$) and technical measurement errors ($\pm 0.5\%$); as each of these sources of variance is independent, the total error is $\pm 3.3\%$ fat, slightly better than the figure usually claimed for densitometry, and of a similar order to that obtained by many of the more sophisticated indirect methods.

Table 3.9. *Correlations between three measures of body fat and (i) underwater weighing (UW), (ii) skinfold data. Measurements made on two groups of young women, one with less than and one with more than 26.5% body fat.*

	Body fat <26.5%		Body fat >26.5%	
Variable	UW weight	Skinfold	UW weight	Skinfold
Body mass	0.40	0.53	0.67	0.80
Neck girth	0.39	0.49	0.36	0.46
Waist girth	0.31	0.57	0.69	0.78

(Based on a report by Murray & Shephard, 1988.)

Circumferences

Because there are technical difficulties in measuring skinfolds accurately, there has been some interest in measuring circumferences (Tables 3.3 and 3.10), either as an independent means of predicting body fat or as a means of adding to the precision of skinfold predictions (Best *et al.*, 1953; Škerlj, 1961; Steinkamp *et al.*, 1965; Katch & McArdle, 1973; Zuti & Golding, 1973; Jackson *et al.*, 1978; Pollock & Jackson, 1984; Sinning & Hackney, 1986; Murray & Shephard, 1988; Forsyth *et al.*, 1988a; Tran & Weltman, 1988). It has been reasoned, in particular, that circumference measurements might be helpful in large-scale field surveys, where individual observers have received only limited training.

Formulae

Best *et al.* (1953) reported a correlation of 0.86 between a ratio of height to abdominal girth and body fat, while Tran & Weltman (1988) noted a SE of 4.4% body fat from a formula based on abdominal circumference (A), buttocks circumference (B), iliac circumference (I) and body mass (M):

$$\text{Body fat } (\%) = -47.4 + 0.579(A) + 0.252(B) + 0.214(I) + 0.356(M)$$

Noppa *et al.* (1979) suggested that in older women (44–66 years), the fat mass (F) could be estimated, but with the substantial SE of ±3.0 kg, by using the formula:

$$F(\text{kg}) = 0.37M + 0.13B + 0.10 \Sigma S - 21.1$$

where M was the body mass (kg), B was the buttocks circumference (cm), and ΣS was the sum of triceps and subscapular skinfolds (mm).

Table 3.10. *Correlations between anthropometric variables and hydrostatic criterion. Cumulative R^2 for selected combinations of variables.*

Single variable	Two variables	Three variables	Four variables
Neck Circumference 0.057	Neck Circumference + body mass 0.446	Neck Circumference + body mass + waist circumference 0.508	Neck Circumference + body mass + waist circumference + 4 skinfolds 0.648
Body mass 0.438	Body mass + waist circumference 0.487	Body mass + waist circumference + skinfolds 0.638	
Waist circumference 0.478	Waist circumference + neck circumference 0.497	Waist circumference + neck circumference + 4 skinfolds 0.635	
4 skinfolds 0.626	4 skinfolds + neck circumference 0.635	4 skinfolds + neck circumference + body mass 0.645	
	4 skinfolds + body mass 0.631		
	4 skinfolds + waist circumference 0.626		

(Average for each of 24 subjects based on data of Murray & Shephard, 1988.)

In similar vein, Weltman & Katch (1978) proposed estimating the total body volume (*TBV*) and thus the percentage of body fat from the equation:

$$TBV = 0.872M + 0.263T - 7.795$$

where M was the body mass (kg) and T was the thigh girth (cm).

Reliability and validity

Observations by Forsyth *et al.* (1988*a*) have substantiated the view that the reliability of moderately trained observers is substantially greater for circumference rather than skinfold measurements. On the other hand, any individual circumference reading has a limited construct validity, as it is determined by a combination of bone dimensions, muscle bulk, skin thickness and local body fat, with the fat accounting for only 10–25% of the total.

Prediction equations such as those of Jackson *et al.* (1978) have recognized this problem, attempting to modify abdominal and gluteal circumferences (which reflect both muscularity and obesity) by forearm and upper arm circumferences (which it is hoped are determined largely by the mass of lean tissue). Nevertheless, the greater construct validity of the skinfold data tends to outweigh the negative impact of poor measurement reliability, so that skinfold predictions are commonly more accurate than those based on circumferences. Thus Jackson *et al.* (1978) found that relative to a hydrostatic criterion, the error of their circumference prediction equation was ±4.8%, compared with ±3.6% for a skinfold prediction. Boileau *et al.* (1981) also noted a larger SE of estimated body density for circumference (0.0078–0.0096 units) than for skinfold readings (0.0064–0.0074 units), while Katch & Michael (1968) found that body density was more closely correlated with the thickness of the triceps skinfold ($r = -0.59$) than with the most predictive circumference (buttocks, $r = 0.52$). The one divergent view is that of Sinning & Hackney (1986); they applied various formulae based on girths and skeletal dimensions to athletes; the systematic errors ranged from 0.7–4.3% fat relative to a hydrostatic criterion, while the SD ranged from 3.2–4.3% fat.

Combined predictions

There remains the possibility of measuring both skinfolds and circumferences, an approach which seems to be gaining ground. Forsyth *et al.* (1988*a*) applied a multiple regression analysis to a combination of skinfold and circumference readings for a large Canadian population, finding that the optimum equation for the prediction of hydrostatic body fat (F) percentage was, in men:

$$F(\%) = 10.2 + 0.11 \, (A) - 1.75 \, (W) + 0.33 \, (C) + 0.21 \, (\Sigma \, S)$$

and in women:

$$F(\%) = -5.1 + 0.68 \, (K) + 0.003 \, (\Sigma \, S)$$

where A was the age in years, W was the wrist diameter in cm, C was the abdominal circumference in cm, $\Sigma \, S$ was the sum of abdominal and triceps skinfolds in mm, and K was the knee circumference in cm. The

Table 3.11. *Relationship between the percentage of body fat as estimated by underwater weighing and other simple indices of obesity. Coefficients of correlation as summarized by Shephard* et al. *(1971).*

Authors	Subjects	Excess body mass	M/H	M/H^2	$H/M^{1/3}$
Von Döbeln (1956)	35M	0.38	0.40	0.46	−0.22
	35F	0.37	0.55	0.38	−0.51
Allen *et al.* (1956)	57M	0.78	0.74	0.65	−0.74
	26F	0.71	0.69	0.71	−0.68
Lesser *et al.* (1963)	8M	0.70	0.61	0.74	−0.66
	8F	0.86	0.85	0.85	−0.84
Goldman *et al.* (1963)	15F	0.66	0.70	0.72	−0.69
	15F after diet	0.52	0.64	0.70	−0.60
	33F	−0.03	0.17	0.12	−0.05
Brozek *et al.* (1963)	10M	0.71	0.72	0.70	−0.65
	10M after diet	0.75	0.76	0.75	−0.70
Shephard *et al.* (1969)	27M	0.29	0.41	0.37	−0.24
	29F	0.50	0.38	0.51	−0.47
Weighted mean*	250	0.56	0.58	0.55	−0.54

*Omitting series of Brozek *et al.* (1963) and of Goldman *et al.* (1963) before dieting, also the 33 girls studied by Goldman *et al.* (1963).

overall correlation with the hydrostatic estimate was 0.82 in men and 0.80 in women. While this compares favourably with alternative formulae (Table 3.6), there remains a need to validate this hybrid equation on a second population sample.

Murray & Shephard (1984) introduced both skinfold and circumference readings into a canonical prediction of hydrostatically determined body fat; their observer was moderately experienced in the collection of skinfold data, and they found that the skinfolds gave a marginally more accurate prediction of body fat than did the circumferences (Table 3.11, see p. 66).

The addition of information on circumferences may be particularly useful in the obese; in such subjects, skinfolds are notoriously difficult to measure, while a large part of the abdominal circumference is attributable to fat (Behnke & Wilmore, 1974; Pollock & Jackson, 1984). Katch & McArdle (1973) reported that in men a combination of arm, forearm and abdominal circumferences matched the best skinfold equation, while in women, these measurements plus thigh girth gave a better prediction of density than any combination of skinfolds.

Weltman *et al.* (1987*b*) found that in obese men (average mass 94.5 kg, body fat > 30%), the percentage of body fat could be predicted with a SE of only 2.5%, using the equation:

$$F(\%) = 0.3146(C,\text{cm}) - 0.1097(M) + 10.83$$

where C is the abdominal circumference, in cm, and M is the body mass in kg.

Ultrasound

High-frequency sound waves pass freely through homogeneous tissues, but a portion of the emitted energy is reflected at any interface between dissimilar tissues, for instance the fascia separating fat from underlying muscle. The reflection can then be converted to an electrical signal.

In the simplest (A-mode) device, the time taken for transmission and return of 2.5 MHz ultrasound pulses is converted to a fat thickness score (Katch & McArdle, 1983), assuming a velocity of 1500 m/sec (a compromise between the anticipated velocity of sound in fat, 1450 m/sec, and muscle, 1570 m/sec). Coefficients of reliability have ranged from 0.78 for the abdomen to 0.97 for the thigh.

Most authors have used a B-mode device. Here the signal is amplified and displayed on an oscilloscope to yield an outline picture of the main tissue structures, including the thickness of the subcutaneous fat layer (Alsmeyer *et al.*, 1963; Katsuki *et al.*, 1965; Haymes *et al.*, 1976; Borkan *et al.*, 1982*b*; Katch, 1983; Volz *et al.*, 1984; Stokes & Young, 1986).

Correlations with carcass fat measurements are generally statistically significant, but in the study of Alsmeyer *et al.* (1963) the coefficients were not particularly large (0.39–0.80 in pigs and 0.46–0.61 in cattle). In contrast, Stouffer (1963) claimed a correlation of 0.986 between ultrasound data and the directly measured thickness of back fat in the pig. However, he also commented that the two methods yielded substantially different absolute values; he attributed the discrepancy to changes which had occurred while the animals were being slaughtered.

Several studies have found correlations of at least 0.80 between ultrasonic measurements and skinfold data in humans (Bullen *et al.*, 1965; Booth *et al.*, 1966; Sloan, 1967). Haymes *et al.* (1976) found test/retest correlations of 0.87 for ultrasonic readings. The correlation with skinfold readings was higher for women than for men. Correlations with soft tissue radiographs were higher for the triceps ($r = 0.88$) than for the suprailiac site ($r = 0.78$); multiple echoes and interfaces were found at the suprailiac site in some subjects. Ranelli & Kuczmarski (1984), likewise, found a correlation of 0.80–0.86 between local ultrasound and skinfold readings, although the correlation with overall body density was only 0.78.

Magnetic resonance imaging

Odedblad (1966) first applied the technique of nuclear magnetic resonance imaging (NMR) to visualize human tissues. The method is based upon the fact that nuclei (particularly hydrogen nuclei) emit electromagnetic waves when they are exposed to a magnetic field.

Separate spectroscopic peaks can be detected which relate to emissions from water and aliphatic lipids, and techniques have also been developed which can selectively suppress the magnetization of water or fat by applying an opposing magnetic field (Frahm *et al.*, 1985). The more sophisticated instruments of this class can thus measure and display water and fat separately or together (Dixon, 1984), although the optimum technique for this purpose has yet to be agreed.

If a whole-body scan is to be undertaken, the subject is positioned inside a very large water- or liquid-helium-cooled electromagnet for up to one hour. Foster *et al.* (1984) and Hayes *et al.* (1985, 1988) have recently used magnetic resonance imaging to provide cross-sectional data analogous to that obtained by ultrasound. An inversion recovery pulse sequence applies an alternating magnetic field at right angles to the static magnetic field; the recovery of alignment of the previous pattern of magnetization is then used to generate an image of the tissues, much as in computed tomography (see the following section). The recovery time for magnetization in the longitudinal axis is several minutes for solids, but only a few seconds for liquids. Contrasts between individual solids depend on differences between recovery times for magnetization in the longitudinal and transverse planes. By allowing an appropriate interval between the initial inversion pulse and application of the 90° pulse, a very good visual contrast can be obtained between muscle and adjacent fat.

Lewis *et al.* (1986) reported that the SD of repeated measurements of total body water by NMR was 2.8%. Cohen *et al.* (1986) noted that caliper readings, ultrasound and MRI readings were well-correlated in men, but in women only the caliper and NMR data showed a satisfactory coefficient of correlation ($r = 0.87$). Moreover, the absolute values did not agree well. Caliper data for the two sexes averaged 6.1 and 8.6 mm, and ultrasound results averaged 6.4 and 12.9 mm, but NMR values averaged 9.5 and 11.5 mm, for males and females respectively. It was suggested that there were fascial planes within the subcutaneous fat, and that the calipers 'pinched off' the more superficial layers of the total subcutaneous fat. Support for the NMR data was found in that they correlated well with figures for metabolic heat production, body heat conductance and cadaver measurements of subcutaneous fat thickness.

Very high capital and operating costs currently limit NMR technology to a few specialized centres. Moreover, while no harmful effects have yet been demonstrated from either electromagnetic or radio-frequency fields, one nagging doubt of the NMR approach remains a possible harmful effect of the magnetic fields upon both subjects and investigators.

Soft-tissue radiography

Given an appropriate duration and intensity of X-ray exposure, the outline of both 'fat' and 'muscle' can be discerned fairly clearly in limb radiographs (Stuart & Reed, 1951; Garn, 1961; Stouffer, 1963; Tanner, 1965; Shephard et al., 1975). Assuming also that the limb has a cylindrical shape, the percentage of fat in the limb can then be estimated from one or more linear measurements on the uniplanar radiograph (Heald *et al.*, 1963).

The method of soft tissue radiography has the advantage that it can be applied to parts of the body where the skin plus underlying fat cannot readily be lifted to allow normal caliper measurements of skinfold thickness. Unfortunately, the radiograph ignores any fat that has been deposited within or between muscles, and in very lean subjects there may be so little subcutaneous fat that the dermal layer cannot easily be distinguished from underlying muscle (Comstock & Livesay, 1963). A further objection to the application of the procedure in human subjects is that except in the rare circumstance where a previous soft-tissue radiograph is available, the subject must be exposed to a small dose of X-irradiation. If the limb is lying relaxed in the supine position, there may also be some distortion of the soft tissues by the counterpressure of the X-ray table; radiographs are preferably taken with the subject standing erect (Reynolds & Grote, 1948). Other technical complications include a distortion of the image by respiratory or other body movements and a lack of parallelism in the X-ray beam. The latter problem can cause a 5–10% magnification of the image (Comstock & Livesay, 1963), although this affects subcutaneous fat and muscle readings to a similar extent. The magnification error can be minimized if the limb is set at a long distance from the X-ray source, with the axis of the body part at a fixed distance from the film. If b is the radius of bone, m is the partial radius of the ring of muscle and f is the partial radius of the ring of fat as seen on the radiograph, the areas of the three tissues can be approximated as:

$$\text{bone} = \pi b^2$$

$$\text{muscle} = \pi(m^2 + 2bm)$$

$$\text{fat} = \pi(f^2 + 2fb + 2fm)$$

While there is a fair correlation between skinfold and radiographic estimates of limb fat (Fletcher & McNaughton, 1987; $r = 0.8$–0.9), a systematic difference in absolute readings would be anticipated, since the radiographs are of uncompressed tissue, whereas the skinfolds are compressed (Bliznak & Stable, 1975). Skinfold readings therefore vary from 65% of the radiographic value in young adults (Garn, 1965) to 84% in older men with less compressible subcutaneous fat (Brozek, 1965).

In the pig, very close correlations ($r = 0.994$) have been claimed between the radiographic estimate of back fat and the direct carcass measurement. In humans, Tanner (1965) reported a reliability of 3–4% for repeated radiographic determinations of fat thickness.

Since the limbs are neither cylindrical nor circular, it is relatively difficult to move from a single cross-sectional measurement of tissue dimensions to a volumetric estimate of total body fat or muscle. Maresh (1963) adapted the formulae of Behnke (1961) to estimate the overall mass of the three main tissue components as follows:

$$\text{Body mass (kg)} = h^{2/3}\{(b^2.Db) + (m^2.Dm) + (f^2.Df)\}/K$$

where h is the stature, Db, Dm and Df are estimates of the densities of bone (1.2), muscle (1.04) and fat (0.9) respectively, and K is an arbitrary age and sex dependent constant in the range 29.7–35.2. Maresh (1963) claimed that this prediction had an accuracy of about ± 1.0 kg for each of the three main component tissues. Nevertheless, it seems overly ambitious to try and predict total body fat or lean mass from a single linear measurement on a uniplanar radiograph. Certainly, the main value of soft tissue radiographs is in examining the local distribution of fat and muscle (Tanner, 1965).

Computerized tomography

Recently, interest has shifted from simple radiography to computerized tomography (CT) of the forearm (Heymsfield *et al.*, 1979; Bulcke *et al.*, 1979; Schantz *et al.*, 1983; Maughan *et al.*, 1984), the thigh (Haggmark *et al.*, 1978) and the abdomen (Borkan *et al.*, 1982a; Ashwell *et al.*, 1985). With the CT technique, a collimated beam of X-ray photons is moved systematically across the body part, and attenuation of the beam as viewed by one or a whole bank of scintillation counters is calculated for each of up to 300 000 'pixels'. A computer then converts the individual attenuation readings into a reconstructed grey cross-sectional image of the part that can be viewed on a suitable visual display unit. A fast (<2 sec) scan is important to a clear image, particularly if the subject is restless. Water gives rise to an arbitrary attenuation of 0 on the CT scale; corresponding figures are for air −1000 units, bone +1000 units or more, fat −50 to −120 units and muscle +30 to +60 units. Thus, by concentrating upon certain levels of attenuation, contrast can be enhanced, facilitating measurements of bone or muscle and fat dimensions. As with soft-tissue radiography, it remains impossible to distinguish muscle from tendon.

To date, CT has found only limited application in body composition studies. The capital cost of the equipment is very high (often more than a million dollars), and operating costs may also require expenditures of US $200 per test. Furthermore, the dose of radiation received by the subject remains a significant issue; for example, Maughan *et al.* (1984) reached 40% of the annually permitted allowance for human volunteers while making six CT scans of the arms. Some authors have used a series of CT scans to estimate the volumes of limb or total body fat (Tokunaga *et al.*, 1983; Maughan *et al.*, 1984; Sjöstrom *et al.*, 1986). As with the simpler radiographic method, it is necessary to assume that the body has a cylindrical or a conical shape and that the observed 'fat' contains some fixed percentage of lipid (for example, 85%). However, neither of these assumptions is completely correct. Thus Heymsfield *et al.* (1979) demonstrated that the arms had an elliptical rather than a conical shape in two thirds of their subjects.

When CT measurements were made on a plexi-glass phantom, the coefficient of variation was only 2.4% (Haggmark *et al.*, 1978). If the objective of the investigator is to determine total body fat, the best prediction is apparently yielded by an abdominal cross-section (Borkan *et al.*, 1983b), and in the interests of reducing irradiation, observations on normal subjects are best restricted to this single site (Kvist *et al.*, 1986). One report found a correlation of 0.99 between a 'whole body' CT value and ^{40}K estimates of body fat (Sjöstrom *et al.*, 1986), while a second study reported the more modest correlation of 0.84 between forearm CT scans and skinfold estimates of percentage body fat (Maughan *et al.*, 1984). In cadavers, Hudash *et al.* (1985) observed a good correlation between local measurements and CT scans of subcutaneous fat, but less accurate estimation of bone dimensions. Likewise, Maughan *et al.* (1984) cautioned that the percentage of muscle found by CT of the forearm did not agree well with the two dimensional estimate based on classical soft-tissue radiography, while Heymsfield *et al.* (1979) found that CT estimates of muscle area were 15–20% smaller than those yielded by anthropometric measurements.

Ashwell *et al.* (1985) have recently noted a close correlation between the ratio of intra-abdominal to subcutaneous fat and the waist to hip circumference ratio; they thus suggest that computerized tomography at the level of the umbilicus has potential application in evaluating 'masculine' and 'feminine' patterns of fat distribution. In men, the proportion of deep to superficial fat, as seen in the abdominal CT scans, increases with age (Borkan *et al.*, 1985).

Dual photon absorptiometry

Mazess *et al.* (1984) have determined the percentage of body fat from dual photon (153Gd) absorptiometry. In their technique, some 1500 picture elements (pixels) are obtained over about 45% of the soft tissue mass of the body. The ratio of attenuation observed for 44 keV and 100 keV photons gives an estimate of the ratio of body fat to lean tissue in any given location, the ratio changing from 1.32 in sectors that are 100% fat to 1.47 in sectors that are 100% lean tissue. The mass-weighted average for the 1500 pixels can estimate body fat with a precision of about 2% (Witt & Mazess, 1978), the main limitation of the method being some exposure to radiation.

Other techniques

Fat tissue absorbs infra-red radiation at a shorter wavelength than lean tissue; Conway *et al.* (1984) have thus suggested that body fat might be estimated by an infra-red interactance method. In the presently used system, near infra-red energy at wavelengths of 916 and 1026 nm is transmitted to the subject via a central bundle of optic fibres, while the peripheral fibres detect reflected energy and return this to the spectro-photometer. Body fat is calculated from the second derivative of interactance at the two wavelengths (Lanza, 1983).

Fat biopsy

Incision of subcutaneous fat has been used in cadavers in order to make direct determinations of local fat thickness. *In vivo* sampling is undertaken in order to estimate the number and size of the fat cells, and thus the subject's propensity to obesity.

The needle of a small syringe is simply plunged into subcutaneous fat, usually over the buttocks (Hirsch & Goldrick, 1964). Washed samples are incubated in buffered osmium tetroxide for 24–72 hours. The number of free cells can then be determined by Coulter counter (Hirsch & Gallian, 1968; Gurr & Kirtland, 1978). A convenient index of cell size is the lipid content (mg/cell):

$$\frac{(\text{Net weight}, \mu g) \times (\text{ratio of lipid to wet weight}]}{\text{Number of cells in sample}}$$

This index assumes an approximate concordance between the diameter of the fat droplet and the size of the cell that contains it. If other information on body composition is also available, the total number of fat cells can then be estimated from the total fat mass, divided by the fat mass per cell.

Comparison with other methods of fixing, staining and counting convinced Hirsch & Gallian (1968) that the osmium tetroxide technique was the most reliable method of examining adipose tissue. They reported a standard error of 5–20% for both cell number and cell size. However, their approach was unable to measure cells containing less than $0.01 \mu g$ of lipid, a shortcoming which may account for reports that fat cell hyperplasia does not occur in adults (Sjöstrom *et al.*, 1971). Problems can also arise from two cells being counted as one larger cell if they enter the Coulter counting chamber simultaneously.

Other authors have used photomicrographs (DiGirolamo *et al.*, 1971; Clarkson *et al.*, 1980; Gurr *et al.*, 1982). These provide not only a cheaper method of analysis, but also a permanent visual record; nevertheless there is then some difficulty in measuring cells smaller than $15 \mu m$, so that Gurr *et al.* (1982) have found it necessary to speak of 'measurable cell number' rather than an absolute cell count. The minimum number of cells that must be counted to provide an accurate description of the tissue lies between 100 (Clarkson *et al.*, 1980) and 300 (DiGirolamo *et al.*, 1971).

An image analysing computer is a third approach now used in hospitals where large numbers of biopsy specimens must be examined (Pallier *et al.*, 1985).

The size of the individual fat cells differs from one body site to another. Thus, the gluteal cells are larger than those from abdominal and subscapular regions (Clarkson *et al.*, 1980), and there are also differences of cell size between deep and superficial fat (Salans *et al.*, 1973). Unless representatives of the smaller deep fat cells are included in the average measurement, the total fat cell number will be underestimated (Björn-torp, 1983*a*,*b*).

Simple epidemiological indices

Mass for height ratios
An increase of body mass subsequent to maturity is usually attributable to an accumulation of fat. It thus seems inherently probable that obesity could be represented by some ratio of total body mass to standing height. The attraction of this approach is its simplicity. Few people object to simple determinations of height and weight, and data on large, representative populations can thus be accumulated quite quickly; indeed, the information is sometimes already available, whether in North American insurance statistics and company records, or in nutritional and anthropological surveys of Third World countries. Moreover, at least in theory, there is little reason for measurements of height and weight to differ systematically between one observer and another.

However, in practice there are a substantial number of obstacles to treating information on height and body mass as more than a rough guide to the degree of obesity in a given individual:

(i)   Much of the already available data has been collected for insurance purposes. Base samples are thus biased towards a favoured socio-economic group.

(ii)   Such data are often very crude. Sometimes the mass of the individual has been ascertained by questioning the subject rather than by weighing, uncertain allowances have been applied for the mass of shoes and clothing (which has rarely been entirely removed), and some subjects have been included twice or more in the data bank.

(iii)   An above average body mass can reflect an accumulation of either lean tissue or fat. It thus provides a poor index of obesity in the athlete (Ross *et al.*, 1986), the growing adolescent (Cronk *et al.*, 1982; Billewicz *et al.*, 1983) and the person with an abnormal ratio of limb length to sitting height (Roche *et al.*, 1981). Likewise, an accumulation of fat can be masked by a concomitant lightening of the bones or an atrophy of the skeletal musculature (Forbes & Reina, 1970).
        Attempts to allow for differences of frame size by expert-rating (Katch & Freedson, 1982) or the measurement of inter-

condylar diameters at the elbow (Metropolitan Life Insurance Company, 1983; Frisancho & Flegel, 1983) remain unconvincing and poorly validated (Katch *et al.*, 1982). Ideally, the search would be for frame-size variables that explained the residual variance of body composition. However, to date, effort has been concentrated on measurements that are easy to perform, have a high coefficient of correlation with body mass and a low correlation with subcutaneous fat (Roche, 1984; Himes & Bouchard, 1985). On these criteria, wrist and ankle measurements have seemed the most useful indicators of frame dimensions. Frisancho (1984) has published norms for body mass, skinfolds and upper arm muscle area on the basis of frame size.

(iv)   In children, the use of weight-for-height standards is equivalent to use of a (M/H^n) standard, where n changes with age (Cole, 1985).

(v)   In the elderly, problems of interpretation arise from a decrease of stature as kyphosis develops and the vertebrae collapse.

Excess body mass

Tables of excess body mass developed by insurance companies have been used as one simple index of obesity (Society of Actuaries, 1959; Metropolitan Life Insurance, 1983) for many years. Others (Parizkova, 1977) have used Broca's index [Ideal mass = $H - 100$ cm; index = (actual/ideal mass)100]. Such calculations are subject to all of the limitations noted above with respect to both the individual subject and the reference criterion.

Mass for height

If data are collected carefully by a trained anthropometrist, simple mass to height ratios correlate quite well with other indices of body fat such as hydrostatic weighing (Roche *et al.*, 1981; Murray & Shephard, 1988; Table 3.12). Billewicz *et al.* (1962) concluded that the ratio of observed to standard weight was not only the simplest but also the most reliable epidemiological index of obesity. Moynihan *et al.* (1986) arbitrarily decided that the threshold for the diagnosis of obesity was a ratio of 1:20, while Wakat *et al.* (1971) went so far as to predict body volume from body mass.

Other observers have claimed that mass to height ratios show poorer correlation with a body fat criterion than do skinfolds or circumferences (Pollock & Jackson, 1984), a conclusion borne out by the averaged correlations of Table 3.12.

Table 3.12. *Coefficients of correlation between percentages of body fat as determined hydrostatically and simple ratios of body mass to height.*

	Coefficient of correlation	
Ratio	boys	girls
Mass/height	0.41	0.38
Mass/height^2	0.37	0.51
Mass/height^3	0.25	0.47

(Based on data of Shephard *et al.*, 1971.)

Other mass/height ratios

The objections raised to mass for height tables and simple mass to height (M/H) ratios apply with equal force to other epidemiological indices of obesity such as body mass index (M/H^2, sometimes described as Quetelet's index or the Kaup index, Van Uytvanck & Vrijens, 1966) and the ponderal index ($H/M^{1/3}$).

There has been much fruitless debate on the relative value of M/H, M/H^2 and $H/M^{1/3}$ as indices of obesity; unfortunately, the basis of validation has commonly been either a lack of correlation of the selected index with H (Billewicz *et al.*, 1962; Khosla & Lowe, 1967; Lee *et al.*, 1981; Norgan & Ferro-Luzzi, 1982) and/or a large coefficient of correlation with M (Lee *et al.*, 1982), rather than a close correlation of the index with a careful assessment of body fat (Shephard *et al.*, 1969; Norgan & Ferro-Luzzi, 1982; Roche, 1984).

The index $M/H2$ was thought to be particularly useful because it was relatively independent of stature (Billewicz *et al.*, 1962; Khosla & Lowe, 1967; Keys *et al.*, 1972; Cerovska *et al.*, 1977; Babu & Chuttani, 1979; Frisancho & Flegel, 1982). However, Knowler & Garrow (1982) have disputed the need for height independence. Norgan & Ferro-Luzzi (1982) suggested further that the addition of age to the index improved predictions of the percentage of body fat.

Benn (1971) suggested that it was preferable to estimate fatness from a ratio M/Hn, where the exponent n optimized the relation to mass for a particular age, sex and racial group. However, Baecke *et al.* (1982) showed that a power function which satisfied the criteria of Benn (1971) offered no advantage over the traditional H^2 as a means of standardizing body mass data. This was confirmed by Frisancho & Flegel (1982) in a sample of 16 459 'black' and 'white' adults and by Garn & Pesick (1982) in a meta-analysis of data from 58 468 subjects.

Normal $M/H2$ values are in the range 19–27 kg/m2, depending on the age of the subject (Van Wieringen, 1972; Rolland-Cachera *et al.*, 1982). DiGirolamo (1986) set the threshold of obesity at an arbitrary $M/H2$ value of 25 in men and 27 in women; massive obesity was associated with scores of 30–40, and morbid obesity with a score >40. Garrow (1983) accepted 25 as an appropriate $M/H2$ threshold for the diagnosis of obesity in men, but he proposed a threshold of 30 for women. In the UK, regional differences in average $M/H2$ values have been described, with the largest proportion of 'obese' subjects in Wales, and the smallest proportion on the south-eastern outskirts of London (Rosenbaum & Skinner, 1985). In adult men, Revicki & Israel (1986) made the damaging criticism that the standard error of a fat estimate based on $M/H2$ was ±5.3%, only marginally better than the basic variance of their sample (±7.3%).

Finally, we may note that physiologists have commonly used $M0.75$ as an index of active metabolic mass. The basal metabolism of animals ranging in size from a mouse to an elephant appears to be correlated with this index (Brody, 1945; Kleiber, 1975).

Linear regression equations

Mellits & Cheek (1970) proposed calculating body water from regression equations based upon standing height and total body mass. Their equations, with the corresponding SE, were as follows:

Water (l)
$$= 0.507M + 0.013H + 0.076 \text{ (SE 0.50) in females } <110.8 \text{ cm}$$
$$= 0.252M + 0.154H - 10.31 \text{ (SE 1.65) in females } >110.8 \text{ cm}$$
$$= 0.465M + 0.045H - 1.927 \text{ (SE 0.60) in males } <132.7 \text{ cm}$$
$$= 0.406M + 0.209H - 21.99 \text{ (SE 2.52) in males } >132.7 \text{ cm}$$

If the assumption is made that lean tissue contains a fixed percentage of water (e.g. 72 or 73%), these equations could in principle be used to estimate body fat, as discussed below; however, in practice, the error of the body water estimate is too large to make this a useful calculation.

General assessment

The overall assessment of the various simple epidemiological indices is that excess mass and the various other possible manipulations of standing height and body mass all show rather similar coefficients of correlation with an independent estimate of body fat made by underwater weighing (Tables 3.11 and 3.12; Womersley & Durnin, 1977; Satwanti *et al.*, 1980; Shephard *et al.*, 1971; Colliver *et al.*, 1983). While the accuracy of such results is sometimes sufficient for epidemiological assessments of obesity, they lack the precision needed for assessments of body composition in the individual.

Estimates from lean mass

The estimation of lean body mass is discussed in Chapter 5. At first inspection, it might appear that the estimation of lean mass would offer a useful indication of body fat (F), since:

$$F \text{ (kg)} = \text{Body Mass (kg)} - \text{Lean Mass (kg)}$$

However, the fallacy in this reasoning becomes apparent when numerical values are introduced:

$$F \text{ (kg)} = 70 \text{ kg} - 60 \text{ kg}$$

If lean mass is determined with a precision of (say) 10%, then there is necessarily a minimum of 6 kg error in the estimation of 10 kg of body fat. Provided there is no systematic error in the technique, a lean mass determination could still indicate the average percentage of body fat in a population, but it would not be well suited to determinations of fat percentage in the individual. In the population sense, there remains the further problem that many approaches to the measurement of lean body mass require the use of complicated apparatus, best reserved for specialized applications and quite small numbers of subjects.

4　Determination of body water

The total volume of body water includes the fluid content of the blood, extra- and intra-cellular fluid, and the water contained in various relatively isolated 'transcellular' compartments such as the cerebro-spinal fluid, the synovial joints, the eyes, and the lumen of the gastrointestinal tract. Several of these compartments have a colloid-like nature, and their fluid content therefore does not behave as a simple aqueous solution.

In the present context, total body water is important partly because of the relatively fixed relationship between this volume and lean tissue mass, and partly because an alteration in the water content of the various fluid compartments can have a significant impact upon both their mass and their physical dimensions (Girandola et al., 1977b; Svoboda & Query, 1986).

The total body water accounts for some 60% of body mass in young adult men, 50% in women, and 80–90% in the neonate; both sexes show a progressive decline of body water over adult life, as lean tissue is replaced by fat.

Principles of fluid volume estimation

The earliest estimations of total body water were based upon the analysis of cadavers. As already noted in Chapter 2, the water content of dead tissue is quickly affected by the drying of the exposed body parts.

Most recent estimates of either total body water or the volume of individual fluid compartments are based on the principle of indicator dilution. A marker substance is ingested or injected into the blood stream, taking due care (by a thorough rinsing of the beaker or syringe) to ensure that all of the intended dose of the marker has been administered. After a suitable period of equilibration (which varies with the marker substance), concentrations of the marker are determined in a biological specimen, usually of blood or urine. If the substance is administered by mouth, problems can arise from gastric retention of a part of the marker, and if equilibrium samples are collected from the blood stream it is necessary to allow for loss of the marker substance that has arisen by hepatic metabolism or urinary excretion.

The ideal marker would be one that is cheap, non-toxic and easy to administer, chemically inert but not radioactive, rapidly and evenly distributed through the body fluid compartments, slowly metabolized and excreted, yet easy to analyse in biological specimens.

In practice, most of the various available marker substances fall short on one or more of these criteria. They pass, some more and some less readily, through the barriers that separate the several body fluid compartments, and thus yield differing equilibration volumes. For example, inulin is water soluble, but does not penetrate cells. Its distribution thus provides an approximation to extracellular fluid volume. Intravenous injections of inulin and thiosulphate do not enter the gastrointestinal tract, and inulin also fails to penetrate connective tissue (Goodford & Leach, 1962); such substances thus yield substantially lower figures for extra-cellular fluid volume than certain alternative markers of the extra-cellular space.

The chloride volume is often regarded as a measure of functional extracellular water, although because the chloride ion penetrates both red cells and glandular cells, the volume indicated by this marker is slightly larger than the true anatomical extracellular volume (Weil & Wallace, 1960; Gamble, 1962).

In some cases, the volume estimation is complicated further by the metabolism or excretion of the marker substance. Blood concentrations of a metabolizable tracer such as N-acetyl-4-amino-antipyrine or ethanol can be measured repeatedly once equilibration has occurred (a period of some three hours for antipyrine, rather less for ethanol), and a semi-logarithmic plot of the data can then be extrapolated back to zero time (Soberman *et al.*, 1949; Gruner & Salmen, 1961; Loeppky *et al.*, 1977). Where excretion occurs (as with inulin and urea), a possible alternative is to calculate the distribution volume from the quantity of marker excreted in unit time, this value being divided by the corresponding drop in plasma concentration (Poulos *et al.*, 1956).

Accurate analysis of the blood concentration of a marker such as sodium ions requires application of a correction for differences of water content between the plasma and the interstitial fluid (Forbes, 1987). Because physical space is occupied by protein molecules, only some 93% of the serum volume is water. In the case of the usually studied cations, it is also necessary to make an allowance for differences in equilibrium concentrations across cell membranes. For example, in the case of Na^+, the Donnan equilibrium factor (the ratio of sodium ion concentrations on the two sides of the membrane) is about 0.95, and the protein volume correction is 0.93. After allowing for these opposing influences, the concentration of sodium ions in the serum is about 140 mE/l, while that in the interstitial fluid is 144–145 mE/l:

$$\text{Interstitial } [Na^+] = 0.95 \text{ Serum } [Na^+]/0.93$$

Measurement of extracellular fluid volume

The value obtained for extracellular volume depends on the completeness with which the extracellular fluid compartments are penetrated by the marker substance. The largest results are obtained with substances that have a low molecular mass (sodium or chloride ions, radioactive sulphate ions, thiosulphate and mannitol), while lesser volumes are reported when sucrose, raffinose or inulin are used as tracers.

Chloride

A 'gold standard' of extracellular volume is provided by the distribution of chloride ions within the body. Taking the distribution volume of injected radioactive chloride as 100%, Swan *et al.* (1954) noted corresponding figures of 61% for inulin, 71% for raffinose, 75% for sucrose, 81% for thiocyanate and 82% for radiosulphate and mannitol. In contrast, the sodium space is slightly larger than the chloride space, since about a quarter of the skeletal sodium depot is rapidly exchangeable (Friis-Hansen, 1965).

Unfortunately, it is not convenient to use radio-isotopes of chlorine when measuring the extracellular volume in humans; the half-lives of the isotopes ^{34}Cl and ^{38}Cl do not pose any health hazard, but they disintegrate too rapidly to allow equilibration with body fluids, while the half-life of the remaining alternative ^{36}Cl is unacceptably long (10^6 years).

Nevertheless, changes in extracellular fluid volume can be assessed from changes in serum chloride concentration, provided that the volume change is either known or assumed. For example, if a litre of urine is eliminated, then:

$$V_2 \text{ (ml)} = V_1 - 1000 \text{ (ml)}$$

and

$$(V_1 - 1000)[Cl_2] = (V_1.[Cl_1] + 1000[Cl_u])$$

where V_2 is the new extracellular fluid volume, V_1 is the initial volume,

$[Cl_1]$ and $[Cl_2]$ are the corresponding chloride concentrations, and $[Cl_u]$ is the concentration of chloride in the urine.

Bromide

The most popular absolute method for the determination of extracellular fluid volume is based on the equilibration of stable bromide or its most readily available and suitable isotope (^{82}Br). The stable bromide can be analysed by fluorescent excitation or by high pressure liquid chromatography (Kaufman & Wilson, 1973; Miller & Cappon, 1984). The latter technique is very sensitive, allowing the use of extremely small doses of radioactive material.

The bromide space is not identical with the chloride 'gold standard'. Relating bromide volumes to the chloride space, both being determined two and a half hours after injection of the corresponding marker, stable bromide gives a 7% larger space, and the ^{82}Br isotope a 2% larger space (Gamble *et al.*, 1953; Cheek & West, 1955); these differences apparently reflect a more ready penetration of the stomach and a less ready penetration of the brain by the bromide marker.

Sodium

Exchangeable sodium amounts to 70–75% of total body sodium (Ellis *et al.*, 1976; Kennedy *et al.*, 1983). The exchangeable fraction can be determined using the radioactive isotopes ^{24}Na or ^{22}Na. Exchange occurs rapidly with extracellular fluid, the gastrointestinal tract and the urine, aqueous, cerebrospinal and articular fluids. However, the analysis is complicated because there is a slow, partial and incomplete equilibration with intracellular sodium (particularly that fraction located in bone).

Potassium

Radio-isotopes of potassium can be used to measure exchangeable potassium. Analyses of the radio-isotopes are relatively easy and accurate, but one disadvantage is the exposure of the subject to radiation (Forbes & Perley, 1949; Corsa *et al.*, 1950).

Other extracellular markers

Among other potential markers of the extracellular fluid volume, thiocyanate is easy to administer and is non-radioactive. However, it tends to penetrate the red cells and the method has a questionable reliability (Lavietes *et al.*, 1936).

Inulin has also been widely used to measure the extracellular space, although it often fails to achieve equilibrium, and thus substantially underestimates the true volume (Benedict *et al.*, 1950).

Measurement of body water

Some of the substances used to estimate total body water suffer from problems of rapid excretion (for example urea, antipyrine and ethanol). With the first two of these markers, there is a requirement for repeated blood sampling in order to plot an excretion curve, although in the case of ethanol it is possible to base estimates on the use of a non-invasive 'breathalyser'.

Urea

Urea can be estimated in the urine, without recourse to intravenous sampling. It is non-radioactive, but suffers from problems of uneven distribution through the body fluids, and equilibrium is difficult to achieve (McCance & Widdowson, 1951).

Antipyrine

Antipyrine is relatively uniformly distributed throughout the body fluids, and excretion is a less serious problem for this marker. It is metabolized fairly slowly, is non-toxic, and is easy to analyse. The main disadvantage is that it binds to protein, thus affecting equilibration (Soberman *et al.*, 1949).

Ethanol

Ethanol has sometimes been used as a marker, but the difficulty is that small doses are rapidly metabolized, while if larger quantities are given the subject becomes intoxicated (Loeppky *et al.*, 1977).

Isotopic forms of water

The best and most logical approach to the estimation of total body water is to follow the dilution of an isotope of water itself. Three possible candidates are $H_2^{18}O$ (Schoeller *et al.*, 1980), deuterated (^2H_2O) and tritiated (H^3HO) water. Ideally, the marker should undergo no absorption, combination or destruction during the period of equilibration.

$H_2^{18}O$

Isotopically marked $H_2^{18}O$ has been measured in timed samples of exhaled gas (Trowbridge *et al.*, 1984); other factors being equal, the quantity of radioactive material exhaled in unit time is proportional to the concentration in the blood stream. There may be a small (2%) systematic error in this method, due to the integration of radioactive oxygen into materials other than water, particularly during periods of rapid growth (Whyte *et al.*, 1985).

Deuterated water

Deuterated water meets most of the criteria for an ideal marker (Von Hevesy & Hofer, 1934; Freeman *et al.*, 1955; Edelman *et al.*, 1952; Ljunggren, 1957; Moore *et al.*, 1963; Hytten *et al.*, 1966b; Krzywicki *et al.*, 1974; Haschke, 1983). It is easily administered, is non-toxic, and despite a heavier mass than normal water, it is quickly and evenly distributed within the body (Schloerb *et al.*, 1950; Pinson, 1952), and it is not toxic to humans at concentrations of less than 20% (Katz, 1960). The one limitation arising from attaching the marker to the hydrogen atom is the exchange of some 5% of the isotope with non-aqueous hydrogen, particularly the hydrogen found in carboxyl and hydroxyl groups. This leads to a corresponding overestimate of body water (Friis-Hansen, 1965; Culebras *et al.*, 1977; Sheng & Huggins, 1979). Freeman *et al.* (1955) and Ljunggren (1965) noted that the deuterium space was 6–8% larger than that estimated by antipyrine, Schoeller *et al.* (1980) found a 3% overestimate relative to $H_2^{18}O$, and Krzywicki *et al.* (1974) reported a 4% overestimate relative to hydrostatic weighing.

The oral dose of the deuterated water ranges from 1 g/kg (Haschke, 1983) to a fixed amount of 10 g (Lukaski & Johnson, 1985). Equilibration is reached within 3 hours of ingestion (Schloerb *et al.*, 1950; Hurst *et al.*, 1952; Faller *et al.*, 1955; Consolazio *et al.*, 1963), and given the ratio of total body water to the normal volume of urine flow, the error introduced by a neglect of urinary losses is usually less than 0.5% over this time (Schloerb *et al.*, 1950; Pascale *et al.*, 1956). Equilibrium concentrations of deuterated water can be determined in urine, plasma or saliva specimens (Mendez *et al.*, 1970; Haschke, 1983). If plasma is analysed, it is necessary to apply a 7–8% correction for plasma solids, as discussed above.

Other minor criticisms of the deuterated water method are a limited supply of deuterium and the need for special instrumentation to analyse blood or urine specimens. We have used a mass spectrometer for this purpose (Solomon *et al.*, 1950; Shephard *et al.*, 1973), but alternative possibilities include a falling drop technique, photo-disintegration, infra-red spectroscopy, gas chromatography, thermal conductivity and freezing point determinations (Stansell & Hyder, 1976; Buskirk & Mendez, 1984). The SD of a single deuterium estimate of body water is about ±2% (Yang *et al.*, 1977). Unfortunately, this error becomes compounded when serial measurements are required (for instance, when assessing the loss of body fat by dieting).

Tritiated water

Tritiated water can also be given orally. It is a soft β-emitter and can thus be analysed by a scintillation counter, at the expense of exposure to some 30 mrem of radiation. For this reason, tritiated water has been a some-what more popular marker than its deuterated counterpart (Siri, 1956*b*; Leibman *et al.*, 1960; Lesser & Zak, 1963; Heald *et al.*, 1963; Boling, 1963; Lewis *et al.*, 1986).

Tritiated water reaches a good equilibrium with all of the body compartments within about 90 minutes of ingestion, and it is metabolized only slowly. Yasamura set the precision of tritium estimates of total body water at 1%, although Sheng and Huggins (1979) noted a 4–15% overestimate relative to values obtained by desiccation. As with deuter-ium, this probably represents an exchange of the isotope with body parts.

Electrical conductivity and impedance measurements

In the future, it may be possible to estimate body water from measure-ments of the electrical conductivity or impedance of the tissue (Kushner & Schoeller, 1986), since these variables are largely dependent upon the fluid content of the tissue (see Chapter 3).

The technology has been developed, but to date, impedance methods have been applied mainly to the estimation of body fat. Van Loan & Mayclin (1987) found only a moderate correlation between total body water (*TBW*) and impedance ($r = -0.65$), and on this basis they pro-posed a correction to the equations proposed by the makers of an impedance measuring device:

$$TBW = 9.987 + 0.000723H^2 + 0.282M - 0.0153R$$
$$- 2.33S - 0.153A$$

where H is the stature (cm), M is the body mass (kg), R the total body impedance (ohms), S the sex (M = 0, F = 1), and A is the age (years). In contrast, Klish *et al.* (1987) reported a close correlation between total electrical conductivity and an ^{18}O estimate of total body water in infants.

Stability of fluid volume estimates

In the context of physical anthropology, the practical value of body water determinations depends to a substantial extent upon the stability of relationships between such data and the mass of lean tissue. Problems of data interpretation can arise both from systematic errors in the measurements and from changes in the hydration of the individual body compartments (Mendez & Lukaski, 1981; Durnin & Satwanti, 1982; Schoeller *et al.*, 1985).

The regulation of body hydration and the interchange of water between the various fluid compartments occurs partly as an active transport process, in association with the pumping of mineral ions across cell membranes by an ATPase enzyme which is dependent on Na^+, K^+ and Mg^{++} ions, and partly as a passive process involving the diffusion of water through cell walls or leakage at inter-cellular junctions.

With regard to the inherent biological stability of body water, there are both short- and long-term concerns. The assumption is often made that human lean tissue contains 72 or 73% water. A study of only six cadavers was enough to indicate an appreciable inter-individual variation, from 67–78% of lean mass (Mitchell *et al.*, 1945; Forbes *et al.*, 1953; Widdowson & Dickerson, 1964). It could be argued that these differences of hydration arose either as the subjects were dying, or during the subsequent dissection. However, substantial variation is also seen when *in vivo* determinations are made. Delwaide & Crenier (1973) used tritiated water dilution to show that water content accounted for 59.8 ± 2.4% of total body mass in men, but only 51.9 ± 1.8% of total body mass in women; given an SD of some 2%, about 5% of the total population could thus have a water content that deviated by 2 l or more from the standard anticipated for their sex.

Total body water apparently decreases from around 61% of body mass at the age of 12 years and 65% at age 18, to 55% at 25, and 47% at 85 years (Heald *et al.*, 1963; Friis-Hansen, 1965; Yoshikawa *et al.*, 1978). A progressive increase of body fat stores and a loss of lean tissue make important contributions to this change. Other factors decreasing total body water in an older person are a replacement of more specific cells by fibroblasts, and a redistribution of body water from the intracellular to the extracellular compartment (Yoshikawa *et al.*, 1978). Cell hydration is decreased due to a progressive impairment in the regulation of sodium ion transport at the cell membrane (Timiras, 1972; Debry, 1982) and/or a deterioration of renal function (Rowe, 1985). Yoshikawa *et al.* (1978) suggested that intracellular water declined from 42% of body mass in a young adult to as little as 33% at the age of 75 years. Longitudinal comparisons of body composition are complicated by these changes in the amount and distribution of body water.

Short-term variations of fluid balance are superimposed upon the long-term trends. The overall daily turnover of water is as large as 7%, and the rate of exchange between individual compartments is even more rapid. For example, a heavy meal may stimulate the secretion of several litres of fluid into the stomach and intestines, decreasing the total volume of body water as indicated by markers that do not enter the gastrointestinal tract. Substantial diurnal variations of both total body water and its distribution occur (Peterson *et al.*, 1959; Moore-Ede *et al.*, 1975) in response to such varied influences as exercise, heat and cold exposure, ingestion of fluid or salt and the depletion of intramuscular glycogen reserves. Thus, if a subject drinks a substantial volume of water or dilute saline in preparation for a prolonged bout of exercise, the body fluid content may be temporarily increased by 0.5–1 l. On the other hand, participation in a football game or a marathon race on a hot day can readily cause a secretion of 3–4 l of sweat; this is met in part through a transient depletion of plasma and extracellular fluid volumes, and in part through a release of bound water as intramuscular stores of glycogen are metabolized (Shephard, 1982*a*). Cold exposure provokes a diuresis, thereby decreasing plasma volume, and if the subject is breathing hard, there may also be a loss of up to 3 l of fluid per day in the expired air (Shephard, 1982*a*). Variations of tissue water content over the menstrual cycle have already been noted (Byrd & Thomas, 1983; Svoboda & Query, 1986). In certain classes of athlete such as wrestlers, body water may also be deliberately manipulated by the self-administration of diuretics (Steinberg & Zaske, 1985) and androgens (Nathan *et al.*, 1963).

Many of the potential short-term sources of variation in body water can be eliminated if the investigator carefully standardizes the conditions of measurement. For example, the subject should avoid strenuous exercise or exposure to extremes of heat or cold on the day prior to testing, and the investigator should further confirm that there has been an overnight fast, with avoidance of food and fluid ingestion until all data has been collected.

Nevertheless, the combined effects of analytic error and unavoidable short-term biological variation leave body water measurements with a minimum coefficient of variation of 3–5%.

Application to determinations of gross body composition

In order to calculate the percentage of body fat or lean mass from any of these estimates of body water, it is necessary to assume a fixed hydration of lean tissue. A commonly assumed figure of 73% water, although Osserman *et al.* (1950) have proposed the use of a figure of 71.8%; in their sample, the latter value yielded a similar average body fat to that found with antipyrine dilution or underwater weighing.

It is important that a suitable allowance be made for long-term related changes in tissue hydration through the adoption of appropriate (but necessarily imprecise) age-specific factors that relate fluid volumes to lean tissue mass.

5 *Determinations of lean tissue*

Lean body mass is commonly viewed as the difference between the total body mass and the mass of fat (Chapter 3), although strictly speaking the figure thus obtained is a 'fat-free' mass. Lean tissue includes also some 4% of 'essential' fat (Lohman, 1981). The main constituent of the lean compartment is muscle, although with many methods of determination body water and bone are also included in the estimate.

On the basis of carcass analyses of both animals and humans, Forbes & Bruining (1976) suggested that fat-free skeletal muscle accounted for some 49% of the total fat-free tissue. Gross dissection of human cadavers likewise suggested to Clarys *et al.* (1986) that muscle comprised 54.0% of the adipose-tissue-free mass in man and 48.1% in women.

This chapter examines potential methods of estimating total and regional lean tissue and/or fat-free mass.

Creatinine excretion

Creatinine is a metabolite of creatine, and as about 98% of body creatine reserves are found in muscle, the daily excretion of creatinine is approximately proportional to muscle mass (Cheek, 1968; Grant *et al.*, 1981; Heymsfield *et al.*, 1983). Creatinine excretion is quite closely correlated ($r = 0.92$) in turn with total nitrogen excretion (Kachadorian *et al.*, 1972).

The daily excretion of creatinine amounts to 18–32 mg in a sedentary man and 9–26 mg/day for a sedentary woman. Figures are greater for athletes who are involved in power sports. Day to day intra-individual variation ranges from 2–30% in different subjects (Boileau *et al.*, 1972; Bistrian *et al.*, 1975; Moran *et al.*, 1980). In young adults, one factor contributing to this variation is the response to prolonged bouts of exercise; thus, a 10% increase of creatinine excretion has been reported after three hours of vigorous marching (Srivastava *et al.*, 1957) and a 60% increase of serum creatinine has been observed immediately following participation in a 90 km ski race (Refsum & Strømme, 1974, but not Castenfors *et al.*, 1967).

Old people with relatively poorly developed muscles may show a marked creatinuria after a relatively mild bout of exercise, probably because they have more difficulty in resynthesizing creatine phosphate. Obese subjects also tend to have a high rate of creatinine excretion relative to the observed activity level (Konishi, 1967), possibly because they must perform more work in displacing a heavy body whenever they do attempt to exercise. Particularly large losses of creatinine are seen if there has been muscle breakdown, whether this is associated with stringent dieting, malnutrition, or an exercise such as running downhill (which requires eccentric contraction of the muscles involved). Other variables potentially distorting the anticipated rate of creatinine excretion include severe infections, menstrual variations in tissue hydration (Lohman, 1986) and chronic renal disease.

Creatinine-based determinations of lean body mass (*LBM*) can be upset if the subject eats meat or fish, as these foods often contain not only a large amount of creatine, but also pre-formed creatinine (Lykken *et al.*, 1980). On an *ad libitum* diet, Forbes & Bruining (1976) established the relationship:

$$LBM \text{ (kg)} = 0.0291 \, Cr + 7.38$$

where *Cr* is the daily excretion of creatinine in mg. Miller & Blythe (1952) had earlier proposed a rather different equation:

$$LBM \text{ (kg)} = 0.00632 \, Cr + 20.97$$

Much presumably depends upon the quantity of meat that is eaten. Forbes & Bruining (1976) found that on a meat-free diet, the prediction equation changed to:

$$LBM \text{ (kg)} = 0.0241 \, Cr + 20.7$$

The practical implication of the large positive intercept is that the ratio of muscle mass to creatinine excretion varies with the rate of creatinine excretion. Given an excretion of 1 g/day, the ratio is about 17–20 kg/g, but if the rate of excretion is doubled, the ratio drops to about 16 kg/g.

Even on a consistent diet which has been meat- and fish-free for three days, some variation of creatinine excretion remains. Forbes (1987) has set the intra-individual coefficient of variation for a careful laboratory measurement at 4.8%. Fortunately, this figure seems free of systematic bias, and if several independent observations are made, the mean estimate of lean mass does not differ significantly from that obtained by anthropometry or use of a whole-body counter (Roessler & Dunavant, 1967; Turner & Cohn, 1975; Forbes & Bruining, 1976).

Given the general proportionality between muscle dimensions and height, Bistrian & Blackburn (1983) developed a table showing the expected creatinine excretion rates in relation to height; this information was used as a means of judging muscle wasting. Bistrian & Blackburn (1983) showed a non-linear increase of excretion from 1.29–1.89 g/day in men, and from 0.78–1.17 g/day in women, both over a 35.5 cm span of inter-individual differences in height.

Chinn (1967) proposed the calculation of the dry mass of muscle protein from a combination of creatinine excretion (Cr, g/day) and total body potassium (K, mE):

$$\text{Muscle protein} = 4.26\,Cr - 1.24\,\text{K}$$

However, the average value obtained in a young adult (3.5 kg of dry mass) was substantially less than the anticipated figure. This formula has therefore not gained wide acceptance.

The collection of one or more 24-hour urine specimens is a cumbersome requirement of any creatinine-based approach to the determination of lean mass. Moreover, a few minutes of error in the timing of the final urine collection can cause a significant error in the 24-hour urine volume. Total plasma creatinine determinations, therefore, have been suggested as a simple and useful alternative to the collection of 24-hour urine specimens (Shutte *et al.*, 1981). If body hydration is carefully controlled, the coefficient of correlation between the two data sets is quite close ($r = 0.82$), and a single plasma creatinine determination has the precision of three urinary measurements. Each mg of plasma creatinine is equivalent to approximately 0.88 kg of striated muscle. A rough estimate of muscle mass can be made on the basis of an assumed plasma volume, but for accurate analysis, a simultaneous measurement of plasma volume must be made.

Meador *et al.* (1968), Kreisberg *et al.* (1970) and Picou *et al.* (1966) have all developed methods of determining muscle mass based upon the feeding of isotopically labelled creatine to their subjects. However, these techniques are unsuitable for routine use, and the interpretation of data is complicated by the fact that 6–7% of the labelled material equilibrates with skin, kidneys and liver rather than muscle.

3-Methyl-histidine

Histidine residues are methylated during the formation of actin and myosin (Asatoor & Armstrong, 1967), but they undergo no subsequent metabolism during catabolism. About 90% of the amino acid 3-methylhistidine (3 mh) is found in muscle, with smaller amounts in the gastro-intestinal tract (3.8%) and elsewhere (Haverberg *et al.*, 1975; Nishizawa *et al.*, 1977; Elia, *et al.*, 1979; Wassner & Li, 1982). The excretion of 3 mh has some potential as a marker of muscle protein breakdown and therefore muscle mass (Young & Munro, 1978; Mendez *et al.*, 1984).

In general, up to the age of four years, the intramuscular concentration of 3 mh increases but thereafter seems relatively constant at 3.3–3.9 mol/g of fat-free muscle mass, with little variation from one striated muscle to another (Tomas *et al.*, 1979). However, a programme of muscle training could alter the fractions of actin and myosin, and thus the amount of 3 mh, that are found in skeletal muscle. Factors influencing the proportion of histidine residues that are methylated remain uncertain. It is also debatable whether the contribution of other tissues to the total excretion of 3 mh can be ignored (Millward *et al.*, 1980; Wassner & Li, 1982); indeed, the turnover rate of 3 mh is much higher for the intestines (24%/day) than for muscle (1.4%/day). Nevertheless, muscle probably accounts for some 75% of the total excretion of this substance (McKeran *et al.*, 1978; Afting *et al.*, 1981; Harris, 1981), and perhaps for this reason the intra-individual coefficient of variation is much greater for 3 mh than for creatinine excretion (Forbes, 1987).

Lukaski & Mendez (1980) noted a correlation of 0.89 between the excretion of 3 mh and a hydrostatic estimate of lean body mass, while the correlation of daily 3 mh excretion with a neutron activation estimate of muscle mass was 0.91. A second report by Lukaski *et al.* (1981) found a good correlation between the 3 mh method and ^{40}K estimates of lean mass. Using maximum oxygen intake as an indirect indicator of lean tissue, Cunningham *et al.* (1983), likewise, found correlations of 0.94 and 0.89 with creatinine and 3 mh excretion, respectively. Buskirk & Mendez (1984) reported further that 3 mh excretion had a correlation of 0.87 with creatinine excretion, and a correlation of 0.79 with a hydrostatic estimate of muscle mass. These authors suggested that lean body mass could be predicted by using the equation:

$$\text{Muscle mass (kg)} = 0.118\ [3\ \text{mh},\ \mu\text{mol/day}] - 3.45$$

In some circumstances, the excretion of 3 mh may give a clearer picture of muscle mass than an estimate of lean tissue derived from underwater weighing. However, the daily clearance of 3 mh is highly dependent on diet, and subjects must be prepared to accept a regimen of three days without meat. Buskirk & Mendez (1984) noted that excretion dropped from the previous daily average of 513 to 230 μmol on the third day of a meat-free diet. On the other hand, the ratio of 3 mh excretion to other indicators of muscle mass increased under conditions where muscle breakdown was enhanced (Miller *et al.*, 1982).

Other factors that may affect the relationship between 3 mh excretion and muscle mass include the sex, age and maturity of the subject, the adequacy of nutrition, hormonal status, physical fitness, recent intense exercise and injury (Buskirk & Mendez, 1984).

Total body potassium

Principle

A woman's body contains some 100 g of potassium, while in the more heavily muscled male a typical figure is 150 g. More than 90% of body potassium is found in lean tissue, and about 60% of this total is located in the muscles (Table 5.1). Morgan & Burkinshaw (1983) have suggested that some 9 kg of the fat-free mass contains almost no potassium, while the remainder (a variable fraction of total mass) shows a potassium to nitrogen ratio of 1.81 mE/g.

The naturally occurring radioactive isotope ^{40}K accounts for a fairly consistent 0.012% of body potassium. The ^{40}K isotope has an extremely long half-life (1.3×10^9 years), but nevertheless each gram of body potassium contains enough ^{40}K to yield 1800 disintegrations per minute; 11% of the released radiation appears in the form of high energy (1.46 MeV) gamma rays and 89% as beta rays. An estimation of the gamma radiation normally emitted by the body thus provides a basis for

Table 5.1. *Typical distribution of body potassium in an adult man.*

Tissue	Total body potassium (%)
Muscle	60.0
Skeleton	11.4
Brain	3.3
Fat	3.2
Liver	2.9
Skin	2.7
GI tract	2.2
Lungs	1.2
Other	13.2

(Based on data of Wagner *et al.*, 1966.)

determining total body potassium, and by inference lean body mass (Anderson & Langham, 1959; McNeill & Green, 1959; Harrison *et al.*, 1975; Hager, 1981; Pierson *et al.*, 1982).

There are formidable technical difficulties when using the ^{40}K approach. It is necessary to make specific assumptions about the concentration and distribution of potassium in the body and about differences in natural abundance of the ^{40}K isotope. The investigator must also contend with substantial and variable background radiation, problems of counter geometry and sensitivity, and the statistical nature of isotopic emissions (Smith, T. *et al.*, 1979; Buskirk & Mendez, 1984); the necessary equipment is both sophisticated and expensive. Nevertheless, some authors (Hager, 1981; Pierson *et al.*, 1982) have suggested that potassium estimation is the method of choice when analysing body composition. Certainly, the ^{40}K method is non-invasive. The only minor inconvenience for the subject is being enclosed in the counting chamber for the period of observation, which is about 20–30 minutes when using modern counters.

Extraneous radiation

The radiation from ^{40}K accounts for approximately 16% of current background emissions. The body must therefore be protected from extraneous sources of radioactivity throughout the ^{40}K measurements. In addition to cosmic radiation, potential external sources of radioactivity to avoid include the muscles of the observer, clothing and jewellery, concrete, wood, nearby cyclotrons and radiotherapy units. Inhalation of atmospheric radon during a bout of vigorous exercise may be sufficient to increase background radiation for some 30 minutes after the vigorous exercise has ceased (Lykken *et al.*, 1983; but not Forbes, 1987).

Measurements of ^{40}K activity must be made in a well-shielded area, either a chimney below which both the subject and couch are moved (the 'shadow shield' method) or a completely shielded room. Enclosure is ideally by materials that have weathered free of contaminants, for example the lead roofs of mediaeval churches and/or steel from pre-World War II battleships. The construction of an entire shielded chamber is obviously cumbersome and costly; Boddy *et al.* (1971) have claimed a standard error of only 2.7% for their shadow shield counter.

Counting emissions

Emissions are usually monitored by plastic or liquid scintillation counters, grouped appropriately around a chair (Talso *et al.*, 1960; Martin *et al.*, 1963; Anderson, 1963; Burkinshaw & Spiers, 1967), or by thallium-activated sodium iodide crystal detectors mounted above and below the measuring couch (Forbes & Hursh, 1963; Harrison *et al.*, 1975; Shephard *et al.*, 1985*b*). When using the latter arrangement, the couch is propelled between the counters during the 20–30 minute period.

Some counters have a high sensitivity to radiation, but only a moderate energy resolution, while others show the opposite characteristics. Multichannel pulse-height analysers can be used with the sodium iodide crystal technique in order to distinguish the 1.46 MeV photopeak from other types of emission (particularly those due to ^{137}Cs, a body burden of radioactivity which humans accumulated during above ground nuclear tests of the 1950s).

Calibration

In animals, the intensity of the observed emissions can be calibrated using cuts of meat (Kulwich *et al.*, 1961; Pfau *et al.*, 1961) or by the sacrifice of representative animals (Kirton & Pearson, 1963; Martin *et al.*, 1963). In humans, the average counting rate attributable to ^{40}K is converted to an equivalent mass of potassium, using a calibration factor based upon the ingestion of known amounts of another (synthetic) isotope such as ^{42}K (Corsa *et al.*, 1950), ^{24}Na, ^{83}Rb or ^{137}Cs. The ^{42}K is particularly suited to this purpose, as it has a photopeak of 1.53 MeV. Total exchangeable potassium can be estimated by administering as little as 10 μC of ^{42}K (Boling, 1963), but nevertheless the administration of a radio-isotope is regarded as undesirable in children. The alternative is to construct a phantom, filled with some potassium-containing substance such as rice. The average potassium content of rice is 67 mE/kg, similar to the figure found for the human body (Forbes & Hursh, 1963).

If ^{42}K is administered to calibrate the system, it is assumed that the synthetic isotope is distributed uniformly and completely throughout the body, although in thin subjects (where the ^{40}K data are the most reliable), the emissions from ^{42}K are only some 89% of those for ^{40}K (Miller & Remenchik, 1963). Emissions from ^{40}K and ^{42}K are compared with those emanating from a bottle containing first a known mass of normal potassium (which necessarily contains its quota of ^{40}K), and later an amount of radioactive material which is equal to that ingested by the subject during calibration of the system (Lloyd *et al.*, 1979; Lloyd & Mays, 1987):

$$\text{K content of human (g)} = \text{Human counts } ^{40}\text{K}$$

$$\times \frac{\text{Bottle K(g)}}{\text{Bottle } ^{40}\text{K}} + \frac{\text{Bottle counts } ^{42}\text{K}}{\text{Human counts } ^{42}\text{K}}$$

Body size effects

The standard calibration factor for a 30 minute observation period (for example, 22 counts/g of potassium, measured over a 0.28 MeV width band centred on the 1.46 MeV ^{40}K peak) must be adjusted to allow for body size effects (which include both the inverse square law relating radiation intensity to the distance from source, and local absorption of gamma rays by the overlying tissues). Although the two size effects tend to cancel one another out, there is an error of at least 3.5% in the third term of the above equation, if calibration data from a 68 kg subject is applied to the results for a person weighing either 45 or 90 kg (Miller & Remenchik, 1963; Hawkins & Goode, 1976; Smith, J. *et al.*, 1979).

There is also some error due to differences in the respective energies of the photopeaks for ^{40}K (1.46 MeV) and ^{42}K (1.53 MeV). If the average path length is 20 cm, differential absorption of the two forms of emission introduces a discrepancy of 2.7%.

Precision

Errors resulting from extraneous sources of radioactivity, size effects and problems of calibration are compounded by an unavoidable 2% coefficient of variation due to statistical fluctuations in both background and muscle radioactivity. Thus, the overall intra-individual error is normally around 5% (Boddy *et al.*, 1972; Harrison *et al.*, 1975).

Using the most elaborate forms of instrumentation (measuring systems with up to 54 counters) on selected population samples, a coefficient of variation as low as 3–4% has been claimed (Cohn & Dombrowski, 1970; Shukla *et al.*, 1973; Pierson *et al.*, 1974; Burkinshaw *et al.*, 1981), with equal accuracy against a phantom. However, such reproducibility of the data does not necessarily imply validity when the method is used in humans. It is unclear how far results are influenced by inter-individual differences of body form. For example, a large proportion of the total body potassium of a distance runner is stored in the thighs and legs, while a swimmer may have unusually well-developed muscles in the shoulder girdle. In obese subjects, an additional error may arise from the screening effect of a thick layer of subcutaneous fat. However, experiments with a pig carcass (Forbes & Hursch, 1963) suggest that body-build is not a major problem in the interpretation of body potassium figures. Exercise or heat exposure may cause further problems, because there is a tendency for a shift of potassium-containing blood to the superficial blood vessels as these become dilated (Lane *et al.*, 1977; Londeree & Forkner, 1978; Thomas *et al.* 1979).

Estimation of lean body mass

As noted in the previous chapter, there are further difficulties in translating a given whole body potassium mass into an equivalent muscle mass. Forbes & Hursch (1963) assumed, on the basis of cadaver dissections, that a fixed ratio could be used:

$$LBM \text{ (kg)} = \text{mE K}/68.1$$

However, in practice, the muscle potassium content as obtained from autopsies is influenced by the blood content of any specimens that are examined; it usually leads to an underestimate of body muscle, and thus a gross overestimate of body fat relative to underwater weighing (Myhre & Kessler, 1966). Moreover, futher investigation has shown that there are age, species and probably sex variations in the potassium content of lean tissue (Anderson & Langham, 1959; Myhre & Kessler, 1966; Delwaide & Crenier, 1973; Cureton *et al.*, 1975; Forbes, 1987; Fig. 5.1). As Anderson (1963) has pointed out, these differences reflect the fact that the body comprises at least three compartments with respect to potassium; however, the specific values that Anderson (1963) has cited for muscle (total potassium 87 mE/kg), other lean tissue (49 mE/kg) and adipose tissue (15 mE/kg) remain somewhat at variance with those presented by Forbes & Hursh (1963, Table 5.2).

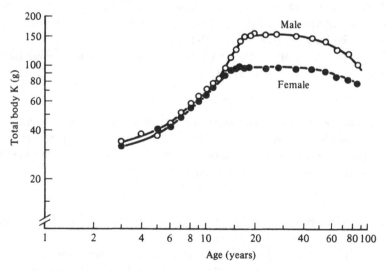

Fig. 5.1. Changes of body potassium content with age (based upon data accumulated by Forbes, 1987).

Table 5.2. *Sex differences in body potassium content (mE) of young adults per kg of total body mass.*

Men	Women	Author
57.0	44.5	Meneely *et al.* (1962)
54.4	44.4	Anderson (1965)
55.1	42.3	Oberhausen & Onstead (1965)
55.6		Krzywicki & Consolazio (1968)
	44.1	Novak (1970)
49.5*		Cohn & Dombrowski (1970)
56.0	44.5	Boddy *et al.* 1971
55.4	45.9	Delwaide & Crenier (1973)

*Heavy subjects

In an old person where muscle is well conserved but there is also some osteoporosis, aging may increase the ratio of potassium to lean mass. More usually, the potassium content of lean tissue decreases as muscle is replaced by connective tissue (Von Kriegel & Airsherl, 1964).

Individual values depend further upon whether the person is well-trained or obese, which inevitably influences the relative proportions of muscle to other lean tissues. Leusink (1974) found that typical values dropped from 69.0 mE/kg of lean mass in the well-trained to 61.6 mE/kg of lean mass in untrained subjects. Likewise, Morgan & Burkinshaw (1983) indicated ratios ranging from 49 mE/kg for a fat-free mass of 30 kg to 62 mE/kg for a fat-free mass of 80 kg.

Body potassium also varies with height. Wagner *et al.* (1966) thus proposed calculating a muscle index:

$$\text{Muscle index} = \text{body K}/140\ (H/174)^3$$

where 140 ġ is the body potassium content of a person with a height of 174 cm, and H is the height (cm) of the individual under consideration. Ellis *et al.* (1974) suggested that body potassium (K) could be predicted from body dimensions and age, using the formula:

$$K = kM^{1/2}H^2$$

where k was an age factor ($5.52 - 0.014A$ in men, $4.58 - 0.010A$ in women, A being the age in years).

The ^{40}K method is finally vulnerable to short-term changes in the average water and mineral content of the body, due for instance to rigorous dieting, or the 'making of weight' by wrestlers and boxers.

Neutron activation

Principle
Neutron activation is a rather sophisticated and expensive technique that has been used to determine several body constituents, including sodium, nitrogen, calcium, chloride and phosphorus (Cohn *et al.*, 1974; Cohn & Parr, 1985). In essence, the subject is uniformly bombarded with a known dose of fast neutrons – for instance, a 50 mrem dose of 14 MeV neutrons (Burkinshaw *et al.*, 1981). These are captured by the element of interest. An unstable isotope is formed and this emits gamma radiation which can be measured in a whole-body counter (Almond *et al.*, 1984).

Whole body nitrogen determinations
In the case of nitrogen determinations, either ^{13}N (Burkinshaw *et al.*, 1981) or ^{15}N (Vartsky *et al.*, 1979) can be produced, depending on the technque that is chosen. The emission from the latter is so prompt that counting proceeds simultaneously with irradiation (Beddoe & Hill, 1985).

The standard neutron activation method is hampered by interference from the formation of other compounds; thus, the laboratories at the University of Toronto have exploited the characteristic rapid gamma rays which are emitted when nitrogen nuclei capture neutrons (McNeill *et al.*, 1979*a,b*; Mernagh *et al.*, 1977; Shephard *et al*, 1985*b*). Subjects are exposed to neutrons from multiple Pu-Be sources. An excited state of ^{15}N is formed, followed after as little as 10^{-16} seconds by decay to the stable isotope ^{15}N; 10.8 MeV gamma rays are emitted during the process – these are then detected by means of sodium iodide crystal detectors.

Calibration and accuracy
As an internal standard, measurement is also made of the 2.2 MeV gamma radiation induced by neutron activation of total body hydrogen. For purposes of absolute calibration, a phantom containing a known amount of urea in water is used. A correction of 1% per cm is applied relative to a standard trunk thickness of 22 cm. The correction allows for differences in absorption of the 2.2 and 10.8 MeV rays.

The subject receives a radiation exposure of 50 mrem. Although Vartsky *et al.* (1979) and Ellis *et al.* (1982) have made optimistic claims of ±3.0–3.5% accuracy for the method, in our experience the test/retest reproducibility of the data is ±6% (Shephard *et al.*, 1985). Chemical analysis of pig and rat carcasses suggests an absolute validity of ±8% for whole body nitrogen estimates (McNeill *et al.*, 1979*a*; Preston *et al.*, 1985).

Estimation of lean mass

In order to convert the data for total body nitrogen to an estimate of lean body mass, it is necessary to assume an average figure for tissue nitrogen content. The data of the International Commission on Radiological Protection (ICRP) shows a value of 30.1 g/kg for tissues other than fat, or (if the nitrogen count is to be converted to lean mass), a figure of 31.9 g/kg.

Cohn *et al.* (1980) assumed a muscle nitrogen content of 30 g/kg, and a figure of 36 g/kg for the remaining tissues other than fat. By simultaneous measurements of ^{40}K, they sought to partition lean tissue into muscle and non-muscle components. Burkinshaw *et al.* (1978) have proposed the following equations for this purpose:

$$\text{Muscle mass (kg)} = [19.25(K) - N]/38.34$$
$$\text{Non-muscle mass (kg)} = [N - 8.45(K)]/20.2$$

where K and N are the total body potassium and total body nitrogen, respectively, both being measured in g.

Garrow (1982) has queried whether there is a consistent K/N ratio for non-muscular tissues that can be used in making such calculations. Forbes (1987) also comments that the K/N ratio varies from 5 mE/g in the brain to 0.45 mE/g in the skin. A further difficulty when using information for both elements is that the combined errors of the potassium and the nitrogen measurements amount to about 18%, leaving occasional subjects with a negative muscle mass! The average muscle mass reported by Burkinshaw *et al.* (1978) after using this approach (6.7 kg in women and 20.0 kg in men), also seems inappropriate.

A simpler alternative is to use the total nitrogen as a measure of lean tissue, as there is little nitrogen in fat. Shephard *et al.* (1985*b*) pointed out that there are some difficulties in making even this calculation, since the ratio of total body nitrogen to lean tissue mass inevitably varies with the muscularity of the individual, the state of body hydration, the protein and mineral content of the skeleton, and possibly age. In women, nitrogen balance also varies over the course of the menstrual cycle (Calloway & Kurzer, 1982). In their data, Shephard *et al.* (1985*b*) found an apparent increase from the ICRP figure of 30 g of N per kg of lean mass to about 35 g/kg in older female subjects.

Because of these various problems, neutron activation estimates of total body nitrogen can occasionally give rise to odd results. Thus, reports from Toronto and elsewhere (Archibald *et al.*, 1983; Vaswani *et al.*, 1983) show instances of subjects who appeared to gain nitrogen while losing weight by stringent dieting!

Total body carbon
Neutron activation can also be used to measure total body carbon. The body is bombarded with fast neutrons, and measurement is made of the 4.43 MeV gamma rays. Since carbon is the main energy reserve of the body, and the main source of energy is fat, there is a general relationship between total body carbon and body composition (Kyere *et al.*, 1982).

Total body chlorine
A futher potential application of neutron activation is the determination of total body chlorine. Knight *et al.* (1986) found good agreement between such observations and determinations of total body nitrogen.

Neutron resonance scattering
A final variant is neutron resonance scattering. In this technique, target nuclei are excited by gamma rays, and as they return to a normal state a characteristic gamma emission is observed. This allows the estimation of certain elements such as Fe, Cu and Mn, which are not susceptible to normal neutron activation techniques (Cohn & Parr, 1985).

Dual photon absorptiometry

Local measurements of body composition can be derived from the attenuation of X-rays or gamma rays. Although the method is not widely used for this purpose, Mazess *et al.* (1984) have pointed out the possibility of determining lean tissue mass by dual photon absorptiometry (Chapter 6).

The local estimate has on occasion been extrapolated to the whole body. In the technique described by Gotfredsen *et al.* (1986), the radiation source was ^{153}Gd, and photon peaks were measured at 44 and 100 keV. Some 40–45% of the scanned tissue is free of bone, and the composition at these locations (fat versus lean tissue) must be considered as representative of the remaining 55–60% of the body.

Gotfredsen *et al.* (1986) suggested that body composition data thus obtained correlated quite well ($r = 0.8$) with estimates based on skinfold measurements; an accuracy of ±2.5% body fat was claimed at the cost of a radiation dose of 5 mrad.

Total body water

Methods of estimating total body water have been discussed in detail in Chapter 4. Since a major part of body water is contained in the lean tissue, such figures can be interpreted in terms of lean mass. However, a constant hydration of the body must be assumed. This creates problems if subjects have been engaged in vigorous exercise, exposed to heat, cold or underwater exploration. Aging and severe malnutrition may also cause departures from the anticipated average level of hydration.

Nuclear resonance

Certain aspects of NMR technology have already been discussed in Chapter 3. Some atomic nuclei have spin as well as mass and a positive charge. Thus, if the body is exposed to a suitable static magnetic field, the orientation of the nucleus can be affected in atoms with an odd number of protons or neutrons. Subsequent exposure to an alternating magnetic field of the same frequency but applied at right angles to the static field, induces specific resonances. These disturbances persist for a few milliseconds after the radio-frequency pulse has stopped, with the formation of images that can be interpreted in terms of the chemical constituents of the body (Mallard *et al.*, 1980; Bottemley, 1983; McCully *et al.*, 1988).

Commonly, the area under a spectral peak or the rate of decay of a peak reflects the quantity of a given metabolite in the tissue scanned by the NMR probe (Meyer *et al.*, 1982). However, the NMR technique also appears to be suitable for determining the distribution of water, sodium and phosphorus. The method is non-invasive, and has considerable promise, allowing the estimation of the cross-section of individual muscles (Narici *et al.*, 1988). Because of high cost, practical application of the technique to studies of lean body mass is just beginning (Lewis *et al.*, 1986). One practical difficulty limiting the use of NMR in children is that the subject must remain still for the 10 seconds per slice needed to obtain a satisfactory image (Roche, 1987).

Derivation of lean mass from body fat

Lean tissue mass can be estimated as the difference between total body mass and fat mass. Indeed, this is an important argument for converting skinfold readings to a prediction of body fat. However, problems can arise from inter-individual differences in the density of lean tissue (Schutte *et al.*, 1984).

Methods of estimating body fat are discussed in Chapter 3.

Anthropometry

Anthropometric measurements have on occasion been used to estimate total lean mass. Bugyi (1972) claimed that in children, the lean mass (kg) was approximated closely by the equation:

$$LBM = 2.514 \, (B,cm) \times (H,m)$$

where B was the sum of styloid diameters at the two wrists.

Crenier (1966) proposed the following equations for adults:

$$LBM(\text{men}) = 0.846T + 0.469H + 1.44A - 0.394B - 109.50$$
$$LBM(\text{women}) = 0.935T + 0.173H - 27.73$$

where T was the lean (skinfold corrected) thigh circumference, H was the height, A was the lean arm circumference, and B the acromial diameter, all measured in cm. A precision of 3% was claimed for these formulae.

Steinkamp *et al.* (1965) proposed two relatively complex multiple regression formulae for estimating the lean body mass of men and women respectively:

$$LBM(\text{men}) = M - [0.894W + 2.53S$$
$$+ 1.003C - 0.353A - 35.69]$$
$$LBM(\text{women}) = M - [0.675B - 5.687D + 1.85A - 39.36]$$

where M was the body mass (kg), W was the waist circumference (cm), S was the 'arm' skinfold (cm), C was the iliac crest circumference (cm), A was the arm length (cm), B was the biacromial diameter (cm) and D was the wrist circumference (cm).

Forbes (1987) suggested that lean body mass was proportional to the simple cube of height. However, the practical value of this relationship was compromised by age and sex differences in the constants to the equation.

Regional assessments of muscular development

The potential of examining the regional development of tissues such as fat and muscle by anthropometry, CT and soft tissue radiography has already been discussed in Chapter 3.

Since muscle is the main component of limb bulk, all of these methods work relatively well in estimating muscle cross-section, muscle force (Van Uytvanck & Vrijens, 1966) and even working capacity while using the limbs (Szakall, 1943).

The main difficulty arises from the complex shape of the muscles,

Table 5.3. *Upper arm circumference and estimated upper arm muscle circumference.*

	Females		Males	
Age (years)	Upper arm circumference (cm)	Upper arm muscle circumference (cm)	Upper arm circumference (cm)	Upper arm muscle circumference (cm)
1– 1.9	15.6	12.4	15.9	12.7
10–10.9	21.0	17.0	21.0	18.0
13–13.9	24.3	19.8	24.7	21.1
16–16.9	25.8	20.2	27.8	24.9
19–24.9	26.5	20.7	30.8	27.3
25–34.9	27.7	21.2	31.9	27.9
35–44.9	29.0	21.8	32.6	28.6
45–54.9	29.9	22.0	32.2	28.1
55–64.9	30.3	22.5	31.7	27.8
65–74.9	29.9	22.5	30.7	26.8

(Based on 50th percentile of data of US Health and Nutrition Examination Survey I, as reported by Frisancho, 1981.)

Table 5.4. *Upper arm muscle area and upper arm fat area (cm²).*

	Female		Male	
Age (years)	Muscle (cm²)	Fat (cm²)	Muscle (cm²)	Fat (cm²)
1– 1.9	12.2	7.1	12.8	7.4
10–10.9	23.0	11.4	25.8	9.8
13–13.9	31.3	16.3	35.5	11.0
16–16.9	32.5	20.1	49.5	10.8
19–24.9	34.1	21.7	59.1	14.1
25–34.9	37.8	25.5	62.1	17.5
35–44.9	37.8	29.0	64.9	17.9
45–54.9	38.6	32.4	63.0	17.4
55–64.9	40.5	33.7	61.4	16.5
65–74.9	40.2	30.6	57.2	16.2

(Based on 50th percentile of data of US Health and Nutrition Examination Survey I, as reported by Frisancho, 1981.)

which limits the interpretation of a single cross-section, and makes it impossible to calculate accurate total muscle volumes for the limbs. Frisancho (1981) and Bishop *et al.* (1981) analysed a large sample of US data (8204 males, 10 893 females), calculating age and sex specific norms of upper arm circumference and the corresponding areas (Tables 5.3 and 5.4) according to the equation:

$$C_m = C_t - 3.142T$$

where C_m is the muscle circumference, C_t the total circumference and T the thickness of the triceps skinfold, all measured in cm. The authors admitted that their formula assumed a cylindrical shape for the arm, and that it ignored sex-related differences in humeral diameter and skinfold compressibility.

6   *Estimation of bone mass*

General considerations

This chapter looks at possible methods of estimating bone mass. This quantity is of importance in assessing the maturation of the skeleton in the growing child, in determining the combined responses to nutrition and habitual physical activity in the adult, and in monitoring the extent of osteoporosis in an older individual.

Because bone is a relatively dense tissue, knowledge of a person's bone mass may also contribute to a more accurate interpretation of overall body density (Boileau *et al.*, 1981; Lohman *et al.*, 1984) and thus to a more precise estimation of other body constituents. The weight of the dry, fat-free skeleton may be up to 20% greater in negroid than in 'white' subjects (Trotter & Hixon, 1974; Vickery *et al.*, 1988), and the bones of negroid subjects may also have a higher content of protein and/or non-osseous mineral; the density of lean tissue in this population is thus 1.110–1.113 rather than the standard figure of 1.100 (Schutte *et al.*, 1984; Vickery *et al.*, 1988). If the figure of 1.100 is used in estimating body fat, a negroid person may thus be assigned a totally unrealistic target weight.

Anthropometry

Matiegka (1921) suggested that skeletal mass could be determined from diameters measured at the humeral and femoral condyles, the wrist and the ankle:

$$\text{Skeletal mass (g)} = 1.2 \, D^2 \, H$$

where H was the standing height, and D was the average of the four diameters.

It is quite easy to take caliper measurements of diameters at such key external landmarks as the condyles of the humerus and the femur, and (unlike Matiegka) to correct such diameters for the particular thickness of overlying skin and fat observed in any given individual. However, there remain substantial inter-individual variations in the anatomy of bones, so that conversion of a locally measured diameter to an average cylindrical radius (Shephard *et al.*, 1988) can never be more than a gross approximation, even for a single bone.

Review of radiographs suggested to Shephard *et al.* (1988) that in adults the average effective radius of the bones in the arm was about 21% of the corrected intercondylar diameter, while in the leg the figure was about 23.5%. The volume of bone in the cylindrical part of the limb could then be estimated from the multiple of this effective radius and limb length. Although a useful approach for examining the regional composition of the arms and legs, their method did not allow any assessment of the volume of bone in the hands, the feet, the hips, the vertebrae, the rib cage or the skull. Moreover, even the answer obtained for the limbs was a volume, and in order to calculate bone mass it would have been necessary to assume an average sex and age specific figure for bone density. A typical figure for the skeleton as a whole is about 1.236, but values range from 1.164 to 1.570 in the various bones of the body, with much inter-individual variation in the magnitude of these regional differences.

Bone anthropometry remains of greatest value as a simple method of converting total limb volumes to estimates of limb muscle volumes. Nevertheless, the Society of Actuaries has recently introduced intercon-dylar dimensions as a means of estimating body frame size and thus of improving the interpretation of population 'weight for height' tables (Metropolitan Life Assurance, 1983).

Radiographic densitometry

Simple radiological assessment (Khairi *et al.*, 1976), while of some clinical value, can provide no more than a semi-quantitative indication of bone density. One technique is based upon the continued integrity of the vertebral column. Each vertebra from T3 through to L5 is assigned a score of 1 if normal, 2 if attenuated by more than 15%, 3 if wedge-shaped, or 4 if totally prolapsed (Marcus, 1986). Nevertheless, most radiologists acknowledge that they are unable to diagnose a mineral loss of less than 30–35% on a routine radiograph.

The mineral content of bone can be gauged somewhat more precisely from quantitative densitometric measurements of the shadow seen in carefully standardized radiographs (Merz *et al.*, 1956; Trotter *et al.*, 1959; Griffith *et al.*, 1973; Dequecker, 1976). Positioning of the body part, the voltage of irradiation and the duration of exposure along with the film processing conditions must all be rigidly controlled if quantitative results are to be obtained. Specific technical problems can arise from changes in film sensitivity and a non-linear change in bone shadow densities with increasing radiation exposure (Mack *et al.*, 1949). Griffith *et al.* (1973) give a detailed account of procedures. In essence, densitometer readings taken over the images of bone and soft tissue are compared with the images of aluminium and polystyrene wedges (Nordin *et al.*, 1962) that have similar radiographic densities to bone mineral and water, respectively. The two equations are:

$$d(\text{bone mineral}) = \frac{0.58\ d(\text{Al}) . t(\text{Al}) - ma(st) . d(st) . t(st)}{d(\text{bone mineral}) . t(\text{bone mineral})}$$

where *d* is the physical density and *t* is the thickness of the soft tissue or the calibrating aluminium (Al) wedge, *ma* signifies the mass absorption coefficient, and *st* soft tissue.

$$d(\text{soft tissue}) = \frac{0.2024\ d(\text{plastic}) \times t(\text{plastic})}{ma(\text{soft tissue}) \times t(\text{soft tissue})}$$

where 0.2024 is the mass absorption coefficient of the plastic wedge at 54 kVP.

A series of analyses of bone density made on beef scapulae showed a systematic difference of 2.8% relative to the results obtained by bone ashing. In two embalmed human feet, errors were 5.2 and 15.7%, while in an unembalmed specimen of a human calcaneus, the error was 11.4% (Griffith *et al.*, 1973).

The radiographic picture that is obtained *in vivo* is greatly influenced by shielding from overlying tissue. Unfortunately, lower energy photons are scattered to a greater extent than those of high energy, and such factors as fluid shifts and muscle atrophy can also distort the amount of scattering. The simple approach of radiographic densitometry has thus been largely superseded by such methods as photon absorptiometry and neutron activation.

Photon absorptiometry

Technique
With the usual technique of photon absorptiometry, a ^{125}I source passes low and uniform energy (27 keV) gamma rays through a short length of a bone such as the radius, the ulna or the calcaneus (Cameron & Sorenson, 1963; Mazess *et al.*, 1964; Cameron *et al.*, 1968; Mazess, 1971; West, 1973; Manzke *et al.*, 1975; Dunn *et al.*, 1980; Lohman *et al.*, 1984*b*).

However, Atkinson *et al.* (1978) used [241]Am to scan the femur of growing boys. Transmitted radiation, inversely proportional to bone mineral content (Mazess *et al.*, 1964) is measured by a suitably collimated sodium iodide crystal detector. Both source and detector are moved across the limb, and the apparatus is subsequently standardized by measuring the equivalent transmission of radiation through a phantom (Smith *et al.*, 1976a).

Positioning of the counter is critical to the accuracy of results, particularly if repeat measurements are intended. Some authors have even mounted a cast around the limb (Smith *et al.*, 1976a) and/or have applied tissue-equivalent material to form flat and parallel surfaces above and below the limb. Smith *et al.* (1981a,b) prefer to make their measurements one-third of the distance from the olecranon to the head of the ulna, arguing that there is minimal variation in either bone mineral content or width with minor displacements about this site.

Expression of results

The standard method of examination cannot take account of the shape of the bone or its dimensions in directions other than that which has been scanned. It is therefore usual to estimate bone mineral in g/cm from the absorption of energy that is seen in a given cross-section.

If a figure for bone width is also introduced, the density can be calculated in g/cm^2. Most authors do not attempt to estimate the traditional units of bone density (g/cm^3), although there have been occasional attempts to assign a shape, such as an oval, to the bone under examination. Klemm *et al.* (1976) made volume determinations for the calcaneum by taking several radiographs in the plane of absorptiometric measurement.

It is also possible to make predictions of total bone mineral from standard photon absorptiometry (Mazess, 1971; Cohn & Ellis, 1976), although such calculations can be misleading, because the regional data is not necessarily representative of bone density in the body as a whole.

Accuracy

Lohman *et al.* (1984b) found a 7-day test/retest reliability of 0.06 g/cm in adults, and 0.04 g/cm in children. Because the annual mineral loss is relatively small, a substantial number of subjects must thus be followed to demonstrate even a 50% change in the rate of calcium loss over a longitudinal experiment (for example, 56 test and 56 control subjects followed for 2.7 years, Smith *et al.*, 1976a,b).

Compton scattering

Another option is to exploit the Compton scattering of higher energy (662 keV) photons. The intensity of scattered radiation is directly proportional to the absolute mass density of the irradiated region (bone, minerals and inter-trabecular soft tissues).

Dual photon absorptiometry

Dual photon absorptiometry allows estimation of total body bone mineral content. Because of the importance of vertebral mineral loss in senile osteoporosis, this technique is frequently applied to local bone density determinations on the lumbar spine (Roos & Skoldborn, 1974).

Roos & Skoldborn (1974) originally used a combination of ^{241}Am and ^{137}Cs, but in the dual photon technique used by Price *et al.* (1977), Madsen (1977), Krølner & Pors-Nielsen (1980) and Peppler & Mazess (1981), a ^{153}Gd source (emitting 44 and 100 keV photons) is located 10 cm below the scanning table. A narrowly collimated scintillation detector collects 8000 picture elements (pixels) at a radiation exposure cost of 1–3 mrem. Some 4500 of the pixels are recorded over the body, and in 3000 of these bone is present.

Comparisons against excised skeletons have suggested that the technique has an error of only 1% (Mazess *et al.*, 1984). There is also a close correlation ($r = 0.97$) with neutron activation measurements of total body calcium (Mazess *et al.*, 1981). In 18 normal subjects (4M, 14F) aged 25–60 years, the average bone mineral mass was found to be 6% of lean mass, but there was also an 11% coefficient of variation above this average, emphasizing that variations of bone structure from one person to another are a major source of error when making hydrostatic determinations of body composition.

One small disadvantage of dual photon absorptiometry is that other factors being equal, the total radiation dose is twice that incurred during single photon absorptiometry.

Neutron activation

Technique

Neutron activation (Nelp *et al.*, 1970; Chestnut *et al.*, 1973; Harrison *et al.*, 1975, 1979; Aloia *et al.*, 1978*a*; Shephard *et al.*, 1985*b*) is a specialized technique, available in only a few major hospital centres. When estimating body calcium, the trunk and thighs are bombarded with neutrons from 12 Pu-Be sources to a total dose of 0.4 rem. The stable isotope ^{48}Ca (some 2 g of which is found in the average person) captures the neutrons and becomes converted to the unstable isotope ^{49}Ca. Ths latter emits 3.1 MeV gamma rays as it decays with a half-life of about 8.8 minutes. Data are based on emissions from a third to a half of the bone in the body (Harrison *et al.*, 1975).

Neutron activation also converts phosphorus to ^{28}Al; this can be measured concurrently with the ^{49}Ca, allowing a calculation of calcium/phosphorus ratios. Gamma emissions are measured by transferring the subject to a whole-body counter immediately after irradiation.

Calibration

In order to estimate the calcium content of the trunk and thighs, appropriate corrections must be applied for the absorption of both neutrons and gamma rays by the overlying body tissues. These are estimated from the attenuation of gamma rays from a thorium source, or from standard figures based upon the individual's height and body mass. There is a residual error of some 5% for each 1 cm departure from standard body thickness.

In the experience of the Toronto laboratory, calcium readings have a reproducibility of ±5.5% in human subjects and ±4% in phantoms, although Cohn *et al.* (1972) have claimed an 'accuracy and precision' of 1.0% in an anthropometric phantom. The cost to the body is an exposure to some 300 mrem of radiation (about the equivalent of a standard chest radiograph).

Data interpretation

It should be stressed that a part of the observed gamma emission is attributable to calcium which is located in sites other than bone, particularly muscle. Results must also be considered in relation to body size. One method of interpreting the data is to calculate a relative calcium content (the so-called 'calcium bone index'). For this purpose, the observed calcium value is divided by that predicted for a normal person of equivalent height (Harrison *et al.*, 1979), according to the equation:

$$\text{Ca count} = 207 \, (H)^3 + 17$$

An alternative approach is to match the observed calcium reading to a predicted value, based on age, sex, height and a potassium estimate of lean body mass (Cohn *et al.*, 1976).

Broader application of data

It may finally be enquired whether the neutron activation data can be combined with other information to yield more precise estimates of body fat and lean tissue.

In terms of the ability to predict a skinfold estimate of lean body mass, Mernagh *et al.* (1986) found that the introduction of figures for total body nitrogen and calcium into their multiple regression equations only marginally improved the coefficient of correlation relative to a simple ^{40}K estimate of lean body mass (the respective multiple *r* values being 0.69 and 0.71).

Other techniques

Chapter 4 has discussed the use of computerized tomography (Jensen *et al.*, 1980) and ultrasound to visualize skin, muscle and bone.

Radioactively labelled calcium (Reeve *et al.*, 1978) can be used to assess bone turnover, although the rate of exchange with calcium depots is a relatively slow process. Finally, the rate of excretion of calcium-binding proteins may, in the future, yield a useful index of the rate of bone breakdown (Price *et al.*, 1980).

7   Regional differences of body composition

Much of the variety of human form reflects not overall differences of body composition, but rather inter-individual differences in the regional distribution of fat, muscle and bone. Tanner *et al.* (1959) noted from limb radiographs that the mass of the three tissue components apparently varies relatively independently, both in athletes and in sedentary adults. Nevertheless, the local width of the bone cortex must be somewhat related to the size of the muscles acting upon it, while the mass of the limb must also have some influence upon the size of the muscles moving it.

Regional distribution of body fat

General considerations

Inter-individual differences in the regional distribution of body fat are seen with respect to both the relative proportions of deep and superficial fat, and variations in the distribution of subcutaneous fat over the body surface. In particular, a centralized or centripetal pattern of fat deposition can be distinguished from a peripheral accumulation of fat; on the other hand, soft-tissue radiographs suggest that within any given individual there is a close correlation between the accumulation of fat in the upper and lower limbs (Tanner, 1964).

The pattern of regional fat distribution seems relatively stable within a given person over periods as long as five years (Garn *et al.*, 1988), but is influenced by age, sex, nutritional status, habitual activity patterns and possibly ethnic background.

Ramirez (1987) applied a principal component analysis to data on the thickness of four skinfolds (triceps, subscapular, suprailiac and medial calf). As in the earlier analysis of Shephard *et al.* (1969*b*), a 'general fatness' factor accounted for a large part of the variance in the skinfold readings (69.0% in men, 75.4% in women). The second component (13.2 and 13.9%, respectively, of the total variance) distinguished trunk from limb fat, the third (6.2 and 8.9% of the total variance) upper versus lower body fat, and the fourth (5.2 and 8.5% of the total variance) medial versus lateral fatness.

Bogin & MacVean (1981) suggested that the functional significance of accumulated fat varies with its location; in particular, a centralized deposition of fat helped to protect vital organs against extremes of temperature. This hypothesis may be true in animals, but in humans thermal protection depends much more upon clothing and other techno-logical innovations than upon the regional accumulation of subcutaneous fat.

Age

The influence of growth and development upon the distribution of body fat is considered further in Chapter 8.

Childhood

Young children have relatively little body fat. At birth, the accumulation is greater on the limbs than on the trunk. As growth progresses, the developing child shows characteristic periods of robustness and linearity, with corresponding changes in the ratios of limb length to circumference (Hampton *et al.*, 1966).

Adults

Reports from several early investigators (Pett & Ogilvie, 1956; Young *et al.*, 1963; Shephard *et al.*, 1969*b*; Parizkova, 1977) showed almost no increase of triceps skinfold readings over the entire span of adult life. Since people generally become fatter as they get older, this implies a selective regional deposition of subcutaneous fat.

Shephard *et al.* (1969*b*) commented 'the greatest age-related gains occurred in the subcutaneous tissue of the lower part of the trunk (waist, abdominal and suprailiac folds), and the reported changes in mean skinfold thickness depend on how far this positive information is "diluted" by readings from folds that do not increase in thickness with age'. Likewise, Cronk & Roche (1982) found that until the age of 18 years, the thickness of the triceps skinfold of male subjects exceeded that of the subscapular region, but that between the ages of 18 and 50 years the reverse was true. As subjects become older, there may even be a reduction of fat in the extremities, concurrent with the deposition of fat over the trunk.

Skewing of the data

Shephard *et al*. (1969*b*) commented on age and region related differences in the skewing of skinfold data.

Their subjects were typical city dwellers. In children, skewness and kurtosis were most marked in the distribution of readings for the chest and subscapular folds, the suprailiac region and the chin. However, in older adults no folds except the chin (in men) and the suprailiac (in women) showed a skewing of the data.

The authors speculated that in children the skewing was due to a small proportion of obese subjects within a generally slender population; in older age groups the great majority of the population tested had become somewhat obese, so that the pattern of data distribution again conformed to a normal curve.

Sex

Essential reserves of fat

Some of the inter-individual differences in the patterning of superficial fat have an obvious relationship to the sex of the individual, particularly the deposits of fat found in and around the breasts of the mature female, which apparently constitute an essential reserve of nutrients.

Katch *et al*. (1980) suggested that because of the energy demands of pregnancy and lactation, there was much less potential for intra-individual variations in the amount of stored fat in women than in men. Support for this view is found during exposure to cold, when women lose much less fat than men (Chapter 10). Garn *et al*. (1988) have also commented that, possibly on account of the greater total amount of body fat and the higher proportion of superficial fat, the regional distribution of fat is more stable in women than in men.

Childhood

Some sex-related differences in the amount of hip fat can be detected between boys and girls within a few hours of birth, although sex differences in fat patterning become progressively more obvious as puberty is approached (Parizkova, 1977).

At puberty, there is also a sex-specific change in the ratio of superficial to deep body fat; for example, an average skinfold reading of 8 mm corresponds to 19.3% of body fat in a boy aged 9–12 years, 21.0% of body fat in a girl of the same age, but only 16.0% of body fat in a 13–16-year-old of either sex (Parizkova, 1977). Whereas the developing female normally deposits fat on the hips and over the proximal parts of the limbs, in young male subjects any increase of subcutaneous fat occurs mainly over the trunk; occasionally, male adolescents may be disturbed by a temporary regional distribution of fat that is inappropriate to the gender (Mayer, 1972).

Principal component analyses

Principal component or factor analysis confirm the sex-related differences (Shephard *et al.*, 1969b; Table 3.5). Typically, women accumulate

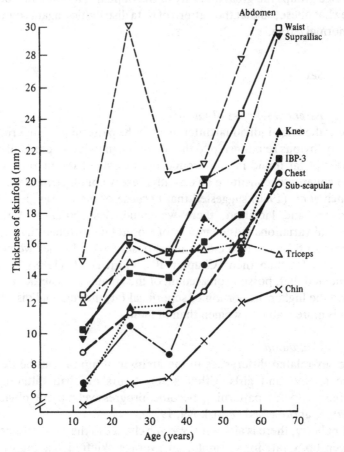

Fig. 7.1. Influence of age upon skinfold thickness of female subjects (based on data of Shephard *et al.*, 1969b). The average of the three folds recommended to the Human Adaptability Project (IBP-3; triceps, subscapular and suprailiac) provides a convenient summary of the data.

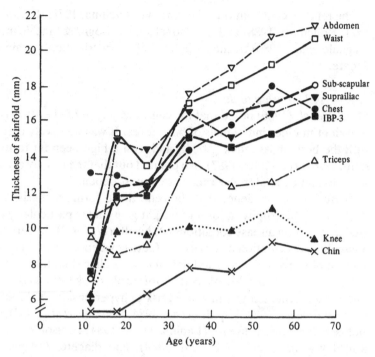

Fig. 7.2. Influence of age upon skinfold thickness of male subjects (based on data of Shephard *et al.*, 1969*b*). The average of the three folds recommended to the Human Adaptability Project (IBP-3; triceps, subscapular and suprailiac) provides a convenient summary of the data.

fat around the hips and the proximal parts of the limbs as they become obese, while men deposit fat on the trunk, both the chest and abdomen (Figs 7.1, 7.2).

Lapidus *et al.* (1984) noted that in women waist/hip circumference ratios ranged from 0.59–1.00. Krotkiewski *et al.* (1983) found average abdominal/buttocks circumference ratios of 0.987 in obese men, but only 0.833 in obese women. In terms of skinfolds, Mueller & Stallones (1981) and Mueller *et al.* (1984) found that the ratio of subscapular to thigh readings gave the clearest indication of a centripetal distribution of body fat, but unfortunately such measurements are not always possible in large scale field surveys. If modesty decrees that such measurements must be limited to the upper half of the body, the distribution of fat (central or centripetal) can be identified correctly from the ratio of triceps plus subscapular to subscapular folds in 72% of subjects (Kaplowitz *et al.*, 1987).

The relative contributions of inheritance (Malina, 1971; Robson *et al.*, 1971; Bouchard, 1988) and nutritional status (Bogin & MacVean, 1981; Dugdale *et al.*, 1980) to such a pattern of fat distribution remain under debate.

Relationship to disease

Mueller *et al.* (1986) noted that the centralized type of obesity was more prevalent in older individuals of both sexes. It was positively correlated with the body mass index (M/H^2) and with a high deep fat to total fat ratio. Garn *et al.* (1982, 1987) also pointed out that the usually calculated skinfold and circumference ratios are fatness dependent.

There is some evidence that the centralized form of obesity reflects adipocyte hypertrophy, a form of weight gain that is particularly associated with certain disease conditions (Ashwell *et al.*, 1982). In particular, a 'masculine' distribution of superficial fat (as evidenced by a high ratio of waist to hip circumference, abdominal to suprailiac skinfolds, or subscapular to thigh skinfolds) is associated with cardiac risk factors (Blair *et al.*, 1988) and a demonstrably increased risk of hypertension (Kalkoof *et al.*, 1983; Blair *et al.*, 1984; Lapidus *et al.*, 1984; Williams *et al.*, 1987), stroke and ischaemic heart disease (Lapidus *et al.*, 1984; Larsson *et al.*, 1984; Reichley *et al.*, 1987; Bouchard, 1988), and diabetes (Vague, 1956; Albrink & Meigs, 1964; Feldman *et al.*, 1969; Szathmary & Holt, 1983; Joos *et al.*, 1984).

It has further been demonstrated that a centralized pattern of fat distribution is associated with insulin resistance; the individuals concerned show high serum insulin levels and a poor glucose tolerance (Evans *et al.*, 1984; Bouchard, 1988). Possibly, the fat patterning serves as an indicator of blood levels of hormones that modify both fat mobilization and cardiac risk (Vague *et al.*, 1971; Evans *et al.*, 1983). In support of this view, Frisancho & Flegel (1982) noted a correlation between early maturation and a centripetal distribution of fat.

Needle biopsy indicates that the hypertrophied fat cells are capable of a high rate of lipolysis, with a correspondingly high output of free fatty acids, which could make an important contribution to the development of glucose intolerance, hyperinsulinaemia and hypertriglyceridaemia (Kissebah *et al.*, 1982; Williams *et al.*, 1987). There are suggestions of important metabolic differences not only in the overall characteristics of hypertrophied fat cells, but also between fat depots in the abdominal and gluteal regions (Smith, 1983). The proximity of 'central' fat depots to the portal circulation may be a further factor increasing susceptibility to metabolic diseases.

Deep versus superficial fat

Computed tomography now provides a relatively simple non-invasive method of determining the amount of intra-abdominal fat (Borkan *et al.*, 1982*a*; Dixon, 1983; Tokunga *et al.*, 1983). Use of this technique has revealed an association between a centralized distribution of body fat and a high ratio of intra-abdominal to subcutaneous fat (Ashwell *et al.*, 1985). In women, the average intra-abdominal/subcutaneous ratio is 0.31, but in men, the proportion of intra-abdominal fat is generally higher (Dixon, 1983; Grauer *et al.*, 1984). In those subjects with a centralized fat distribution the ratio may rise as high as 0.61.

Nutritional status

Malnutrition and over-nutrition are considered further in Chapter 10. Moderate under- and over-nutrition both lead to changes in the distribution of subcutaneous fat.

Under-nutrition

In poorly nourished groups, 'essential' trunk fat is apparently conserved, while little fat is retained over the arms (Bogin & McVean, 1981; Johnston *et al.*, 1984).

Perhaps because of a low level of nutrition and a resultant alteration in the distribution of body fat (Eveleth, 1979), the generally applied skinfold prediction equations (Chapter 4) have proven unsatisfactory in several populations, including Indian women (Satwanti *et al.*, 1977) and older men in New Guinea (Norgan, 1982). However, the same formulae have been used relatively successfully in other populations from a number of developing countries (Shephard, 1978*a*).

Over-nutrition

If food intake exceeds energy expenditure, the thickness of subcutaneous fat is increased at characteristic sex-specific sites (the hips and thighs in women, the abdomen and chest in men, Björntorp, 1983*a*). The local accumulation of fat can subsequently be reduced by an appropriate exercise and/or dietary reducing regimen.

Després *et al.* (1985) found a 22% decrease of average trunk folds, but only a 12.5% decrease of extremity folds, when initially sedentary (but not particularly obese) young men participated in a rigorous 20 week physical training programme.

It has been suggested that the differential fat loss which occurs during exercise and/or dieting reflects regional and sex-specific differences in the sensitivity of the adipocytes to catecholamines (Smith *et al.*, 1979; Lafontan *et al.*, 1979; Lafontan & Berlan, 1981).

Superficial versus deep body fat

The proportion of subcutaneous to deep body fat generally increases as a person becomes more obese (Edwards *et al.*, 1955; Allen *et al.*, 1956; Durnin & Womersley, 1974). Conversely, long-distance runners (who have a low total body fat) tend also to have extremely little subcutaneous fat (Després *et al.*, 1985).

Interpretation of the relative amounts of superficial and deep body fat becomes complicated in severe protein deficiency malnutrition; oedema increases total body water, thickens subcutaneous tissue and alters the compressibility of the skinfolds. In extreme cases of malnutrition (Dean, 1965), 85% of total body mass may be attributable to water.

Habitual activity patterns

Various entrepreneurs have promoted fat-mobilizing devices that supposedly trim fat from selected regions of the body such as the thighs or the buttocks, without significant energy expenditure on the part of the subject. For example, unfortunate obese or figure-conscious young women have been persuaded to spend long hours attached to devices that induced vibration in the undesired deposits of subcutaneous fat. Given that fat is mobilized by lipases acting upon individual adipocytes, the only possible basis for such a regional loss of fat would be the development of an irritant reaction to the vibrating belt. If a subject were to engage in prolonged and vigorous exercise having induced a local hyperaemia in this manner, then it is just conceivable the affected region might receive a larger dose of circulating catecholamines, with a potential for an enhanced local activation of lipases.

The same type of phenomenon might occur in the opposite sense if a particular region of the body was kept unusually still during the pursuit of a particular sport (Table 3.9). Thus, comment has already been made on the relative fatness of the knee folds in whitewater paddlers (Sidney & Shephard, 1973), while older cyclists have been found to have thick triceps folds (Mittleman *et al.*, 1986).

A further possible influence of physical activity upon skinfold thicknesses arises from a local stretching and thus a thinning of the subcutaneous tissue as the underlying muscle hypertrophies. Krotkiewski *et al.* (1979) have suggested that unilateral isokinetic strength training can cause a significant local reduction of skinfold thickness, without any corresponding change in the size of individual fat cells.

Ethnic background

Genetic influences

The influence of genetic factors upon body composition is considered in detail in Chapter 9. Bouchard *et al.* (1985) and Bouchard (1988) have suggested that genetic factors determine a substantial proportion of the variance in both overall obesity, and also the distribution of subcutaneous fat between the trunk and the extremities. In particular, several reports have suggested differences of fat patterning between 'black' and 'white' populations, with the black subjects showing a greater amount of fat on their shoulders and backs, and the 'whites' more fat on the abdomen and thighs (Robson *et al.*, 1971; Watson & Dako, 1977; Harsha *et al.*, 1980; Vickery *et al.*, 1988).

Centripetal distribution

West *et al.* (1974) concluded that North American Indians were particularly prone to a centripetal distribution of fat. Others, also, have commented on populations where there seems to be a high proportion of subjects with a centralized distribution of body fat, for example the Dogrib Indians in the North West Territories of Canada (Szathmary & Holt, 1983). Ramirez & Mueller (1980) described a characteristic patterning of fat in Tokelau (Polynesian) children.

It is less clear that such ethnic differences have a genetic basis. After studying fat distribution among the Dogrib Indians of northern Canada, Szathmary & Holt (1983) suggested that over-nutrition and the stresses associated with acculturation to a 'white' urban lifestyle might well account for a concentration of centripetal tendencies in the populations concerned. In support of the latter hypothesis, Ramirez (1987) noted that the proportion of fat on the trunk and upper body was much higher in Tokelau Islanders who had migrated to New Zealand than in non-migrants.

Ratio of superficial to deep fat

The constancy of the relationship between subcutaneous and deep body fat is a critical issue when using skinfold equations developed on one population to predict the total percentage of body fat in a different population.

Controversy concerning this question has not been helped by uncertainties regarding an appropriate 'gold standard' for total body fat (Chapter 3). Some authors have maintained that there is no such thing as a generalized skinfold equation. Even if a formula is carefully validated on a second sample of a given population, it cannot necessarily be applied to a different population (Satwanti *et al.*, 1977; Eveleth, 1979; Norgan, 1982). This argument apparently has some justification if the groups that are to be compared differ widely in average body composition (for example, the sedentary versus the highly active, or the malnourished versus the well-nourished and/or obese).

Thus, Shephard *et al.* (1973) commented on the apparent discrepancy between the very thin average skinfolds of a Canadian Inuit population, and the substantial total percentage of body fat estimated for the same subjects using a deuterium dilution method. Unfortunately, there remain two possible explanations of their findings. On the one hand, the Inuit may have had a much higher proportion of deep body fat than their 'white' counterparts, but on the other hand there may have been a systematic error in their deuterated water estimates of total body fat. Simple carbon monoxide estimates of blood volume suggested that the hydration of the Inuit had been decreased by vigorous exercise in cold and dry air, and the abnormally low tissue water content would have biassed the deuterium estimate of lean mass in a downward direction.

Parizkova (1977) maintained that ethnic differences in superficial versus deep fat distribution were small, even when groups with very different morphological features were compared (for example, Taiwanese and Czechoslovakians).

Regional distribution of muscle

An unusual regional development of muscle may reflect either inheritance or local training. We shall discuss the influence of local muscle size upon aerobic power and athletic performance in these contexts.

Muscle development and aerobic power

The regional distribution of muscle has an important influence upon the subject's ability to undertake specific tasks, particularly those requiring fatiguing endurance exercise. The force developed by a muscle is proportional to its cross-sectional area (Shephard, 1982*a*), with the consequence that large muscles in a given region allow the same force to be developed at a lower figure per unit of cross-section; this in turn facilitates perfusion of the part, lessening fatigue and increasing aerobic performance.

Empirical results show that the power which can be developed when operating a normal cycle ergometer is more closely correlated with the local volume of the leg muscles than with the total body mass or the lean body mass (Davies, 1971). Because the arm muscles are smaller than those in the legs, the aerobic performance when using an arm ergometer is linked yet more closely to the local volume of active muscle (Shephard *et al.*, 1988).

Regional muscular development in athletes

Inter-individual differences in the regional distribution of muscle have been studied most fully in athletes, where (through a combination of competitive selection and training) there is an unusual development of the muscles contributing to success in a particular sport.

Sommer (1985) noted that because there was difficulty in learning complex motor skills after the age of 14, some quite young children were now being subjected to intensive unilateral training in special sports schools. He argued that the resultant muscle asymmetry and imbalance had undesirable effects upon the development of other body parts.

Local muscular training tends to have more effect on the arms (which normally get little exercise) than on the legs (which get some exercise from normal walking, Tanner, 1964). The classical study of Tanner (1964) examined 137 track and field competitors at the Rome Olympic Games. He noted the perils of using simple ratios to compare arm muscle width with calf or thigh muscle width, given that most linear regressions had a significant intercept. By applying a simple grid as a form of covariance analysis (Fig. 7.3), he demonstrated that throwers had arms which were much more muscular than their calves, when compared with track athletes. The muscles of the throwers were also much heavier than those of the track competitors when related to bone width (Fig. 7.4).

Eiben (1980) compared the local muscular development of female volleyball, basketball and handball players. The volleyball players were well adapted to a relatively stationary game. He noted that their bodies were not very muscular, with the exception of the upper arms, which were powerfully muscled. Basketball was a sport that demanded more movement than volleyball, and perhaps in consequence the basketball players had much more muscular extremities. Handball players were characterized by very powerful shoulders and upper arms. Eiben also commented upon differences of body form among the various categories of fencers. The foil fencers were short and light, with little by way of unusual muscular development. Epée fencers had muscular arms, but relatively slender legs, while sabre fencers had robust upper and lower extremities.

Maughan *et al.* (1983) used a CT scan technique to compare marathon runners, sprinters and non-athletic subjects. The marathon runners had

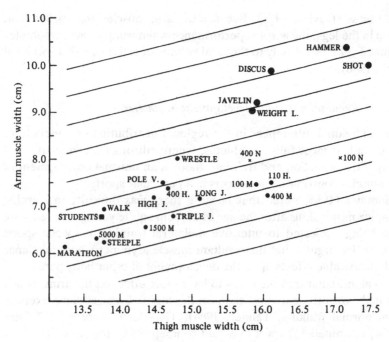

Fig. 7.3. Relationship between muscle cross-section of arms and of thighs in selected groups of Olympic athletes. Note large arms of throwers and weight-lifters (heavy circles) relative to runners. The diagonal grid indicates the slope of the relationship within a given athletic discipline. (From J. M. Tanner (1969), by permission of the publishers, George Allen & Unwin.)

much the same thigh dimensions and muscle strength as control subjects, but the sprinters had a 10% greater muscle cross-section and an 18% advantage of muscle strength. Nelson & Craig (1978) likewise noted a 6% increase in thigh and calf circumference when a young man cycled across the United States.

Maas (1974) examined the influence of hand dominance, finding that the left arm was larger in left-handed subjects. Likewise, Gwinup *et al.* (1971) demonstrated significant muscle hypertrophy of the dominant forearm in tennis players.

One practical consequence of unusual regional development is a corresponding shift in the distribution of body potassium (Buskirk & Mendez, 1984; Garrow, 1982); this can have consequences for the measurement of lean tissue mass by the ^{40}K method (Chapter 5).

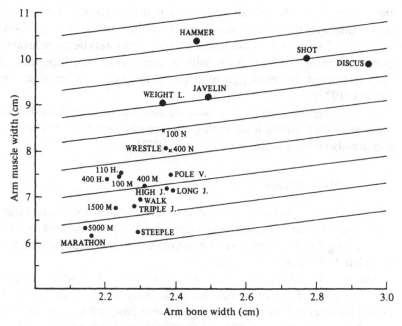

Fig. 7.4. Relationship between muscle width and bone width in the arms in selected groups of Olympic athletes. Note that throwers and weight lifters (heavy circles) have large muscles relative to their bones. The diagonal grid indicates the slope of the relationship within a given athletic discipline. (From J. M. Tanner (1964), by permission of the publishers, George Allen & Unwin.)

Regional distribution of bone

The regional development of bone is here discussed in relation to growth, ethnic differences, athletic selection and rigorous athletic training.

Growth

The bones in specific regions of the body grow rapidly at individually characteristic stages in the course of a person's development. Thus, some 90% of the growth in the skull and the brain occur in the first five years of life, and development in this region is virtually complete by 10 years of age (Fig. 1.7).

In the upper limbs, the hands develop earlier than the forearms, and the forearms are in turn consistently closer to their final adult size than are the upper arms. A similar pattern of regional growth (feet > lower legs > thighs) is seen during development of the lower limbs (Harrison *et al.*, 1964). Particularly at adolescence, there may also be asynchrony of growth, leaving the individual with a short waist and hips that are disproportionately developed relative to the shoulders, or vice versa (Mayer, 1972).

As sex differentiation occurs, the male develops broader shoulders and narrower hips than the female; the arms (particularly the forearms) and legs are also longer in the male (Mayer, 1972).

Ethnic differences

The relative proportions and shape of the various bones are largely inherited characteristics, although dimensions may be modified by vigorous training if this is undertaken before closure of the epiphyses has been completed (Shephard *et al.*, 1978).

At one time, anthropologists attached considerable adaptive value to both overall and regional differences in body form. Thus Bergmann (1847) commented 'Within a polytypic warm-blooded species, the body size of the sub-species usually increases with decreasing temperature of its habitat'. J. A. Allen (1877) noted an increased linearity of body form in parts of the world where the environmental temperatures were high. The population of such regions was reputed to show a large limb to trunk length ratio, with a small relative limb diameter. Both authors were in essence postulating that the tropical residents had maximized their body surface area as a means of facilitating heat dissipation during heavy work (see Chapters 10 and 11). In support of this idea, both West Africans (Schreider, 1957) and Australian aborigines (Wyndham, 1966) were shown to have very long arms, and studies within the US demonstrated that blacks had longer extremities with relatively narrow hips (Krogman, 1970; Malina, 1973; Malina *et al.*, 1987; Martorell *et al.*, 1988). On the other hand, critics of the hypothesis have noted that groups with widely differing stature and relative limb length (such as the African Hoto and Twa) thrive under apparently similar geographic and environmental conditions (Hiernaux, 1966).

Moreover, in groups such as the Lapps, the Eskimos and the Japanese, stature, limb circumferences and the ratio of leg length to sitting height, for long regarded as fixed racial characteristics, have changed quite rapidly with acculturation to a sedentary, urban lifestyle (Shephard, 1978*a*; Tanner *et al.*, 1982). Finally, the adaptive value of a linear body form remains quite debatable. Convective heat exchange is certainly increased by a linear body form, but in the desert this increases heat gain, adding to the thermal load imposed by bright sunlight. Moreover, at night the large surface area associated with a linear body form causes a rapid loss of heat by both radiation and convection, creating substantial cold stresses for groups such as the Kalahari bush dweller and Australian aborigines.

Interestingly, a small body has also been suggested as having adaptive value in dense jungle conditions (Roberts, 1953; Newman, 1960). It has been argued that a diminutive form not only allows a person to crawl through dense vegetation, but also (by reducing energy demands) helps survival when food is in short supply. However, the classical pygmy populations of the African jungle, like the Lapps and the Eskimos, are currently undergoing a rapid increase of average height that is out-dating many of these anthropological 'laws' and hypotheses (Ghesquière, 1971).

Athletic selection

Athletes tend to be selected for a particular sport on the basis of unusual regional bone dimensions (Kohlrausch, 1929; Telkka *et al.*, 1951; Correnti & Zauli, 1964; Tittel, 1965). Certain ethnic groups have an advantage in specific athletic events for this reason. North American negroes commonly have longer legs than their 'white' counterparts (Tanner, 1964). On the other hand, Orientals have classically shown a shorter leg to trunk-length ratio than 'white' subjects, although the recent secular trend to an increase of stature (which has been particularly fast in the Orient, Asahina, 1975) is now eliminating some of these inter-population differences.

It is a significant advantage for weight-lifters and gymnasts to have a low centre of gravity, and thus a low limb to trunk-length ratio (Maas, 1974), while the converse is true for high-jumpers. Sprinters also tend to have short legs. While this offers some mechanical advantage as far as the natural frequency of limb oscillation is concerned, a part of the association may be indirect; muscular individuals tend to mature early, and premature closure of the epiphyses may leave them with short legs. It is also well recognized that tall people have legs of above average length in relation to their height (Rother *et al.*, 1973). Among the throwers, arm length appears to be long in relation to leg length, while walkers have an unusual hip width in relation to trunk length.

Tanner (1964) has used a covariance grid to demonstrate these relationships (Figs 7.5 and 7.6). Limb radiographs show further that the throwers have very wide arm bones relative to the dimensions of their

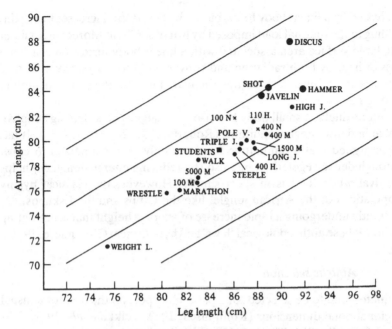

Fig. 7.5. Relationship between arm length and leg length in selected groups of Olympic athletes. Note that the throwers (heavy circles) have long arms relative to their legs. The diagonal grid indicates the slope of the relationship within a given athletic discipline. (From J. M. Tanner (1964), The Physique of Olympic Athletes, by permission of the publishers, George Allen & Unwin.)

legs, although it is unclear whether this is a response to traction of the active muscles on the periosteum of the arm bones or merely the competitive selection of individuals with very large upper limbs (Tanner, 1964). Other authors have also noted regional differences of bone structure in relation to limb dominance and participation in specific categories of sport. Ruff & Jones (1981) summarized data from cadavers and archaeological collections showing that the right upper limb was typically 1–3% longer and 2–4% heavier than the left, while the left lower limb was 1% heavier and longer than the right. The asymmetry is greater in men than in women; it apparently varies from one racial group to another (Trotter & Gleser, 1952; Latimer & Lowrance, 1965) and in both sexes it declines with age. While an effect of limb dominance seems the most plausible exlanation of both the asymmetry and the effects of age, race and sex, Ruff & Jones (1981) have speculated that there might also be a compensatory redistribution of bone to compensate for overall effects of aging. Schultz (1926) noted femoral asymmetry in the foetus

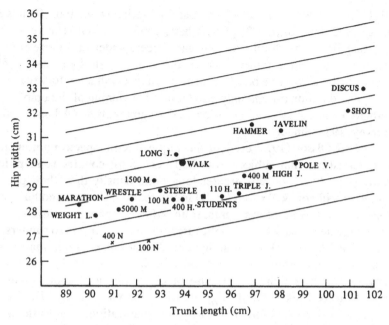

Fig. 7.6. Relationship between hip width and trunk length in selected groups of Olympic athletes. Note that in comparison with track athletes, the race walkers (heavy circle) have broad hips relative to their trunk length. The diagonal grid indicates the slope of the relationship within a given athletic discipline. (From J. M. Tanner (1964), by permission of the publishers, George Allen & Unwin.)

from the age of 4 months, while Plato *et al.* (1980) noted greater cortical thickness in the right second metacarpal, independent of dominance. However, Nilsson & Westlin (1971) were unable to detect any asymmetry of femoral bone mineral content in non-athletes.

Effects of athletic training

While people with an appropriate body build undoubtedly gravitate to the corresponding type of sport, there is also some evidence that athletic participation can itself lead to overall growth and strengthening of the more frequently used bones, with a strengthening of the local trabecular architecture. As a result, one frequently reported review noted hypertrophy of the playing arm in tennis players (Steinhaus, 1933). More recently, Toriola *et al.* (1987) commented that Nigerian basketball players had a broader humeral intercondylar diameter than volleyball players.

Studies of tennis (Buskirk *et al.*, 1956; Gwinup *et al.*, 1971; Jones *et al.*,

1977; Huddleston *et al.*, 1980) and baseball (King *et al.*, 1969; Tullos & King, 1972; Watson, 1973) players have confirmed that relative to the opposite limb, the dominant arm has longer, wider and more robust bones, with greater mineralization and density. In some cases (for example, the elbow of a baseball thrower), a local inflammatory reaction to repeated minor trauma may also lead to a widening of the epiphysis, with its eventual demineralization and fragmentation (Adams, 1968; Harvey, 1986; Singer, 1986).

Dalén & Olsson (1974) used the technique of X-ray spectrophotometry to demonstrate that subjects who had been cross-country runners for many years had a 6–19% greater long bone density than their peers. Likewise, Nilsson & Westlin (1971) found that the femoral and humeral bone mineral content was greater in athletes than in controls, the advantage decreasing from weight-lifters through throwers to runners.

Aloia *et al.* (1978*a*) found an increased total bone mass in marathon runners. On the other hand, there have been suggestions that excessive pressure on the epiphyses, because of either physical labour (Wurst, 1964; Kato & Ishiko, 1966) or gymnastics (Shephard *et al.*, 1978*b*) can cause a local retardation of growth and maturation. For those who recommend vigorous exercise to the young, the problem still remains, as suggested by Steinhaus (1933), of determining an intensity of activity that will stimulate rather than retard normal bone development.

8 Developmental changes, growth and aging

The processes of growth and maturation reflect the combined response of the body to genes, hormones, nutrients and other environmental influences. Anthropometric data, therefore, represent the phenotype – the extent to which genetic potential has been realized in a given environment.

Growth-promoting hormones serve mainly as a 'linear amplifier' of tissue synthesis; depending upon favourable environmental conditions, some growth is likely even in the absence of growth hormones. At the opposite extreme, the illegal administration of growth hormones to an athlete may hasten maturation, but stunts ultimate development by precipitating a premature closure of epiphyses in the long bones.

The form and relative proportions of the various body constituents change progressively over the course of both growth and aging (Table 8.1). Possible techniques of adjusting data for size effects have been explored in Chapter 1.

Table 8.1. *Relative size of various organs in the newborn child and adult.*

Organ	Percentage of total	
	Newborn	Adult
Muscle	25	40
Fat	10–15	15–25
Skeleton	13	14*
Bone	4	7.0
Cartilage	9	1.6
Skin	15	7
Viscera	27–35	21–31
Heart	0.7	0.5
Liver	5.0	2.6
Kidneys	1.0	0.5
Brain	13.0	2.0

*Includes marrow and periarticular tissue.
(Based on data compiled by Forbes, 1987.)

General and regional patterns of growth

General growth patterns

Body size and body mass increase rapidly over the first two years of childhood. There is then steady growth for six or more years, a sharp acceleration before and during puberty, and a final deceleration in late adolescence (Tanner, 1962).

Longitudinal observations on children in Norway and in West Germany (Rutenfranz *et al.*, 1984) showed that the median age of peak height velocity was 14.5 years in Norwegian boys, 12.8 years in Norwegian girls, 14.6 years in German boys and 13.9 years in German girls. However, vital capacity (and by implication, the dimensions of the thoracic cage) does not peak until about 24 years of age.

Most boys end their growth in height 3–4 years after reaching their peak height velocity, while girls complete their growth in height 2–4 years after the greatest observed height velocity. Total body mass and lean body mass increase somewhat more than predicted from the increase of height during the final 3–4 years of growth.

Normality of growth

Normality of growth is usually assessed relative to standard curves for height, body mass, and body mass in relation to height (Chapter 1), making due allowance for regional differences of nutrition and growth, in order to avoid imposing unrealistic standards on Third World populations. Account must be taken of secular trends towards earlier maturation and an increase of adult size. Previous generations failed to realize their genetic potential as a result of poor environmental conditions. However, in many societies, environmental factors are now near optimal for the growing child, and secular trends to earlier maturation and an increase of adult size are coming to an end.

Some anthropometric data such as body mass show significant skewing (Jéquier *et al.*, 1977), and for such data interpretation should be based on percentiles rather than mean values. The expected range of normal values (5th to 95th percentiles) can be narrowed substantially by including information on the size of the parents and the skeletal maturity of the individual (Weech, 1954; Garn & Rohmann, 1966). It remains unclear whether regular vigorous physical activity slows or accelerates growth (Parizkova, 1974; Shephard *et al.*, 1978*b*; see Chapter 10).

Developmental age

Several authors (Bouchard *et al.*, 1968; Beunen *et al.*, 1972; Kemper & Vershuur, 1974) have argued that the morphological characteristics of the individual are more closely correlated with developmental than with chronological age.

Developmental age is most accurately determined from radiographs or assessments of sexual maturity. A reluctance to expose the growing child to either unnecessary X-rays or sexual assessments has led to a search for alternative indices of maturation, including the eruption of the teeth (Shephard *et al.*, 1978*b*) and observations of body shape in partially clothed subjects (Janos *et al.*, 1985). Janos *et al.* (1985) claimed that the sum of age equivalents based on stature, body mass and Conrad's plastic index (the sum of biacromial distance, lower arm girth and hand circumference) showed a correlation of 0.92 with the radiographic assessment of maturity.

Prediction of adult form

The prediction of adult body form has considerable practical importance when selecting child athletes for special sports schools (Mészaros *et al.*, 1984). Bar-Or *et al.* (1974) attempted to predict the aerobic power of boys two years hence, using an equation based upon lean body mass, penis development and height. Tanner *et al.* (1956) found that 62% of the variation in adult size could be predicted from the standing height at an age of three years, the respective linear equations being

$$\text{Adult height} = 1.27\,(H, \text{cm}) + 54.9 \text{ for boys}$$

and

$$\text{Adult height} = 1.29\,(H, \text{cm}) + 42.3 \text{ for girls}$$

Weech (1954) proposed a somewhat similar formula, based on the height H at age 2 years and the averaged stature of the two parents P:

$$\text{Adult height} = 0.545\,(H, \text{cm}) + 0.544\,(P, \text{cm}) + 37.7 \text{ for boys}$$

and

$$\text{Adult height} = 0.544\,(P, \text{cm}) + 0.544\,(P, \text{cm}) + 25.6 \text{ for girls}$$

Bayer & Bayley (1959) developed tables that allowed an assessment of skeletal age to be incorporated into the prediction.

Sex differences

Probably because of the influence of testicular androgens, girls are on average marginally smaller than boys during early childhood. However, since they also reach puberty earlier, girls become heavier (and sometimes taller) than boys between the ages of 12 and 14 years. The female advantage of body mass is particularly marked in societies where the adolescent girl remains physically active (for instance, in Inuit populations, where teenage girls carry smaller members of the family on their backs throughout much of the day, Rode & Shephard, 1973a).

Growth of body mass

Girls show a rapid increase of body mass at 12 or 13 years of age, associated with an accumulation of fat in the breasts and around the hips. In boys, the pubertal growth spurt occurs one to two years later; it lasts longer than in the girls, reflecting in part an increase of physical dimensions and in part, an increase of muscle mass per unit of stature. The timing of the pubertal spurt is influenced by socio-economic conditions, being later in rural areas and in isolated populations such as the Inuit (Rode & Shephard, 1973a).

Mirwald & Bailey (1986) noted that in 61 of 75 boys and 17 of 22 girls the peak height velocity either coincided with or preceded the peak velocity for body mass. In boys, the two events were quite closely correlated ($r = 0.89$), but in girls, gains of overall body mass were distorted by a substantial accumulation of subcutaneous fat, and the corresponding coefficient of correlation ($r = 0.45$) was relatively weak.

Development of individual organs

At birth, the brain, viscera and skin together account for a larger proportion of the total body mass than in the adult (Table 8.1). On the other hand, the relative muscle and fat content of the neonate is less than would be anticipated in a mature individual.

The large percentage of body mass which is attributable to skin in the newborn, reflects in part a large relative surface area and in part, the high water content of the skin (Forbes, 1987). Water accounts for some 83% of body mass at term, the high extracellular water content facilitating the transport of nutrients to the developing cells (Forbes, 1987). Based upon analyses of body water, nitrogen and ash, Moulton (1923) concluded that many species reached 'chemical maturity' when about 4–5% of their lifespan had elapsed. Human body composition generally stabilizes by the age of 3–4 years, although the precise age varies with the constituent (Spray & Widdowson, 1950); for instance, potassium concentrations stabilize earlier than calcium levels.

Body composition during gestation and neonatal life

Overall course of growth

All tissues grow very rapidly in the foetus. Over the first few weeks, the relative increase in tissue mass is in excess of 105% per day, and even in the third trimester of pregnancy the growth rate is 1.5% per day, compared with a mere 0.04% per day during the adolescent growth 'spurt'. Hyperplasia gives place progressively to hypertrophy, with a decrease of tissue DNA content (Rozovoski & Winick, 1979). The water content of the various tissues declines, while there is a roughly parallel increase in the proportions of protein and fat.

The absolute rate of tissue growth reaches a maximum about the 36th week of pregnancy. Reasons for the slowdown at term are unclear. Placental inadequacy is unlikely, since the normal foetus has a much larger available mass of placental tissue than that available to twins, even in the final weeks of pregnancy. Possibly, the limiting factor is the maximum potential size of the uterus (McKeown & Record, 1952; Hytten & Leich, 1971). Unusual characteristics of the foetus include a low proportion of dry, fat-free mass in the skeletal tissue, and a significant proportion of metabolically hyperactive brown fat in the adipose tissue (Hull & Smales, 1978; Houdas & Ring, 1982).

Body fat

A high proportion of foetal body fat is subcutaneous (Southgate & Hey, 1976). Skinfold thicknesses increase by about 0.18 mm/week between the 26th and the 40th weeks of pregnancy (Whitelaw, 1979). Oakley *et al.* (1977) found thicknesses peaked at 38 weeks for the triceps fold, and 39 weeks for the subscapular fold. Because subcutaneous fat increases with maturation, there is a fairly close correlation between body mass and skinfold thickness at birth (Farr, 1966; McGowan *et al.*, 1975; Oakley *et al.*, 1977). Inter-individual differences in fat stores can be correlated with skinfold thicknesses and thus the nutrition of the mothers (Frisancho *et al.*, 1977).

Brown adipose tissue (BAT) accounts for up to 10% of all body fat at birth (Merklin, 1974); it has a very high heat-generating capacity when stimulated by catecholamines (Shephard, 1985a), and is a site of lipid synthesis (Minokoshi *et al.*, 1988). After birth, BAT accounts for a progressively smaller fraction of total body fat (Hassi, 1977), although some authors have described deposits even in adults after exposure to severe cold (Heaton, 1972; Huttunen *et al.*, 1981).

Body water and minerals

During early foetal development, the cells have a high relative content of water, sodium and chloride, but a low content of calcium (Tables 8.2 and 8.3). As development proceeds, the body becomes enriched with respect to calcium, phosphorus, magnesium, nitrogen, iron, copper and sulphur. Potassium concentrations show little change, while the concentrations of sodium, zinc and chloride diminish (Needham, 1950; Forbes, 1987).

Changes in mineral concentrations during the latter part of gestation (Table 8.2) seem to be associated with a relative reduction in the volume of extracellular fluid. At birth, the chloride to potassium ratio of 1.25 still remains much above the usual adult figure of 0.73 (Forbes, 1987). It commonly takes 2–3 days for infants to establish an adequate intake of fluids and food energy. During this early phase of neonatal life, body mass declines by 5–10%, due to a loss of both water (mainly extracellular) and solids (mainly carbohydrate and fat rather than protein, Anderson *et al.*, 1979; Cheek *et al.*, 1984).

Birthweight is typically regained within a week of delivery, although premature infants require rather longer to make good early losses. The extent of dehydration in the neonate can be gauged from the decrease in skinfold readings between 15 and 60 seconds of compression (Brans *et al.*, 1974); the skinfolds of the dehydrated infant are less compressible than those of a well-nourished child.

During the first few hours of extra-uterine life, blood volumes depend greatly upon such factors as the feeding schedule and the time of clamping of the umbilical cord (Usher *et al.*, 1963). An early clamping of the cord provides a larger initial store of haemoglobin. Over the next few days, the ratio of plasma to total body mass usually increases because of tissue dehydration, but haemolysis of foetal red cells leads to a decrease in haematocrit. Probably because of associated glycogen depletion, there is also a substantial decrease of body potassium stores (Forbes, 1987).

Table 8.2. *Body composition at selected points of foetal life.*

Gestational age (weeks)	Total mass (g)	Lean mass (g)	Fat (g)	Nitrogen (g)	Sodium (g)	Potassium (g)	Calcium (mg)	Iron (mg)	Copper (mg)	Zinc (mg)
25	800	790	10	9.0	0.66	0.97	3.33	0.46	2.34	12.5
30	1480	1400	80	17.2	1.87	1.79	6.77	0.93	4.60	21.0
35	2450	2250	200	33.0	3.18	3.27	13.80	1.89	9.04	35.4

Change in body content of individual elements can be represented by an exponential equation of the type N_2 (mg) $= 346.7\ e^{(0.0186\,days)}$ over the period 24–36 weeks of gestation (Shaw, 1973). The initial rate of increment is greater than suggested by this equation, and the accumulation of individual constituents slows further in the final weeks of pregnancy. (Based in part on data of Shaw, 1973.)

Table 8.3. *Influence of age on body water content.*

Age	Extracellular water (% body mass)	Intracellular water (% body mass)	Total body water (% body mass)
0–1 day	44.5	33.9	78.4
1–30 days	39.7	31.8	74.0
1–3 months	32.2	43.3	72.3
3–6 months	30.1	42.1	70.1
6–12 months	27.3	35.2	60.4
1–2 years	25.6	33.6	58.7
2–3 years	26.7	38.3	63.5
3–5 years	21.4	45.7	62.2
5–10 years	22.0	42.3	61.5
10–15 years	18.7	46.7	57.3

(Based on data reported by Friis-Hansen, 1961.)

Bone

The relative skeletal mass almost doubles, from 1.7% of total body weight in a small foetus, to 3.2% in a foetus nearing maturity (Trotter & Peterson, 1969). Photon absorption measurements suggest an exponential increase of bone mineral content with time (Greer *et al.*, 1983), while radiographs of the humerus in the newborn infant suggest a linear relationship between the cross-sectional area of the bone cortex and total birth mass (Poznanski *et al.*, 1980). However, figures for the arm are unlikely to be representative of the body as a whole; both data sets indicate a decrease of score per unit of body mass as growth proceeds, whereas the total body calcium increases (Table 8.2).

Sex differences

At birth, there are appreciable sex differences, both in the thickness of subcutaneous fat (greater in the female at all of five sites, McGowan *et al.*, 1975) and in overall body mass (the male is some 200 g heavier than the female). The mechanism responsible for early sexual differentiation may be testicular androgen production, which begins relatively early in foetal life (Forbes, 1987).

Body build and composition in childhood and adolescence

Overall body build

While height provides some indication of an individual's growth status, the development of other organs does not conform very closely to the predictions of any height-based proportionality theory (Chapter 1).

Analyses restricted to the adolescent part of the growth span are further complicated by secular changes in the timing of the pubertal growth 'spurt'. The Belgian growth curves of Vajda *et al.* (1977) provide a specific example of this problem. The secular tendency to an increase of adult stature (discussed further in Chapter 9) had led to a progressive increase in the height of Belgian children at all ages, but there had also been a decrease of relative body mass, particularly in children under the age of 10 years. Cross-sectional data for 1840 suggested that the body mass of boys increased as height$^{2.08}$ between the ages of 6 and 13 years, but by 1970 the exponent had increased to 2.58 (Vajda *et al.*, 1977); likewise, in girls, the exponent had increased from 2.34 to 2.68. Plainly the shape of the relationship, and thus the magnitude of the exponent, was affected more by the pubertal growth spurt in 1970 than in 1840.

Body composition methodology

Body fat

Methods for the determination of body fat in the adult are considered in Chapter 3. Data for the newborn infant have been obtained by gas absorption (Mettau, 1978) and by volume displacement (Taylor *et al.*, 1985). With the latter approach, there remains the difficulty of ascribing an appropriate density to lean tissue. Mettau (1978) reported that in a well-nourished full-term newborn infant, the fat mass was given by:

Fat mass (g) = 0.15 (body mass, g) − 133 g

With a total body mass of 3.5 kg, this would be equivalent to approximately 14.4% body fat.

Methods applicable to older children have been reviewed by Lohman *et al.* (1984*a*). Hydrostatic weighing is complicated by difficulties in determining the residual volume while submerged, and some authors have suggested that in healthy children it is more accurate to predict the residual volume from vital capacity than to attempt a direct measurement of the residual volume while submerged (Chapter 3). A further problem arises when converting density readings to equivalent percentages of body fat. Because of incomplete calcification of bone and an atypical water content of the fat-free mass, the density of lean tissue is substantially less than the 1.100 assumed in the usual adult two-compartment model. Indeed, the figure for a child may be as low as 1.085 (Lohman, 1984*a*); the same percentage of body fat would thus yield an overall body density of 1.076 in a 25-year-old man, but the density would drop to 1.051 in a 9-year-old boy.

The relationship between skinfold readings and body fatness also changes progressively over adolescence. Equations that are used successfully to predict body density in adults thus overestimate density by an average of 0.023 g/ml in 8–12-year-old boys, and 0.010 g/ml in girls of similar age (Lohman *et al.*, 1984*b*). There are some specific skinfold prediction equations for children; however, the 'gold standard' used in developing such equations has generally been underwater weighing, with subsequent application of the adult two-compartment model of body composition, so that the results must be viewed with some suspicion. The International Biological Programme Working Party compared skinfold readings with hydrostatic data in 10–13-year-old children, and derived the linear equations:

$$D = 1.0628 - 0.00289 \, (\Sigma \, S - 7.80) \text{ in boys}$$

and

$$D = 1.0511 - 0.00273 \, (\Sigma \, S - 9.82) \text{ in girls}$$

where $\Sigma \, S$ was the sum of three skinfold readings (triceps, subscapular and suprailiac). For reasons discussed in Chapter 3, Durnin & Womersley (1974) suggested the use of a logarithmic formula, using the same three skinfolds:

$$D = 1.1147 - 0.0612 \, (\log_{10} \Sigma \, S) \text{ in boys}$$

and

$$D = 1.1309 - 0.0587 \, (\log_{10} \Sigma \, S) \text{ in girls}$$

A third option, proposed by Parizkova (1977), translated the sum of 10 skinfolds (cheek, chin, thorax I and II, triceps, subscapular, abdomen, suprailiac, thigh and calf) directly into a percentage of body fat:

$$\text{Body fat} = 26.60 \, \log_{10} \Sigma \, S - 31.34 \text{ in boys}$$

and

Body fat $= 29.99 \log_{10} \Sigma\, S - 24.57$ in girls

All of these formulae refer to immediately pre-adolescent children. In younger age groups, it is better to refer skinfolds to tables of average skinfold thickness than to attempt a prediction of body fat using formulae derived from older subjects.

Other body constituents

It is undesirable to use radioactive markers of body water in a growing child, but the deuterium dilution method can be applied to estimate total body water. Nevertheless, if the results are to be interpreted in terms of lean mass, a problem arises with respect to the water content of the lean body compartment, as this generally differs from the 72–73% assumed in adults.

The whole body counter/^{40}K technique is also suitable for children who will accept the enclosed counting chamber, but again there are some practical difficulties in selecting an appropriate figure for the average potassium content of lean tissue.

Normal fat content

The percentage of body fat increases rapidly over the first year of life, rising from an initial 10–15% to 20–25% of body mass. There is a subsequent decline, as the infant becomes more mobile. Thus, Tanner & Whitehouse (1975) noted a decrease in the 50th percentiles for the triceps and subscapular folds between 1 and 8 years of age, while Parizkova (1977) reported that skinfold thicknesses reached a nadir at 6–8 years of age.

Girls show a second rapid increase of body fat content at puberty, as adipose tissue develops in the breast and over other parts of the body such as the hips (Tanner, 1962; Parizkova, 1977). Calculations based upon ^{40}K determinations of lean body mass (Burmeister & Bingert, 1967; Forbes & Amirhakimi, 1970) estimate the fat mass at 5, 10 and 17.5 kg for 7.5, 12.5 and 17.5 years of age, respectively. Frisch (1976) noted that from menarche to the age of 18 years, girls accumulated an additional 4–5 kg of fat. The growing proportion of adipose tissue inevitably leads to problems of data interpretation when variables such as maximum oxygen intake are expressed relative to total body mass.

The age-related trends in overall body fatness are reflected in the thickness of individual folds such as the triceps (Tanner & Whitehouse, 1975; Parizkova, 1977; Fig. 8.1). From early childhood, all skinfolds are somewhat thicker in girls than in boys, but because the arms are larger in

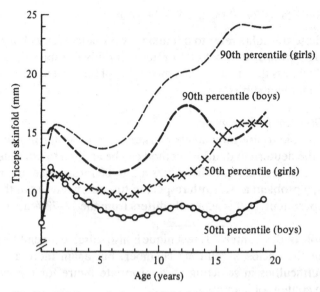

Fig. 8.1. Thickness of triceps skinfold; based on data of Tanner & Whitehouse (1975) for British children.

male subjects, boys may carry about the same mass of superficial fat on their limbs as do girls (Frisancho, 1981).

Boys show a modest increase of subcutaneous fat in the years immediately preceding puberty (Novak *et al.*, 1973; Parizkova, 1977; Tanner & Whitehouse, 1975), but the velocity of increase in skinfold thickness becomes negative coincident with peak height velocity, and there may be an appreciable loss of fat during late adolescence (Garn & Clark, 1976, but not Novak, 1963). The ^{40}K determinations of lean mass indicate that boys have a total mass of 3, 6 and 9 kg at 7.5, 12.5 and 17.5 years of age, respectively.

Adipocyte growth

Knittle *et al.* (1979) examined the number and size of adipocytes at different ages. The cell number increased tenfold from birth to maturity. About half the adult count was reached between the ages of 2 and 4 years, and adipocyte proliferation usually ceased between 14 and 16 years of age. Nevertheless, the normal infantile increase of body fat reflects mainly an increase in the fat content of the individual adipocytes (Fig. 8.2).

There is evidence that fat cells are particularly prone to hyperplasia if

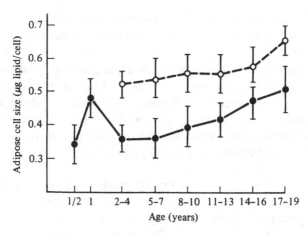

Fig. 8.2. Adipose cell size as a function of age in obese (○) and non-obese (●) children (sexes combined). Based on data of Knittle *et al.* (1979).

over-feeding occurs during early childhood (Hager *et al.*, 1977). Furthermore, training prior to weaning reduces the number of fat cells, at least in animals (Oscai *et al.*, 1974). Women generally have more fat cells than men, and this accounts for most of the greater fat content of the female. However, the increase of percentage body fat with age reflects an increase of fat cell size rather than an increase of cell numbers (Björntorp, 1974).

Once the number of adipocytes has been established at an above-average level, the 'set-point' of body fat stores shows a corresponding increase, and it becomes increasingly difficult to regulate the total amount of body fat to what would be regarded as a 'desirable' level in a subject with a lower fat cell count. Stern & Greenwood (1974) suggested that in those individuals with an early onset of obesity had 4–5 times the normal number of adipocytes, although the fat content of individual cells was usually increased only moderately. Likewise, Björntorp demonstrated a linear relationship between the fat cell count and the body fat content in kg. On the other hand, Knittle & Hirsch (1968) noted that overweight in children was closely correlated with the size of the fat cells; moreover, in starvation the diameter of the fat cells decreases steadily by 0.39 μm/day, coincident with a 439 g decrease of body mass (Faulhaber, 1974).

Body fat and menarche

Frisch hypothesized (Frisch & Revelle, 1970; Frisch, 1983; Frisch & McArthur, 1974) that menarche was triggered when an adolescent reached a critical body mass (48 kg) and percentage of body fat (17%); she further suggested that about 22% of body fat was needed to restore menstruation in an amenorrhoeic individual. The secular trend to an earlier menarche was explained on the basis that girls now became bigger and fatter at an earlier age (Wyshak & Frisch, 1982). It was argued that evolution had led to the selection of females who delayed puberty and/or turned off ovulation when food supplies were short. Suggested mechanisms included a conversion of androgens to oestrogens in fat, together with alterations in the metabolism, binding and excretion of oestrogen as body fat increased (Frisch, 1983).

However, a critical review of the original work shows that body fat was not measured. Rather, it was predicted very indirectly from the relationship of body water to height and body mass. Subsequent research has stressed that the Mellits & Cheek (1970) equations used by Frisch lead to a systematic overestimate of body fat in thin women (Johnston, 1985). The Frisch hypothesis is thus largely discredited, despite research supporting the concept that androgens are converted to oestrogens in adipocytes (Kley *et al.*, 1980). The body mass at menarche can vary widely from 28–97 kg (Scott & Johnston, 1982). Women with a body mass of less than 48 kg are known to have become pregnant (Garn & LaVelle, 1983), while regular patterns of menstruation have been described in athletes having as little as 8% body fat (Carlberg *et al.*, 1983).

Body fluids

Total body fluids

Most investigators have used the deuterium method in examining the total fluid volume of growing children. Total body water decreases from about 78% of body mass at birth to around the adult figure of 60% at one year of age (Friis-Hansen, 1961). Readings then increase slightly to three years of age, before returning to the adult value (Table 8.2). The data of Heald *et al.* (1963) suggest for boys aged 12–18 years a decrease of about 0.6% per year to a final figure of 71.9% in late adolescence. Young *et al.* (1968) studied girls aged 9 to 17 years, finding some decrease of water content after menarche, with the adult value being reached at about 15 years of age.

As noted above, variations in the water content of lean tissue have a significant influence upon the interpretation of body density measurements.

Extracellular fluid volume

Several authors have made simultaneous estimates of extracellular fluid volume (using the bromide dilution technique) and total body water (using the deuterium dilution method). Pooling the information for children of various ages (Lohman *et al.*, 1984a), it appears that extracellular fluid accounts for a progressively diminishing fraction of total body water over the early part of growth. The extracellular fraction settles to about 41% of total body water by the time that the latter has expanded to a total volume of 10 l (Fomon *et al.*, 1982; Forbes, 1987). Changes of fluid distribution occur equally in male and female subjects, and are not affected significantly by obesity (Cheek *et al.*, 1970; Shizgal *et al.*, 1979; Egusa *et al.*, 1985).

Blood volume

Blood volumes have generally been calculated from plasma volumes and venous haematocrit readings. Unfortunately, the venous haematocrit bears a somewhat variable relationship to whole body haematocrit, particularly in the newborn, where there is an SD of ±0.03 about the average ratio of 0.87 (Mollison *et al.*, 1950).

Depending on stature, the adult man has a total blood volume of about 5500 ml (2200 ml red cells and 3300 ml plasma), while the female has 4100 ml (1500 ml red cells and 2600 ml plasma). Hawkins (1964) accumulated data showing that the blood volume increased from 7% of body mass at birth to 8% in the child and 8.5% in a young man, but dropped back to 7% in young women, 7.5% in older men, and 6.5% in older women. Shephard (1956) further reported that the oxygen carrying capacity of unit volume of blood increased from an average of 129 ml/l at 3 years of age to 186 ml/l at 15 years.

Mineral content

Potassium

The body potassium content increases from an average of 49 mEq/kg (1.91 g/kg) at birth to about 68 mEq/kg (2.65 g/kg) in the adult male, and 64 mEq/kg (2.50 g/kg) in the adult female (Ziegler *et al.*, 1976; Fomon *et al.*, 1982; Haschke, 1983; Cohn *et al.*, 1984). Data on young children are relatively limited, although from a search of the literature, Lohman *et al.* (1984a) concluded that concentrations for both boys and girls were in the range 2.5–2.7 g/kg.

There is an associated decrease in the potassium to nitrogen ratio, from about 2.0 at birth to 1.8 in a young adult, and 1.7 in an older individual.

As might be anticipated, the age-related decrease in the proportion of extracellular water is accompanied by a progressive decrease in the chlorine to potassium ratio, from about 1.24 at birth to 0.72 in the adult. Possible explanations include a dramatic decline in secretion of the mineral-regulating hormone aldosterone, with a resultant decrease of sodium retention (Weldon *et al.*, 1967; Sippell *et al.*, 1980), or a decrease in the surface area to volume ratio of individual cells as they hypertrophy.

Iron

The total iron content of the body (found mainly in blood and myoglobin) develops largely in parallel with lean mass, although pubertal boys have a high plasma mass of transferrin and alpha-2 macroglobulin (Koch & Röcker, 1980). At puberty, males are accumulating about 1.2 mg of iron per day, and females about 0.8 mg. Even taking account of menstrual iron losses, the male need of iron may exceed that of the female.

At maturity, the female body contains about 64% of the iron found in a male. Since adipose tissue contains relatively little blood, all of these percentages are reduced in obese individuals (Allen *et al.*, 1956).

Calcium

The total body calcium of boys doubles between the ages of 12 and 18 years, while the body mass increases by a factor of only 1.7 (Christiansen *et al.*, 1975). Mazess & Cameron (1972) found that the calcium content of the radius (expressed in g/cm) doubled from 8 years of age to adulthood; adjusting for differences in bone width, the gain was still 62% in males and 48% in females. Klemm *et al.* (1976) examined the os calcis, and found that at 6 years of age, the bone mineral content was 33% of the adult figure, while the density (g/ml) was 67% of the adult value; maturation of the os calcis was largely completed at 15 years of age – the boys having reached 90% and the girls 80% of the adult density.

Other elements

The body content of most other elements (for instance, magnesium, zinc, nitrogen and phosphorus) increases relative to body mass over the course of maturation, but copper stores increase somewhat less than the gain in overall body mass (Forbes, 1987). Moulton (1923) noted that the total ash content of the fat-free body rose from 3–4% in an infant to 9% in an adult male.

Lean tissue

Methodological issues

Information on the growth of lean body mass is based in part upon ^{40}K determinations and indirect calculations from hydrostatic estimates of body fat (Forbes, 1978; Fomon *et al.*, 1982). We have noted already the technical problems caused by age-related changes in body potassium, lean tissue hydration and bone density.

Local information on the course of muscle growth has also been accumulated by anthropometry and soft-tissue radiography (Tanner, 1962; Maresh, 1966). Unfortunately, anthropometry provides only a measure of muscle plus bone, and neither approach can distinguish and exclude intramuscular fat deposits from the total volume described as 'muscle'.

Further indications of muscle development have been taken from the rate of urinary creatinine excretion at different ages (Viteri & Alvarado, 1970), although the interpretation of such data is also complicated, since there is a decrease of muscle to creatinine ratio as body size increases, perhaps because the synthesis of muscle is biochemically more efficient in a child than in an adult.

Growth pattern

The body composition moves progressively towards the adult picture as childhood progresses; growth of the lean fraction can be seen in figures for water, potassium and protein content and overall density. At birth, the relative water content of the body is 110% of the adult figure, and at the age of 10 years it is 104%. Likewise, the potassium content is 74% and 99%, the protein content is 73% and 93%, and the density 96% and 98%, respectively, of the adult figures (Forbes, 1987).

The mass of skeletal muscle increases from less than a kilogram at birth (about 25% of total mass) to about 28 kg (40% of total mass) in an adult male. The rapid development of lean tissue which is characteristic of the foetus continues for a year or more after birth, but then the body settles down to a slower growth rate until the adolescent 'spurt', when there is again a substantial increase in the accretion rate for lean tissue, particularly in boys. The energy cost of synthesizing the new tissue (12–17 kJ/g) is lower than in adults, but, bearing in mind the substantial water content of the tissue that is formed, the net conservation of energy is less than 25% of the cost.

Sex differences

Males have a slightly greater relative content of muscle than girls at birth, and the male advantage increases with the onset of the adolescent growth 'spurt', so that the adult female has less than 60% of the muscle found in an average man, with a corresponding difference in their rates of creatinine excretion (Clark *et al.*, 1951) and in their respective muscular strengths (Asmussen, 1964).

Throughout childhood, the lean tissue mass is slightly greater in boys than in girls, and this advantage is much enhanced at puberty. Between the ages of 10 and 20 years, a boy gains some 33 kg of lean tissue, while a girl gains only 16 kg (Forbes, 1987).

Forbes (1987) has related the developing sex differences to the male advantage of testosterone production (from the time of puberty, this amounts to about 6 mg/day, giving a cumulative difference in dosage of more than 20 g of testosterone from the age of 13 years to maturity).

Issues of proportionality

Lean mass shows a general relationship to stature, and the peak height velocity coincides fairly closely with the peak velocity for increase of lean body mass (Parizkova, 1977). However, in some studies (for example, Berg & Bjure, 1974), the maximum oxygen intake of adolescents is related more closely to height[3] than to lean mass, perhaps because of the rapid development of non-cellular tissue in skeleton and tendons.

Muscle cell hypertrophy and hyperplasia

During post-natal growth, the number of muscle fibres generally remains constant, although there is a substantial increase in fibre size and an increase in the number of muscle nuclei, the latter being derived from satellite cells (Malina, 1978).

Prolonged and rigorous training may also lead to a splitting of fibres (Gonyea, 1980). Cheek (1968) estimated the total number of muscle cells in the body from the DNA content of biopsy specimens and urinary creatinine excretion, on the assumption that each muscle cell had a single nucleus containing 6.2 pg of DNA. His calculations showed the cell count increasing from 0.2×10^{12} in the newborn infant to 3×10^{12} in the adult male and 2×10^{12} in the adult female.

Other authors have argued that there may be an increase of DNA content (Buchanan & Pritchard, 1970; Bailey *et al.*, 1973; Hubbard *et al.*, 1974), but that the cell number becomes relatively fixed by a few months after birth (Goldspink, 1972; Edgerton, 1973; Zak, 1974; Korecky & Rakusan, 1978). Thereafter, any increase in muscle volume reflects a hypertrophy of existing cells, without change in the relative proportions of slow and fast twitch fibres (Thorstensson, 1976). Typically, cell size is approximately doubled over the course of growth.

Bone

Growth patterns
The skeleton continues to grow for rather longer than other tissues, often not reaching its maximum size until as late as the third decade of life (Trotter & Peterson, 1970). Both the amount of bone and its chemical characteristics change over the course of development. In the long bones, the process is ultimately halted by a closure of the epiphyses, while in the skull, growth is severely limited once the sutures between individual bones have closed.

Chemical composition
The total body calcium remains approximately proportional to the third power of stature throughout the period of growth (McNeill & Harrison, 1981). In the young child, a large proportion of the total skeleton is in the form of cartilage, but most of the cartilaginous material becomes calcified as growth proceeds. Human milk contains only about a fifth as much calcium as cows' milk. Thus, at the time of weaning, a formula fed infant may have almost twice the body calcium of a breast-fed infant (Stearns, 1939; Greer *et al.*, 1981).

Maturation of bone is associated with a decrease in the water content and an increase in the calcium and sodium content of cortical bone (Forbes & McCoord, 1963). The marrow, also, is filled initially with the precursors of red and white cells, and these are progressively replaced by mature cells and fat.

Crystals of hydroxyapatite ($[Ca_{10}(PO_4)_6(OH)_2]$) lie in close contact with the underlying collagen, and are bounded by a thin film of water of hydration. Certain mineral ions can be adsorbed onto the surface or even incorporated into the crystal lattice, more readily in younger men than in older tissue. During the 1960s, this phenomenon attracted considerable interest due to the atmospheric testing of nuclear weapons and the resultant 'fall-out' of radioactive ^{90}Sr.

Skeletal age
A determination of skeletal age contributes to the interpretation of anthropological data (Szabo *et al.*, 1972), aerobic power (Shephard, 1971), motor performance (Beunen *et al.*, 1987*a*,*b*) and track perform-

152   *Body composition in biological anthropology*

ance (Cumming *et al.*, 1972), although in some cases the proportion of the variance described by the skeletal index is relatively small after due allowance has been made for the effects of calendar age (Shephard *et al.*, 1978b).

Sexual and racial differences
Until puberty, the cross-sectional area of cortical bone is only marginally greater in boys than in girls, but by the age of 18 years the figure for female subjects is no more than 80% of that for males (Gryfe *et al.*, 1971; Christiansen *et al.*, 1975; Garn *et al.*, 1976; Ringe *et al.*, 1977).

Forbes (1987) suggests that in the US there are also some inter-racial differences, the bone cortex being thicker in black children, and thinner in Spanish American children. It remains unclear whether these differences are inherited or due to environmental factors.

Effects of habitual physical activity
Regular physical activity increases both the mineral content and the density of bone, but in most studies it does not appear to have either accelerated or delayed maturation (Cerny, 1969; Kotulan *et al.*, 1980; Novotny, 1981).

In the Trois Rivieres study (Shephard *et al.*, 1978b), the difference in months between the radiographic and calendar ages of pre-adolescent children as measured at the wrist was given by the formula:

$$\text{Delta age} = 4.56 \, (\text{Sex}) + 0.39 \, (\text{Milieu}) - 3.59 \, (\text{activity}) - 13.0$$

where male = 0, female = 1, urban = 0, rural = 1, and the added activity programme = 1. This equation implies that there was earlier maturation in females, but that some delay was caused by an added hour of classroom exercise per day. However, any effect of physical activity upon bone maturation must have been purely local, as the corresponding equation for development of the lower jaw and teeth is:

$$\text{Delta age} = 1.33 \, (\text{Sex}) + 4.53 \, (\text{Milieu}) + 1.47 \, (\text{activity}) - 0.28$$

Regulation of development
The rate and ultimate extent of growth are determined partly by inheritance (Chapter 9), acting through neuro-hormonal regulators, and partly by environmental factors such as nutrition, physical activity and psychosocial constraints (Chapter 10). Given the ability of exercise to influence

blood levels of growth hormone, androgens and thyroid hormones (Shephard, 1982*b*) and also the potential influence of genetic factors upon lifestyle, the distinction between inheritance and environmental influences is far from clear cut.

Growth hormone

One important regulator is the growth hormone secreted by the pituitary gland. This compound promotes the incorporation of amino acids into tissue protein, and encourages the development of muscle and bone rather than fat (Campbell & Rastogi, 1969). Nevertheless, it is not essential to tissue hypertrophy (Goldberg & Goodman, 1969); apparently, it functions mainly as a linear amplifier of stimuli received from other regulators of the growth process.

Thyroid hormone

Adequate amounts of thyroid hormone are essential to normal growth. Thyroxine activates many enzyme systems, and in particular is essential for RNA production and therefore protein synthesis.

Androgens

Androgens are a third factor influencing growth. The blood levels of several hormones (dehydroepiandrosterone, androstenedione, androsterone and somatomedin) all increase in the period preceding adolescence, and given the known anabolic response to such hormones (Kochakian, 1976; Forbes, 1985), it has been suggested that they could contribute to both early sexual differentiation and the modest relative increase of lean body mass observed prior to puberty.

The main effect of the androgens is an increase of cell size, but they also favour the development of muscle relative to fat, and accelerate closure of the epiphyses in the long bones (Ezrin *et al.*, 1973). Androgen levels also strongly influence the formation of haemoglobin (Krabbe *et al.*, 1978; Shephard *et al.*, 1977). On the other hand, there is a five month discrepancy between the age of maximum testosterone secretion and the age of maximum bone mineralization. It is thus unlikely that androgens play a major role in controlling bone development.

The pubertal growth spurt is associated with modest increases in blood levels of dehydroepiandrosterone (30%) and somatomedin (40%), together with a major increase of serum testosterone levels (Krabbe *et al.*, 1984). The last is much larger in boys (500%) than in girls.

Oestrogens

Oestrogens have a weak androgenic effect, stimulating an increase of cell size while restricting cell multiplication. On the other hand, progesterone has an antiandrogenic effect.

The increase of body fat in girls seems linked to rising serum oestradiol levels, and oestrogens also have an influence upon bone mineralization, at least in the female (Chapter 11). Oestradiol increases extracellular fluid volumes in muscle and liver, while decreasing intracellular fluid volumes (Cole, 1952).

Other hormones

Other hormones can also influence growth. Insulin increases protein synthesis (Martin & Wool, 1968). Glucocorticoids promote protein synthesis if the subject is in positive energy balance, but lead to a selective loss of protein from slow twitch fibres if energy is in short supply (Elson, 1965; Mansour & Nass, 1970). Prolonged administration of cortisone can interfere with normal growth (Loeb, 1976).

Possible trophic factors

There has been vigorous debate concerning the factors endowing individual muscle fibres with specific metabolic characteristics (Buller *et al.*, 1960; Kow *et al.*, 1967; Robbins *et al.*, 1969; Lentz, 1971). Gutmann (1976) argued for the importance of mechanical loading and a trophic factor transmitted along the nerve axons, and McArdle & Sansone (1977) suggested the importance of information transmitted by the afferent nerves, while Brown (1973) maintained that fibre differentiation was dependent upon the pattern of muscle activity that occurred.

The current consensus seems that the motor nerve is important only in terms of the type of contraction that it initiates, and given that the muscle fibre can influence its innervation (Purves, 1976), a key role is played by the genetic information stored in the muscle nuclei (Colling-Saltin, 1980).

Ethnic differences

It is well recognized that in many populations from underdeveloped nations, the growth spurt occurs a few months later than in urban 'white' society (Rode & Shephard, 1973*a*; Shephard, 1982*b*), and that the adult size of the underdeveloped groups is also smaller than in developed nations. However, it has proven extremely difficult to disentangle the respective roles of physical activity, diet and genetic factors in determining ethnic differences of growth.

One study of the Alaskan Eskimo adolescent found the samples were short for their age, and also light in relation to their height (Nobmann, 1981). In some instances, this led to a bad self-image. The children wanted to be both taller and lighter, although they were already underweight. In contrast, Jamison (1976) found the North Western Alaskan Eskimos to be short, but above the average weights for 'white' children. Postl (1981) noted that between the ages of 1 and 7 years, Canadian Inuit children were short but above the 50th percentile in terms of body mass; he further commented that the additional mass reflected lean tissue rather than fat.

In older Canadian Inuit children, Rode & Shephard (1973a) found a continuation of these same trends, with an exaggeration of the normal period when the girls were heavier than the boys between the ages of 11 and 14 years. They supported the view that the body mass of the arctic populations was increased by a life that demanded hard physical work, with a development of muscle rather than body fat.

Body composition from young adulthood to old age

Methodological considerations

Cross-sectional studies of aging are confounded by such factors as secular changes in diet, patterns of habitual activity, and other variables (Shephard, 1987). Moreover, a diminishing proportion of subjects volunteers for testing in the older age categories, and a rising prevalence of disease makes it difficult to distinguish the normal processes of aging from disease-related changes of body form and composition. Some authors, unwisely, have compared free-living young adults with the hospitalized elderly, where body composition has inevitably been distorted by enforced inactivity and intercurrent disease.

The data obtained from longitudinal surveys is generally thought to be more reliable than cross-sectional results, although the consistency of longitudinal values may be compromised by a change of measuring techniques or of personnel. Prolonged investigations tend to attract fit, health-conscious volunteers, and there may be a progressive elimination of the sick as a survey progresses. In some instances, the prospect of a repetition of testing may also encourage a surge of physical activity or dieting in the weeks immediately preceding the test measurements, but in other instances the inherent effect of aging becomes exaggerated by a progressive age-related decrease of habitual activity as the survey continues.

Whether using a cross-sectional or a longitudinal survey design, measurements of body composition tend to become less accurate in an older person for purely technical reasons. For instance, the interpretation of hydrostatic data is compromised by a decline of bone density (Trotter *et al.*, 1960), skinfolds become more compressible (Chapter 3), and ^{40}K measurements of lean tissue are affected by changes in the ratio of potassium to body water (Bruce *et al.*, 1980; Cohn *et al.*, 1980; Pierson *et al.*, 1982). Again, if bone density is estimated by photon absorptiometry measurements over the radius and ulna, the observer must take into account possible differences in the rate of calcium loss in other parts of the skeleton. Estimates of body composition that are based on total body water should be less affected by age, as should measurements involving the uptake of fat soluble gases. Neutron activation is also little affected by aging, with the exception that an increase of subcutaneous fat may increase the screening of gamma emissions.

Several surveys have illustrated differences in the apparent rate of decline in stature between longitudinal and cross-sectional surveys (Miall *et al.*, 1967; Borkan *et al.*, 1983a; Shephard & Rode, 1985). The secular trend normally accounts for a cross-sectional decrease in stature of about 0.8–1.0 cm/decade, although in some populations that are undergoing rapid acculturation it can amount to 2–3 cm/decade (Shephard, 1987; Chapter 9). Selective migration also leads to cross-sectional changes of stature with age. The decrease of height that is observed thus tends to be less in a longitudinal study than in cross-sectional comparisons. Nevertheless, aging does lead to increasing kyphosis, a narrowing of intervertebral discs, and sometimes a collapse of vertebrae, these tendencies being exaggerated by osteoporosis and mechanical trauma to the vertebral column (for instance, frequent operation of snowmobiles over rough ice, Shephard *et al.*, 1984b).

Age-related changes of standing height are of practical importance, in that a person's stature is frequently used to standardize other measures of body composition.

Overall body composition

The International Commission on Radiological Protection (1975) has specified the typical composition of the young adult male of average size (Table 8.4). Figures for a young woman are generally about two-thirds of those for a man, although since oxygen, carbon and hydrogen are present in fat, the amounts of these elements may reach three-quarters of the male totals.

Table 8.4. *Characteristics of the 'reference man' as specified by the International Committee for Radiation Protection (1975).*

Gross organ mass		Body content of key elements		
Organ	Mass (kg)	Element	Mass (g)	(mol)
Whole body	70	Oxygen	43 000	1340
Adipose tissue	15	Carbon	16 000	1333
Skeletal muscle	28	Hydrogen	7000	3500
Skeleton	10	Nitrogen	1800	64
cortical bone	4	Calcium	1100*	27
trabecular bone	1	Phosphorus	500*	16
marrow	3	Sulphur	140	4.37
cartilage and periarticular	2	Potassium	140	3.60
Skin	4.9	Sodium	100	4.17
Viscera	12.1	Chlorine	95	2.68
Liver	1.80	Magnesium	19	0.78
Brain	1.40	Iron	4.2	0.075
Heart	0.33	Zinc	2.3	0.035
Kidneys	0.31	Copper	0.07	1.1×10^{-3}
		Iodine	0.01	0.079×10^{-3}

*Values revised from 1000 g Ca and 780 g P to take account of neutron activation studies of Cohn *et al.* (1976a).

Body fat

'Ideal' body fat
As discussed previously, it is difficult to decide upon a 'normal' amount of body fat, given both the skewed distribution of data and the widespread discrepancy in average body fat content between inhabitants of the western world and those people living in developing countries.

A survey conducted in the context of the International Biological Programme (Shephard, 1978a) showed that in many developing countries, the average body mass was 5–15 kg below the actuarial ideal (Table 8.5). One suggestion has been that young men should have no more than 14% and young women no more than 18% of body fat (Amor, 1978). Given the loss of other tissues with aging, percentages of body fat may increase somewhat in an older person, even if there has been no increase in the absolute fat content of the body.

Normal pattern of aging
The typical finding from cross-sectional surveys of the general North American population (Shephard, 1977; Table 8.6) is that body mass

Table 8.5. *Discrepancy between body mass and actuarial standards in selected indigenous populations.*

Population	Excess weight (kg)	Author
Easter Island	0	Ekblom & Gjessing (1968)
Ethiopia		
Addis Ababa		
Workers	−10.7	Areskog *et al.* (1969)
Airforce cadets	−7.1	Areskog *et al.* (1969)
Adi Arkai	−13.0	Anderson (1971)
Israel		
Kurds	−0.4	Samueloff *et al.* (1973)
Yemenites	+2.5	Samueloff *et al.* (1973)
Jamaica	−5.3	Miller *et al.* (1972)
Malaya		
Medical students	−9.1	Duncan (1972)
Nigeria		
Yoruba		
Active	−3.7	C.T.M. Davies *et al.* (1973)
Inactive	−4.0	C.T.M. Davies *et al.* (1973)
South Africa		
Bantu	−4.7	Wyndham (1966)
Tanzania		
Active indigenous	−1.2	C.T.M. Davies & Van Haaren (1973)
Inactive indigenous	−4.1	C.T.M. Davies & Van Haaren (1973)
Trinidad		
Negroes	−1.3	Edwards *et al.* (1972)
East Indians	−3.4	Edwards *et al.* (1972)
Zaire		
Hoto	−6.7	Ghesquière (1971)
Twa	−7.7	Ghesquière (1971)

(Based on data accumulated by Shephard, 1978*b*; see source for details of references.)

increases in early middle age, but remains constant or diminishes as old age is approached. The latter tendency reflects a loss of lean tissue with a constant or even an increasing amount of body fat (James, 1976; Shephard, 1977; Norgan, 1982). Studies by densitometry (Young *et al.*, 1963; Durnin & Womersley, 1974), and by the ^{40}K method (Forbes & Reina, 1970) suggest that the average fat accumulation is as large as 3–4 kg per decade. The onset of moderate obesity in middle age usually reflects a decrease of physical activity (Greene, 1939), and it is associated with an over-loading of existing fat cells rather than adipocyte hypertrophy (Salans *et al.*, 1971).

Table 8.6. *Changes of excess body mass and skinfold thickness with aging.*

Age (yr)	Women		Men	
	Excess[1] body mass (kg)	Average[2] skinfold thickness (mm)	Excess[1] body mass (kg)	Average[2] skinfold thickness (mm)
20–29	8.3 ± 5.3	16.2 ± 3.8	1.7 ± 8.7	11.2 ± 5.3
30–39	1.4 ± 5.3	13.5 ± 5.2	6.4 ± 8.5	16.1 ± 10.6
40–49	6.8 ± 8.4	17.3 ± 5.4	9.3 ± 9.5	14.0 ± 5.8
50–59	4.9 ± 7.2	18.2 ± 5.1	8.8 ± 7.7	15.2 ± 6.7
60–69	4.5 ± 9.5	22.5 ± 7.9	5.1 ± 7.3	15.4 ± 2.7

Mean ± SD.
[1]Relative to *average* 'ideal' mass proposed by Society of Actuaries (1959).
[2]Average of eight skinfolds (see Figs 7.1 and 7.2 for individual sites).
(Based on the data of Shephard, 1977.)

In Britain, a comparison of four studies conducted prior to 1951 and four completed subsequent to 1968 noted a secular trend to an 8 kg increase of body mass in male subjects. Only a small part of this increase was attributable to an increase of stature (Montegriffo, 1971), and most of the gain was presumed to be the result of body fat. In the US there has been little secular tendency to an increase of body mass over the past several decades (Malina, 1979), in part because most of the population were already inactive and obese in the 1940s.

In the absence of vigorous muscular exercise or the administration of androgens, an increase of body mass after the age of 25 years almost invariably reflects an accumulation of fat. This is well illustrated by data on former champion runners (Dill *et al.*, 1967). Those who had gained no more than 5 kg since the date of their final competition had an average body fat content of 15%, but those who had gained 6–12 kg carried 20% fat, and in those who had gained 15–22 kg the average body fat was 30%.

Lean tissue

Much of the variability in total body mass is due to differences of body fat content. Inter-individual variability in the amount of lean tissue is much smaller (Burmeister & Bingert, 1967). Age-related changes in lean tissue may be examined in terms of total lean mass, cross-section of individual muscles and histological changes.

Size effects

Stature accounts for a substantial part of the observed differences of lean mass between adults, lean mass being approximately proportional to the third power of stature (Forbes, 1974):

$$\text{Log } LBM = -4.99 + 3.0 \log H$$

A simple linear ratio rather than a logarithmic equation is frequently used when comparing adults of differing size. Although there are theoretical objections to this approach, it works quite well in practice, particularly if comparisons are made over only a limited range of body sizes. During the early part of adult life, the slope of the linear relationship is about 0.69 kg/cm in men and 0.48 kg/cm in women. Sex differences per unit of stature are small in old people, the slope decreasing to about 0.35 kg/cm in both sexes (Forbes, 1987).

Age trends in total lean mass

Similar age trends in lean body mass are suggested by measurements of ^{40}K (Allen *et al.*, 1960; Myhre & Kessler, 1966; Novak, 1972; Flynn *et al.*, 1972; Rutledge *et al.*, 1976; Noppa *et al.*, 1979), creatinine excretion (Young *et al.*, 1963; Vestegaard & Leverett, 1968; Rowe *et al.*, 1976), and estimates of lean tissue mass derived from body density determinations, the absorption of foreign gases or body water determinations (Brozek, 1952; Young *et al.*, 1963; Myhre & Kessler, 1966; Borkan & Norris, 1977; Lesser & Markofsky, 1979).

A peak of lean mass is reached at about 20 years of age in a man and 18 years in a woman. A plateau is then sustained to about 40 years of age, but thereafter most subjects show an accelerating loss of lean tissue, so that by the age of 80 years cross-sectional studies suggest that the cumulative loss amounts to more than 40% of the young adult value in men and some 20% in women. In later adult life, the decrease of muscle mass is typically greater than the decrease of body mass, since muscle tends to be replaced by fat (Shephard, 1977).

There have been relatively few longitudinal studies of body mass, and available observations have generally begun in early adult life (when subjects tend to be in the plateau phase). Shukla *et al.* (1973) found potassium losses over 12 years of 3.3 and 3.8% per decade in men and women, respectively, while Keys *et al.* (1973) and Tzankoff & Norris (1978) found a decline in basal metabolic rate (also, to some extent, a marker of lean tissue mass) of 3.2 and 3.7% per decade, respectively. However, individual cases have been cited where little loss of lean body mass occurs over an extended period (Forbes, 1987).

Given the accelerating course of the loss, caution must be shown when interpreting results that have been calculated as a percentage loss per decade, particularly if the measurements have extended over a lengthy time span.

Local cross-sectional studies

Ultrasound (Young *et al.*, 1980; Vandevoort *et al.*, 1983) has been used to study local age-related decrease in the cross-section of the quadriceps muscle. The rate of tissue loss seen by this approach is typically 0.5–0.7% per year (Young *et al.*, 1980), a figure generally compatible with cadaver estimates of muscle wasting (Haggmark *et al.*, 1978) and changes of creatinine excretion (Tzankoff & Norris, 1977). In women, there is a fairly consistent relationship between muscle cross-section and peak isometric force at all ages, but in young men the isometric force per unit of muscle cross-section is greater than in older subjects; this may be because the muscles of a young man have a different shape (Young *et al.*, 1985), or it may reflect a more effective coordination of individual motor units.

Histological changes

The histological appearance of the muscles can readily be studied in humans using modern methods of muscle biopsy. Such investigations suggest that the fibre pattern and composition change with aging.

There is a progressive decrease in both fibre number and fibre diameter (Campbell *et al.*, 1973; Hanzlickova & Gutmann, 1973) – ultrastructural proteolytic changes being particularly marked near the fibre surface. Inokuchi *et al.* (1975) examined the composition of the rectus abdominus muscle from adults aged 20–80 years, finding that the fat content increased from 1–5% in the youngest subjects to 40–50% in the oldest. In women, there was also an age-related increase of connective tissue content, from 3% in young adults to 20% in nulliparous and 30% in parous women aged 50 years. Finally, a selective loss of type II fibres has been described in the proximal muscles of the lower limbs (Larsson *et al.*, 1979; Green, 1986).

Biochemical analyses have shown a decreased mitochondrial content (protein per g of tissue), a decrease of stored glycogen, and a lesser activity of ATPase, oxidative and glycolytic enzymes (Fitts, 1981).

It is difficult to disentangle true effects of aging from the consequences of decreased habitual activity. Inactivity and associated reductions of protein synthesis (Booth & Seider, 1979) undoubtedly play a major role in the loss of lean tissue, and aging does not seem to limit the possibility of restoring muscle tissue by an appropriate training regimen (Tomanek &

Wood, 1970; Goldspink & Howells, 1974). There are nevertheless age-related hormonal changes that reduce protein anabolism, including decreased serum levels of somatomedin-C, dehydroepiandrosterone, testosterone (exacerbated by an increase of binding globulin) and oestrogen (Forbes, 1987), Gutmann (1977) has suggested that there may also be an age-related slowing in either the synthesis or the neural axoplasmic transport of specific trophic agents that stimulate muscle growth in a younger person.

Body water

Total body water shows a steady decline after the age of 30 years. Both measurements of gross fluid volumes and tissue analyses suggest that this reflects a decline in the amount of intracellular fluid (Allen *et al.*, 1960; Wilson & Franks, 1975; but not Möller *et al.*, 1979).

Borkan & Norris (1977) reported that the intracellular fraction decreased from 58% of body water at the age of 30 to 52% at 80 years. The sodium space accounts for a correspondingly larger fraction of total body water (Pierson *et al.*, 1982) and the volume of histochemical specimens (DuBois, 1972; Möller *et al.*, 1979).

Bone

There is general agreement that the mineral content of bone decreases with age, although it is less clear how far this is an inevitable accompaniment of aging, and how far it reflects a decrease of physical activity or of pathological change. The interpretation of statistics for skeletal mass or body calcium requires a consideration of body size effects.

Influence of body size

Dissections of cadavers (Borisov & Marei, 1974) suggest that the total skeletal mass (bone plus marrow, but excluding periosteum and ligaments) of a human is approximated by the equation:

$$\text{Log Skeletal Mass} = -1.487 + 2.46 \log H$$

Although the logarithmic relationship is to be preferred, simple linear ratios are often used in making inter-individual comparisons, particularly where differences of size are relatively small. In men, the skeletal mass averages 10.4 kg, changing by 258 g/cm of stature, while in women the average mass is 9.0 kg, with a change of 273 g/cm (Forbes, 1987). Smith *et al.* (1981*a*) have pointed out that although the mass of bone mineral decreases with age, its composition remains relatively constant. However, there is a progressive enlargement of the medullary cavity in the long bones, with an associated weakening of their structure.

The total body calcium of adults is proportional to a little over a third power of height. Both Nelp *et al.* (1972) and McNeil *et al.* (1982) have found height exponents of 3.10. Typical average values for young adults are about 800 g in women and 1200 g in men. Ellis & Cohn (1975) noted that the total calcium increased by 20.5 g for each cm increment of height. The calcium to potassium ratio generally decreases somewhat in tall people (Forbes, 1987), suggesting that tall individuals have more muscle in proportion to bone than do shorter subjects.

Changes with age
Neutron activation studies (Cohn *et al.*, 1980) indicate that a progressive loss of body calcium begins quite early in adult life. The process is more rapid in women (38–51 g/decade) than in men (36 g/decade). The loss of cortical bone can be seen quite clearly in the hand and forearm (Virtama & Helelä, 1969; Ringe *et al.*, 1977; Exton-Smith, *et al.*, 1969; Mazess & Mather, 1974; Christiansen & Rödbro, 1975; Garn *et al.*, 1976), although changes are more marked for the spine than for the limbs (Genant *et al.*, 1982).

In an 80-year-old male, the density of the spine may be only 55% of that in a young adult, while in an 80-year-old woman the figure may have dropped to as little as 40% of earlier values. Adams *et al.* (1970) noted a 10% loss of cortical bone in men and a 15% loss in women over an 11 year period of observation. However, other longitudinal studies (Smith *et al.*, 1976a; Milne, 1985) have found a lower rate of loss than that observed in cross-sectional investigations. Smith *et al.* (1981a) have suggested that the true rate of bone loss is 0.75–1.0% per year, starting at the age of 30–35 years in women and 50–55 years in men.

Biochemical basis
Reasons for the calcium loss remain unclear. While some loss of bone seems an almost inevitable accompaniment of aging, a more rapid loss sufficient to cause 'spontaneous' fractures may be a pathological abnormality (Chapter 11). Further contributing variables are a sedentary lifestyle and an inadequate calcium intake (Heaney, 1986).

Early work suggested that the overall growth and local architecture of bone depended substantially upon the local stresses and strains that had been applied (Wolff, 1892; Howell, 1917). The greater rate of loss in postmenopausal women suggests some role for oestrogen, while the rapidly deleterious effects of bed rest (Krølner & Toft, 1983) and the reversal by weight-bearing exercise support the idea that a progressive decrease of physical activity is also involved (Chapter 11).

Possibly because of deteriorations in renal function, older subjects have lower levels of the active form of vitamin D (1,25 hydroxycholecalciferol). Together with oestrogen lack, this factor may reduce intestinal absorption of calcium, stimulating parathyroid hormone secretion (Heaney *et al.*, 1982). Alternatively, inactivity and lack of oestrogen play a primary role, leading to a diminution of parathyroid hormone and thus a decreased synthesis of 1,25 hydroxycholecalciferol in the kidneys.

Connective tissue

Age leads to an increase in the proportion of connective tissue in 'muscle' and other lean tissues. The increased proportion of connective tissue reflects in part a 'scarring' from disease injury and ischaemia, and in part an atrophy of more active tissues.

There are also important modifications in the characteristics of connective tissue. Structural and functional changes (Balazs, 1977) include a yellowing of the elastin, with formation of pseudoelastin (Shephard, 1987), an increased stability of collagen, associated with greater molecular cross-linking, changes in the ground substance and a thickening of the basement membranes separating connective tissue from other tissues. These various changes lead to a loss of elasticity in the affected tissue.

9 Genetic influences upon body composition

General anthropological considerations

Origin of racial differences

Inherited differences of body build may arise by either genetic drift or natural selection. If a small population colonizes a remote habitat, this group may by chance have an unusual frequency of genes favouring a particular body form (founder effect), and because of limited opportunities for mating, these characteristics will persist in subsequent generations (Harvald, 1976). Moreover, there will be fewer heterozygotes than in a larger community, and some gene combinations with a low initial frequency may disappear from the population by mere chance (a form of 'genetic drift'). However, if a particular body form has favoured survival, there will also be a selective pressure increasing the frequency of any related gene combination within the population.

An environment might be conceived where the homozygote was at a major disadvantage, but the heterozygote gained some advantage. Such a situation would be particularly difficult to detect other than by a very careful examination of gene frequencies. Further, in an isolated population, the apparent advantages of a particular body form might be exaggerated by emergence of unusual patterns of diet and lifestyle within the community under investigation.

Adaptive pressures in specific habitats

Principle

By analogy with classical Mendelian analysis, it is easy to envisage a general relationship between body build and inheritance. Moreover, it seems reasonable to argue that at least in less developed societies, certain body builds confer an advantage in regions where the climate is extreme, or food is in short supply. The hunting of game, the tilling of infertile soil, or an escape from predators might all be facilitated by a particular

165

phenotype, and this could conceivably have had a cumulative survival value over the centuries (Shephard, 1978*a*; Roberts, 1988).

One of the basic premises of the International Biological Programme was that genetically determined differences would be revealed by comparisons of different ethnic groups sharing a common habitat and/or similar ethnic groups colonizing widely differing habitats.

Migrant studies

Several groups of investigators have examined the characteristics of migrants. One country well-suited to such studies is Israel. Here, recent immigration has led to genetically very different populations living under similar environmental and ecological conditions. Particularly on the kibbutzim, physical activity, nutrition and socio-economic conditions are all well controlled and closely comparable from one subject to another.

Glick & Schvartz (1974) compared four groups of Israelis whose parents had migrated respectively from Europe, North Africa, Iraq and the Yemen (Table 9.1). Differences of diet and physical activity were supposedly ruled out, yet the Yemeni were much smaller, with less fat and lean tissue than the other groups. Glick & Schvartz (1974) commented 'ethnic group differences in PWC [physical working capacity] in this study were, therefore, largely a result of genetic differences between these groups . . .' however 'more work is needed to determine relationships between ethnic origin, anthropometric characteristics, and PWC'.

More recent studies from Israel have tended to contradict these findings (Samueloff, 1988). Moreover, Tanner (1966) has pointed out the fundamental weakness of migrant studies – those who chose to migrate are rarely typical of the base population; often, they are the taller and more intelligent members of a particular community.

Table 9.1. *Differences of body composition between four segments of the Israeli population.*

Origin of population	Height (cm)	Body mass (kg)	Body fat (%)	Lean mass (kg)
Europe	173	70.7	14.6	60.4
Iraq	169	62.6	10.6	56.0
North Africa	169	59.4	9.9	53.5
Yemen	165	52.5	7.7	48.5

(Based on data of Glick & Schvartz, 1974.)

Adaptation to heat

A second postulate of the International Biological Programme was that nature might select an ethnic group with an unusual build and physique to colonize a particular environment, in accordance with such anthropological precepts as the Bergmann and Allen 'rules' (see also Chapter 10). Differences of growth patterns might lead to an immediate environmental variance of adult characteristics within a genetically homogeneous population (Anderson, 1967), while exposure to the demands of a given environment over many generations might lead to heritable differences of body form and composition.

Thus Hiernaux (1977) has followed tradition in seeking to explain the small size of the African pygmy as a 'useful' adaptation to the hot and wet conditions of a tropical rain forest. A north/south gradient of thermal stress can certainly be correlated with differences of stature between the Sara (inhabitants of Southern Tchad), the Tomba and the Twa found in the forests of Zaire (Austin *et al.*, 1979). However, this does not prove a causal relationship. Ghesquière & D'Hulst (1988) have pointed out that there is also a west/east gradient in the opposite sense, the Twa showing an average height of 159 cm, but the Bambuti (who live in a cooler and less humid habitat) having an average height of only 144 cm. The concept of thermal advantage from a small body size is explored further in Chapter 10.

Adaptation to physical work

Andersen (1969), Ghesquière (1971) and more recently Ghesquière & Eekels (1984), Ghesquière & D'Hulst (1988) have drawn attention to the obvious differences of build between Bantu negroes (Ntomba) and the pygmoid people (Twa), living side by side in the same jungle environment. Andersen (1969) noted that if data were expressed per unit of body mass, the Twa had a 15% advantage of aerobic power. He thus wrote the 'difference is believed to demonstrate a genetically determined difference in fitness for work'. Likewise, mining supervisors believed that Bantus from the Basuto tribe had a particular tolerance for hard physical work, although a formal study of Bantu from 10 different tribes (Wyndham *et al.*, 1966) revealed no inter-tribal differences in either body mass or aerobic power.

Adaptation to food shortages

Neel (1962) suggested that there was evolutionary advantage in a thrifty, energy-storing and diabetes-prone type of metabolism in regions where food supplies were limited. However, he was quick to point out that the same populations were at a metabolic disadvantage after acculturation to normal city life with abundant food reserves. Likewise, Lasker & Womack (1975) argued that the smaller body size of Central Americans relative to their peers in the US might represent a practical adaptation to sustained food shortages.

In contrast, Rothwell & Stock (1981) suggested that leanness and propensity to increase metabolism conferred advantages in terms of locomotion, reproduction and thermal balance. Plainly, it is difficult to specify a metabolic phenotype that will confer advantage in other than a very specific habitat.

Evolutionary pressures

Such hypotheses of natural selection imply a substantial evolutionary pressure from some adverse feature of the environment, with the prospect of relief of this pressure through adoption of a particular body form or physique.

In practice, it is difficult to assign a numerical value to the importance of such contributions to human adaptability. Successful colonization of a given habitat depends upon the amount of physical work needed for survival, population pressures and available technology (Breymeyer & Van Dyne, 1979). In some, if not all environments, a diversity of physical challenges or the importance of intellect relative to brute strength may well have precluded the emergence of an advantageous body form (Shephard, 1978a). There is finally no guarantee that a genetic variant giving a substantial advantage in one type of situation may not have a negative adaptive value with respect to some other environmental challenge.

Information from genetic markers

It is possible to demonstrate substantial differences of genetic markers when superficially similar peoples are compared from one habitat to another (Dossetor et al., 1971; Simpson & McAlpine, 1976), but the overall physique and body composition seems remarkably similar not only within a given ethnic group, but also from one indigenous population to another. Most of the small documented differences of body build could easily have arisen through differences in nutrition and physical activity rather than through genetic factors.

Nature or nurture?

When looking at factors responsible for the inter-individual variations in human form, the overall impression is of the dominance of immediate environmental factors, rather than long-term variations of genotype. We shall look briefly at ethnic differences of stature, percentage of body fat, and bone density, together with regional differences of body form.

Stature
With regard to stature, the influence of nurture appears to outweigh by far any effect of inheritance. Substantial socio-economic and/or nutritional gradients of overall body size have been described within ethnic groups in India (Malhotra, 1966), Tunisia (Parizkova *et al.*, 1972), Samoa (Greksa & Baker, 1982) and Kinshasa (Ghesquière & Eekles, 1984). Smaller differences of body size have been seen between ethnic groups, but these have generally disappeared if samples have been restricted to well-nourished individuals. In general, the malnourished person is small, but the body form (weight for height) remains relatively normal (Ghesquière & D'Hulst, 1988).

Body fat
In terms of total percentage body fat, the role of nurture again seems dominant. Figures are generally comparable from a variety of developing countries (Table 9.2), all values being considerably less than anticipated for the same or different ethnic groups living in western Europe or North America.

Table 9.2. *Estimates of body fat in selected populations.*

Population	Age (yrs)	Per cent fat Males	Per cent fat Females	Author
Scotland	20–29	15	29	Durnin & Womersley (1974)
	30–39	23	33	
	40–69	25	35	
East Africa	25	12	18	Di Prampero & Cerretelli (1969)
	55	11	18	
Tunisia	25	18	27	Di Prampero (1969)
	55	22	34	
New Guinea	23	9	21	
	35	10	22	
Canadian Inuit	25	15	20	Shephard *et al.* (1973)
	35	12	26	
India	12–35	19	—	Jones *et al.* (1976)
	18–30	—	15	Satwanta *et al.* (1977)
	20–25	—	18	Raja *et al.* (1977)
Colombia (mild under-nutrition)	33	18	—	Barac-Nieto *et al.* (1978)

(Based on data collected by Shephard, 1978a and Norgan, 1982. See these reports for details of references.)

Fat stores of people from developing countries are particularly small when account is taken also of their small body size. However, thinness appears to have a cultural rather than a genetic basis. The amount of subcutaneous fat increases rapidly with migration to large cities (Greksa & Baker, 1982), while both cross-sectional and longitudinal observations illustrate a similar trend to an increase of fat stores within a more traditional habitat as the population becomes acculturated to a western lifestyle (Rode & Shephard, 1984).

Bone calcification
A poor calcification of bone has been seen in Eskimos (Mazess & Mather, 1975; Thompson *et al.*, 1981) and in North American Indians (Orchard *et al.*, 1984; Evers *et al.*, 1985), while a high calcium content and bone density has been observed in US negroes (Thompson *et al.*, 1981; Cohn *et al.*, 1977), in Polynesians (Reid *et al.*, 1986), in the Singapore Chinese and the South African Bantu (Calmers & Ho, 1976). It is difficult to disentangle such issues as activity patterns, calcium intake and lack of vitamin D, from genetic causes or skin pigmentation, although a study of ancient bones suggests that the poor calcium content has been a feature of Eskimo body composition for many centuries. Eskimo bone also has more secondary osteons, suggesting a more rapid turnover of bone (Thompson *et al.*, 1981). Smith *et al.* (1973) further noted a concordance of bone density that was greater in identical than in fraternal twins.

Regional differences of body form
Regional differences of body form and composition have sometimes been attributed to inheritance. Issues discussed include the unusual arm length of African tribes (Ghesquière & D'Hulst, 1988), the relative and absolute shortness of the legs in circumpolar Lapps (Eriksson *et al.*, 1976), the propensity of Europeans and Euro-Americans to deposit more fat on the upper limbs than Afro-Americans, Asiatics or Amerindians (Eveleth, 1979), and the greater thigh development of Negev Israelis relative to those of similar race living in Jerusalem (Samueloff, 1988).

Johnston *et al.* (1974) suggested that differences in arm fat had a strongly genetic basis, but differences in subscapular readings were attributable mainly to environmental factors. On the other hand, a twin study by Brook *et al.* (1975) led to the conclusion that genetic factors had more influence on trunk than on limb fat. The differing dimensions of thigh and calf regions in the data of Samueloff (1988) apparently reflect the involvement of the Negev group in agricultural work, as opposed to the predominant office work of the Jerusalem population.

A final critical comment may be left to Wolanski (1970). He has queried the usefulness of trying to distinguish environmental and genetic influences, pointing to the importance of non-genetic transmissibility through what he terms the 'maternal regulator' – foetal nutrition, lactation and the subsequent environment the mother chooses or is able to offer her offspring. Certainly, the question of nature versus nurture seems too complicated to resolve through a simplistic comparison of peoples or habitats.

Heterosis, environment and secular trends

Many primitive societies were built around very small communities, with much inbreeding. There has, therefore, been much discussion of the importance of an increase in heterosis, or 'hybrid vigour' (Knussmann, 1967) to the observed characteristics of a given population with a possible potential for maximization of human growth potential through an increase in heterozygosity.

Secular trends

It is well documented that the physical characteristics of most populations have shown a progressive change over the past two centuries.

The overall figures for Canada (Shephard, 1986) are fairly typical of western nations, showing an increment in stature of some 1 cm/decade over the past 30 years. Moreover, the increase of size accounts for most of the increase in body mass over the same period. However, within this average, the course of change has varied, being relatively slight for the Toronto area and for the large prairie cities, but much more pronounced in rural Quebec (Landry *et al.*, 1980).

A rapid secular trend has been observed in many of the classically small populations, such as the Eskimos (Jamison, 1976), the Japanese (Nukada, 1975), the Lapps (Skrobak-Kaczynski & Lewin, 1976) and the subarctic Icelanders (Palsson, 1981). Among the Eskimos (Jamison, 1976), a gradient of stature, with an associated accentuation of the 'shortness for body mass' characteristic, has been observed; the tallest people are found among the most acculturated groups on the Alaskan north shore, while the smallest are seen in Eastern Canada (where there has been more pressure to preserve a traditional lifestyle).

Heterosis or environment?

There has been some suggestion in Alaska that the tallest Eskimos have a greater admixture of 'white' genes (Jamison, 1976). Other authors (for instance, Suzuki, 1970, and Nukada, 1975) have also linked the secular trend to an increase of body dimensions with the migration of populations to the cities and resultant increased opportunities for interbreeding (Wolanski, 1977). However, most investigators believe that environmental rather than genetic factors have been the main cause of the secular trend to an increase of physical dimensions.

Certainly, migration to the cities affects not only heterozygosity, but also socio-economic conditions (Harrison *et al.*, 1964; Palsson, 1981), including size of family, size of home, facilities for sleeping, the incidence and severity of disease (Hauspie *et al.*, 1977; Shephard, 1978*a*) and many other aspects of lifestyle (Tanner, 1962; Douglas & Simpson, 1964; Grant, 1964). Particular attention has focussed upon an improvement of nutritional status. Some authors think that an increased protein intake is the key change (Chapter 10), although Martin (1981) has commented that in practice acceleration of growth has been associated with an increase of sugar consumption, a poor protein intake and low haemoglobin levels. A second possible factor is a lessened exposure to climatic extremes (Eriksson *et al.*, 1976).

Studies of athletes

Athletes may be considered as extreme members of a population distribution curve, their physical characteristics reflecting the combined influence of rigorous environmental pressures (prolonged training) and a thorough selection of the human race for favourable variants.

A wide variety of body forms are selected for different sports. Size varies from the petite gymnast to the basketball player with a height of 215 cm. In general, the amount of body fat is small, and in some endurance sports male competitors may have as little as 7–8% body fat (Table 9.3). Likewise, some endurance competitors are quite lightly muscled, while at the opposite extreme the lean mass of some power athletes is almost twice that of the average adult.

The standard deviation of lean mass in a typical sample of middle-aged adults is about 12.5%, so that at first inspection a very careful search of the population might disclose a person at four standard deviations (50%) above the average value for a given population (Shephard, 1978*a*). Unfortunately, it is difficult to be certain of the normality of the distribution curve when dealing with such extreme values, but if we assume a continuation of the normal curve, then any additional advantage beyond a 50% increase of lean mass must be the result of training rather than a genetically determined population variance; the genetic

Table 9.3. *Percentage body fat in various categories of athlete.*

	Percentage fat		
Sport	M	F	Author
Track and field			
100–200 m	7.8	—	Malhotra *et al.* (1972)
	16.5*	—	Barnard *et al.* (1978)
	—	19.3	Malina *et al.* (1971)
400 m	8.1	—	Malina *et al.* (1972)
	12.4	—	Rusko *et al.* (1978)
Long distance	7.1–9.5	—	Costill (1967)
	8.4	—	Malhotra *et al.* (1972)
	7.5	—	Costill *et al.* (1970)
	8.4	—	Rusko *et al.* (1978)
	—	19.2	Malina *et al.* (1971)
	—	15.2	Wilmore & Brown (1974)
	11.2–13.6*	–	Pollock *et al.* (1974)
	13.2*	—	Lewis *et al.* (1975)
	18.0*	—	Barnard *et al.* (1979)
Pentathlon	10.0	—	DiPrampero *et al.* (1970)
	—	11.0	Krahenbuhl *et al.* (1979)
Discus and	19.6	—	Behnke & Wilmore (1974)
throwing events	16.4–16.5	—	Fahey *et al.* (1971)
	16.3	—	Wilmore (1986)
	—	25.0–28.0	Malina *et al.* (1971)
Jumpers and	8.9	—	Malhotra *et al.* (1972)
hurdlers	—	20.7	Malina *et al.* (1971)
Orienteering	16.3	18.7	Knowlton *et al.* (1980)
Water sports			
Canoe/kayak	5.9	9.2	Sidney & Shephard (1973)
	9.4*	—	Sidney & Shephard (1973)
	12.4	—	Rusko *et al.* (1978)
Rowing	16.8	—	De Pauw & Vrijens (1971)
	11.8	—	Wright *et al.* (1976)
Sailing	19.9	—	Niinimaa *et al.* (1976)
Swimming	6.9	—	Shephard *et al.* (1973)
	8.5	19.2	Sprynarova & Pařízková (1963)
	5.0	—	Novak *et al.* (1968)
	—	26.3	Conger & MacNab (1968)
Team sports			
American football	10.0	—	Behnke *et al.* (1942)
	13.8	—	Novak *et al.* (1968)
	13.7–17.6	—	Kollias *et al.* (1972)
	12.5	—	Balke (1972)
	13.9	—	Forsyth & Sinning (1973)
	11.5–19.1***	—	Wickiser & Kelley (1975)
	9.4–18.6***	—	Wilmore *et al.* (1986)
Baseball	14.2	—	Novak *et al.* (1968)
	11.8	—	Forsyth & Sinning (1973)
	12.6	—	Wilmore (1986)

(*continued*)

Table 9.3 (*cont.*)

Sport	Percentage fat M	F	Author
Basketball	7.1–10.6***	—	*Parr et al.* (1978)
	—	20.8	Sinning (1973)
	—	26.9	Conger & MacNab (1967)
Ice-hockey	15.0	—	DiPrampero *et al.* (1970)
	15.1	—	Wilmore (1986)
	13.0	—	Rusko *et al.* (1978)
Rugby football	12.7	—	Williams *et al.* (1973)
Soccer	9.6	—	Ravey *et al.* (1976)
Volleyball	10.2	—	de Rose (1973)
	11.6	—	Zelenka *et al.* (1967)
	13.3	—	Seliger *et al.* (1970)
	9.9	—	Serfass (1971)
	12.4	—	Williams *et al.* (1973)
	—	25.3	Conger & MacNab (1968)
	—	21.3	Kovaleski *et al.* (1980)
Other sports			
Gymnastics	4.6	—	Novak *et al.* (1968)
	—	23.8	Conger & MacNab (1967)
	—	16.8	Sprynarova & Pařízková (1969)
	—	15.5	Sinning & Lindberg (1972)
	—	9.6–17.0	Parizkova (1972)
Racquetball	8.1	—	Pipes (1979)
	8.5	14.0	Hagerman *et al.* (1979)
Alpine skiing	7.4	—	Sprynarova & Pařízková (1971)
	14.1	—	Rusko *et al.* (1978)
	10.2	20.6	Haymes & Dickinson (1980)
Nordic skiing	12.5	—	Niinimaa *et al.* (1978)
	10.2	—	Rusko *et al.* (1978)
	7.9	21.8	Haymes & Dickinson (1980)
Ski jumping	14.3	—	Rusko *et al.* (1978)
Speed skating	11.4	—	Rusko *et al.* (1978)
Tennis	15.2	—	Forsyth & Sinning (1973)
	16.3*	20.3*	Vodak *et al.* (1980)
Weight-lifting	9.8	—	Sprynarova & Pařízková (1971)
	12.2–15.6	—	Fahey *et al.* (1978)
Wrestling	6.9	—	Katch & Michael (1971)
	5.0	—	Pařízová (1972)
	10.7	—	Gale & Flynn (1974)
	8.8	—	Sinning (1974)
	9.8	—	Fahey *et al.* (1975)
	4.0	—	Stine *et al.* (1979)
	14.3	—	Taylor *et al.* (1979)

*Master's competitors.
***Dependent on playing position.
(Based upon data accumulated by Shephard, 1978a and Wilmore, 1986; see sources for details of references.)

contribution cannot explain more than 50% of the difference between the athlete and the average sedentary individual.

Unfortunately, it is very difficult to partition the basic 50% of variance about the population mean between the potential causes of inheritance, habitual physical activity and dietary factors.

Formal genetic studies

Impact of inheritance upon body composition and human performance

Any analysis of the extent and strength of genetic influences upon body composition, human performance and adaptation to a particular habitat is complicated by the polygenic nature of human inheritance. Many of the variables of interest to the processes of human adaptability are biologically complex. Individual genes may make additive, non-additive or interactive contributions to the impact of other genes and of environment upon growth and development. In consequence, most data sets show a continuous distribution of physical characteristics, rather than the bimodal distribution associated with classical Mendelian analysis.

Moreover, the impact of inheritance appears to be flexible, so that in some instances the influences of aging, ill-health and external environment can over-ride the control exercised by the individual's genotype. In humans, the relatively long period of growth enhances the likelihood that there will be a substantial interaction between the immediate effects of environment and the genetic potential of the individual (Malina, 1978). The long life span of humans also complicates the collecting of observations, relative to the task of those who study a simple organism such as the fruitfly *Drosophila* (Roberts, 1988). Finally, while genetic effects can be demonstrated conclusively at the population level of analysis, the predictive power of genetic markers is weak when similar concepts of inheritance are used to characterize the physical characteristics of the individual.

Inherited abnormalities and genetic markers

There are quite a number of relatively rare inherited conditions that influence growth, maturation and adult body form. However, in many of these disorders the specific genetic defect has yet to be determined.

In galactosuria, there is difficulty in metabolizing milk sugars to glucose and glycogen, resulting in a retardation of growth. In the various abnormalities of glycogen mobilization ('Von Gierke's Disease') there is again retardation of growth. Gargoylism is associated with an accumulation of mucopolysaccharides in many organs, shortness of stature and a grossly abnormal appearance. Achondroplasia is a dominant abnormality where cartilage cells mature less rapidly than normal, with short long bones and dwarfism. Morquio's disease (osteochondrodystrophy) is a recessive abnormality characterized by a decreased growth of articular cartilage, with a secondary slowing of metaphyseal development. Albers-Schonberg disease (osteoporosis congenita) has both dominant and recessive forms. The bones in this condition are dense with a small marrow space, and if the child survives it remains small. Osteogenesis imperfecta is a dominant condition where there is a deficiency of osteoblasts; repeated fractures of the long bones lead to dwarfism.

Certain chromosomal defects have been linked to characteristic abnormalities of body build. The most obvious is Turner's syndrome, where an XO chromosome pattern is associated with rudimentary genital organs, dwarfism, webbing of the neck and an increased carrying angle of the arms (Eiben *et al.*, 1977). Girls with a 45-XO chromosome count lack functioning ovaries, and fail to lay down the normal sex-related depots of subcutaneous fat at puberty. Klinefelter's syndrome is another disorder due to an abnormal chromosome count; it is characterized by a 47-XXY chromosome count. Affected males have low serum testosterone levels and an average of 9.4 kg less lean tissue than anticipated (East *et al.*, 1976).

In some medical conditions associated with particular body builds, such as hypertension and ischaemic heart disease, a preponderance of specific gene-linked blood groupings has again been demonstrated (Cruz-Coke *et al.*, 1973).

However, it is less certain that there is a packaged and readily identifiable group of genes determining body composition and form. Certainly, the ability to link body form to genetic markers in the general population remains highly debatable. In a study of the Igloolik population, Shephard & Rode (1973) compared the frequency of various blood groupings in short, average and tall Inuit. Even with this simple physical characteristic, application of chi^2 statistics showed no significant associations with blood groupings. It thus seems most unlikely that a linkage to readily accessible markers could be demonstrated for more complicated phenomena such as the relative proportions of muscle, fat and bone.

Twin studies

Much of the early work on inheritance was based upon comparisons of variance between supposedly monozygous and dizygous twins, with a simple partitioning of variance into genetic, environmental and methodological components (Table 9.4). The usual equation estimated the contribution of heritability (H_{Est}) as:

$$H_{Est} = (o^2dz - o^2dzm) - (o^2mz - o^2mzm)100/(o^2dz - o^2dzm)$$

The equation in essence assumes that the variance for the monozygous twins represents a simple summation of environmental and methodological variance, and that in the dizygous twins this figure is augmented simply by genetic effects.

When making a calculation of heritability in this manner, the zygosity of the twins is often not confirmed by serological tests. Moreover, the premise of a total genetic identity between monozygous twins is open to question, except immediately after fertilization has occurred (Roberts, 1988). Even if the theoretical basis be accepted as reasonably sound under controlled laboratory conditions, in free-living humans substantial problems arise from a greater similarity of environment for the monozygous than for the dizygous pairs, both before and after birth. Furthermore, both types of twin encounter less inter-pair environmental differences than would be anticipated in the general population. Moreover, the constancy of methodological error for the two types of twin is open to question.

When confronted by a family of fat and poorly muscled individuals, there is thus real difficulty in disentangling the influence of inheritance

Table 9.4. *Ratio of dizygous pair variance to monozygous pair variance.*

Stature	Body mass	Arm length	Hand length	Leg length	Foot length	Author
9.05	—	5.02	—	—	—	Dahlberg
4.11	2.23	6.45	4.54	4.53	—	Von Verschauer
8.29	3.24	9.72	5.49	—	5.37	Clark
9.03	6.68	—	—	—	5.77	Vogel & Wendt
10.42	1.36 (n.s.)	7.13	2.06	9.92	5.45	Osborne & De George
13.16	4.35	—	—	—	4.72	Vandenberg & Strandskov

(Based on anthropometric data for twins, as accumulated by Vandenberg, 1962; see source for details of references.)

from the effects of a physically inactive environment that favours obesity and poor muscular development. The whole family may well be habitually inactive and over-eat. In this connection, Garn *et al.* (1979) noted a synchrony in both fat gain and fat loss between spouses (although if one ignores the complication of assortative mating, they would normally be of quite different genetic backgrounds). A further complication is that genetic and environmental factors are not totally independent, but rather tend to interact. In particular, some individuals are genetically predisposed to show adaptations of fat, muscle and cardio-respiratory performance in response to a training programme, while others show little response to a substantial modification of their environment (Bouchard, 1986*b*).

Wilson (1986) pointed out yet more difficulties in interpreting data from twin studies. All twin pairs share some prenatal influences that tend to exaggerate their genetic similarities. Indeed, some 70% of monozygous twin pairs are born with a monochorionic placenta. Varying degrees of vascular anastomosis lead to unequal nutrition of the developing twins, so that at birth there may be greater within-pair variability of body composition for monozygous than for dizygous twins (Naeye *et al.*, 1966). However, by the age of three months, the influence of inheritance is becoming more important than intra-uterine environment, and the body mass of the monozygous twins differs less than that of the dizygous pairs.

Confidence in the twin-pair estimates of heritability have been weakened because the results have often proved rather unstable. For instance, Brook *et al.* (1975) concluded that heritability was marked for both trunk and limb fat in older children, whereas in those under the age of 10 years environmental effects were dominant. Despite such problems, genetic influences have been inferred from twin studies. Genes apparently contribute to the regulation of both bone (Tanner, 1962) and skeletal maturation (Hiernaux, 1966, 1977). Vandenberg (1962) accumulated the results from six twin studies, showing significant differences of variance between monozygous and dizygous twins for a number of anthropometric measurements (Table 9.4). Twin-pair studies have also indicated a genetic contribution to muscular development (Hewitt, 1958; Osborne & DeGeorge, 1959) and body fatness (Hewitt, 1958; Tanner & Israelson, 1963). Identical twins who are reared in the same environment show smaller differences of body mass than do non-identical twins, while adopted children show little or no relationship to the mass of their parents (Norgan, 1982).

Family studies

More recent investigators have studied whole families (Table 9.5), calculating the coefficients of correlation between data for relatives of varying closeness (parent–parent, parent–offspring, full sib–sib, dizygous and monozygous twins). The results have been explored using path analyses and mixed models of inheritance. Such formal analyses allow the variance of body composition data to be partitioned into at least four components – the environmental, and the additive, dominant and inter-active portions of genetic variance (Li, 1961; Cavalli-Sforza & Bodmer, 1971). The additive component reflects differences between homozygotes, summed over the relevant combination of genes, the dominant component reflects deviations of the heterozygote from the value intermediate between two contrasting homozygotes, and the interactive component indicates the extent of interaction between genetic loci. One possible basis of calculation is as follows:

Additive fraction $= 2(r_{p-o})$

Dominant fraction $= 4(r_{s-s} - r_{p-o})$

Genetic fraction $=$ Additive $+$ Dominant fractions

Environmental fraction $= 1 -$ (Additive $+$ Dominant fractions).

Other methods of making the necessary calculations are discussed below. Broad estimates of heritability based upon a variety of familial relationships are currently regarded more favourably than pure twin studies, partly because a larger body of data can be used, and partly because a wider and more representative range of environments can be

Table 9.5. *Inter-class correlations in pairs of cultural and/or biological relatives, computed with residuals of age and sex.*

Variable	Adopted pairs	Unrelated pairs	Cousin pairs	Biological sibs	DZ twins	MZ twins
Sum of 6 skinfolds	−0.01	0.11	0.28	0.27	0.39	0.83
Extremity/trunk ratio	−0.01	0.07	0.06	0.37	0.40	0.80
Body density	−0.14	−0.04	0.18	0.20	0.22	0.73
Subcutaneous/total fat	−0.01	0.12	0.32	0.29	0.15	0.61
Fat-free mass	0.03	0.06	0.28	0.26	0.53	0.93

(Based on data of Bouchard *et al.*, 1985.)

Table 9.6. *Broad heritability estimates for body composition, as calculated from (i) biological siblings and (ii) monozygous and dizygous twins.*

	Heritability estimates	
Variable	Biological siblings	MZ/DZ twins
Sum of 6 skinfolds	0.54	0.88
Extremity/trunk ratio	0.74	0.80
Body density	0.40	1.02
Subcutaneous/total fat	0.58	0.92
Fat-free mass	0.52	0.80

(Based on data of Bouchard *et al.*, 1985.)

introduced into the analysis (Table 9.6). Nevertheless, the theoretical basis of both twin and family analyses remains weakened by significant differences in the extent of similarities for maternal–sibling and paternal–sibling comparisons (Montoye & Gayle, 1978; Bouchard *et al.*, 1981). Presumably, both parents influence their offspring to a differing extent through their differing impacts upon foetal, neonatal and childhood environments (Wolanski, 1969a, 1970).

Pérusse *et al.* (1987, 1988a) recently examined the inter-generational transmission of several variables by examining data collected on 304 French Canadian nuclear families from the Quebec area and a representative sample of 13 804 Canadians tested during the Canada Fitness Survey (Table 9.7). They followed Rice *et al.* (1978) in considering the phenotype P as determined by transmissible factors T (a combination of genetic and environmental influences) and all other environmental influences E:

$$P = tT + eE$$

where t and e are the partial regression coefficients for transmissible and non-transmissible factors respectively. The transmissibility (Table 9.8) is then given by t^2, and the non-transmissible component of the variance $e^2 = 1 - t^2$. A more detailed model (Rice, 1981; Cloninger *et al.*, 1983) includes path coefficients expressing the contribution of each parent to the phenotype of the offspring:

$$P = hA + bB + eE$$

$$A_o = \tfrac{1}{2}A_F + \tfrac{1}{2}A_M + sS$$

$$B_o = bB_F + bB_M + rR$$

where a quantitative trait P is partitioned into additive genetic (A) and

Table 9.7. *Family correlations for selected data collected in the Canada Fitness Survey of 1981.*

Variable	Spouse	Correlation coefficients			
		Parent–offspring	Biological sibs	Uncle (aunt)–nephew (niece)	Grandparent–grandchild
Height	0.43	0.20	0.34	0.10	0.48
Weight	0.16	0.16	0.34	0.10	0.48
Body mass index	0.12	0.20	0.31	−0.11	0.05
Sum of 5 skinfolds	0.15	0.21	0.27	−0.14	0.04
Trunk/extremity ratio	0.05	0.20	0.34	0.11	−0.12
Waist/hip ratio	0.11	0.16	0.23	−0.06	0.07

(Based on data of Pérusse *et al.*, 1988*a*.)

Table 9.8. *Transmissibility (t²) with constraint* r = 0.5.

Variable	Transmissibility (t^2)	Transmissibility for single parents	
		Father	Mother
Height	0.28	0.22	0.33
Weight	0.27	0.25	0.31
Body mass index	0.36	0.38	0.36
Sum of 5 skinfolds	0.37	0.38	0.37
Trunk/extremity ratio	0.37	0.33	0.41
Waist/hip ratio	0.28	0.30	0.26

(Based on an analysis of Canada Fitness Survey data completed by Pérusse *et al.*, 1988*a*.)

cultural (B) factors transmitted from parent to offsping, E is the sum of all other non-transmissible environmental effects, and h, b and e are the path coefficients for a particular phenotype. The respective path contributions of the father (F) and the mother (M) in genetic transmission are taken as $\frac{1}{2}$ (based on accepted notions of diploid autosomal inheritance), while b represents a path coefficient from the parent's to the offspring's (o) cultural value B. The model further allows for the combined effects of assortative mating and a common domestic environment upon the parents m (particularly the tendency to select a spouse of a similar physique) and the shared but non-transmissible environment of siblings (c), dizygous twins (c_{dz}) and monozygous twins (c_{mz}) (see Table 9.4). Note that the transmissible portion of the variance is given by:

$$t^2 = h^2 + b^2 + 2whb$$

where w represents the correlation between genetic and cultural factors arising from assortative mating.

In the French Canadian sample, fat-free mass, body density and subcutaneous fat all showed significant intra-class correlations for sibs and for families, but only body density showed a weak correlation ($r = 0.17$) between spouses. In the national sample, all variables showed a transmissibility substantially greater than zero (Table 9.8), with highly significant components attributable to assortative mating and a common childhood environment. If the transmission of variance had been dependent solely upon polygenic inheritance, the transmissibility would have been $r = 0.5$; however, all variables departed significantly from this expectation, implying also a significant environmental contribution to the transmitted variance. In general, environment accounted for more than

Table 9.9. *Chi square tests of goodness of fit of model where* r *is set at* 0.5, *and likelihood ratios that transmission path coefficient* r = 0.5, *assortative mating coefficient* m = 0, *transmissibility* $\mathrm{t}^{134.2}$ = 0, *and common childhood environment* c = 0.

Variable	Goodness of fit of general model (r = 0.5)	Likelihood ratios		
		$m = 0$ (r = 0.5)	$r = 0$ $t^2 = 0$	$c = 0$ (r = 0.5)
Height	4.8	652	290	93
Weight	n.s.	81	185	106
Body mass index	n.s.	47	303	36
Sum of 5 skinfolds	n.s.	63	319	14.3
Trunk/extremity ratio	n.s.	6.7	266	57
Waist/hip ratio	n.s.	37	169	20

(Based on analysis by Pérusse *et al.*, 1988*a* of data collected in the Canada Fitness Survey.)

50%, and the combined biological and cultural inheritance some 30–40% of the total variance. In contrast to some other studies, the respective contributions of the father and mother were relatively similar (Table 9.9). The authors cautioned that because of the large size of the survey and the use of around 80 testing teams, the error variance may have been larger than in some surveys; since a given team tested an entire family, this could have inflated estimates of transmissibility.

Body size and somatotype

In terms of the inheritance of somatotype, Bouchard (1978) set the genetic effect at about 0.40 for mesomorphy and 0.50 for endomorphy and ectomorphy. In a subsequent analysis of the Canada Fitness Survey data (Pérusse *et al.*, 1988*a*) the total coefficient of transmissibility (which includes both genetic and shared environmental factors) was of a similar order (0.45 for mesomorphy, 0.36 for endomorphy and 0.42 for ectomorphy).

Parents and children tended to cluster in terms of body size – the family resemblance being stronger for skeletal lengths than for skeletal breadths. Part of the familial co-variation in both height and body mass is due to a shared environment; thus, Bouchard (1978) found that coefficients of correlation for height and body mass drop from 0.53 and 0.48 to 0.34 and 0.29 after correction for socio-economic indicators. Use of the TAU multivariate statistical model has suggested an overall coefficient of transmissibility as high as 0.65–0.70 for height, although this estimate may have been inflated by the use of twins (Byard *et al.*, 1984, 1985) or residents of an isolated community (Kansas Mennonites, Devor & Crawford, 1984; Devor *et al.*, 1986), both of which tactics would have reduced the environmental component of the total variance. In the Canada Fitness Survey analysis, the coefficient for height was only 0.28 (Pérusse *et al.*, 1988b). For body mass, transmissibility estimates have ranged from 0.50 (Byard *et al.*, 1984, 1985) through 0.40 (Devor *et al.*, 1986) to 0.27 (Pérusse *et al.*, 1988b).

Bouchard (1978) concluded that a single genetic system with pleiotropic (multi-sited) effects determined skeletal lengths in pre-pubertal children, and a similar single factor determined skeletal breadths. In contrast, Dallaire (1978) suggested that height was determined by specific loci on the short arms of both the X and Y sex chromosomes. The effects of chromosomal abnormalities (see section *Inherited abnormalities and genetic markers*) suggests that size can also be influenced by other genetic loci.

Body fatness

The likelihood that a child will be obese is 40–50% if one parent is obese, but 70–80% if both parents are obese (Garn & Clark, 1976). Typically, family studies suggest that about a third of the inter-individual differences in fatness stems from genetic causes (Mueller, 1983; Savard *et al.*, 1983).

Using the index M/H^2 as a criterion of obesity, Stunkard *et al.* (1986) noted that adopted children continued to resemble their biological rather than their adoptive parents. Bouchard *et al.* (1985) studied 971 individuals ranging in age from 8–26 years, and drawn from more than 400 families. The sample included 80 pairs of adopted children, 120 pairs of unrelated siblings, 95 pairs of cousins, 370 pairs of biological siblings, 69 pairs of dizygous twins and 87 pairs of monozygous twins. Inter-class correlations for these various groups are summarized in Table 9.5. Narrow estimates of heritability were based upon parent–child correlations, after correction for attenuation by the variance of the measuring techniques, while broad estimates of heritability were based on sib–sib correlations, also corrected for attenuation (Bouchard, 1978).

Bouchard (1978) found that the impact of living together upon overall body fatness was much less, both for parents and for adopted children, than had been suggested by some authors. Thus, spouses were not significantly more homogeneous than parents from different families (Bouchard, 1978). In adopted children, a common environment also had little influence upon the extremity to trunk skinfold ratio, the subcutaneous to deep fat ratio, or the fat-free mass (as estimated by hydrostatic weighing). The same was true of biologically unrelated pairs, Bouchard's data contrasting somewhat with the observations of Garn *et al.* (1979), who had noted intra-environmental correlations ranging from 0.12 to 0.29 for the thickness of different individual skinfolds.

First cousins share (on average) an eighth of their genes by descent, but they generally live in different households. Bouchard *et al.* (1985) found a closer correlation between the fatness of pairs in this group than for subjects who were unrelated. Indeed, for most of the variables except the extremity to trunk ratio the coefficients were as great as for the biological sibs (where on average one half of the genome is shared). The correlations for the latter sub-group (0.20–0.37) were of the same order as those reported by other authors (Howells, 1966; Garn *et al.*, 1979; Mueller, 1978; Hawk & Brook, 1979; Savard *et al.*, 1983).

Inter-class correlations showed a further modest increase for the dizygous twins (Table 9.6), probably reflecting the fact that the twins shared a more similar environment than other biological siblings. Again, the findings were of a similar order to those previously reported by Brook *et al.* (1975) – 0.34 for the subscapular and 0.49 for the triceps fold.

In agreement with Brook *et al.* (1975), both body fat and lean mass were closely correlated in monozygous twins. Two possible estimates of broad heritability were provided by (i) twice the correlation between biological sibs, and (ii) twice the difference in correlations between dizygous and monozygous twins (Table 9.6). Such calculations suggested that a substantial part of inter-individual differences in body composition is inherited. However, the estimates derived from the twin studies also seemed unrealistically high, presumably reflecting a greater similarity of environment for monozygous than for dizygous twins.

The Canada Fitness Survey data yielded transmissibility estimates of 0.37 for the sum of five skinfolds (Pérusse *et al.*, 1988*a*), somewhat lower figures than the heritability estimates discussed above, although agreeing with the estimates of Byard *et al.* (1984, 1985) and Devor *et al.* (1986), who had used a similar methodology.

A re-analysis of earlier sibling data, using an alternative form of path analysis, suggested to Bouchard (1985) that while transmission accounted for 40% of the variance in fatness, 35% of this variation was due to sociocultural factors, with only 5% of the variance in skinfolds and 10% of the variance in body mass index being attributable to heredity.

Plainly, caution is needed in partitioning variance between genetic and socio-cultural inheritance in all familial studies. The figures remain rather unstable, and current indications are that the genetic component is quite small.

Skeletal and cardiac muscle

Detailed analyses of heritability have yet to be completed for individual skeletal muscles, but there have been some comparisons of ventricular dimensions among relatives (Table 9.10).

Spouses did not co-vary significantly for any of the ventricular measurements, with the exception of one small negative correlation. On the other hand, there were quite strong correlations between values for parents and their children (Bouchard, 1978). Left ventricular volume per unit of body surface showed a clear gradient of correlation with the closeness of the biological relationship. However, other ventricular measurements did not differ clearly between relations and unrelated co-habitants, suggesting that the main determinant of most aspects of cardiac size is some factor other than inheritance.

Cellular perspective

Genetic control of the cell

It is well recognized that genetic factors play a major role in hyperplasia and cell differentiation. Hyperplasia involves a replication of cellular DNA, with a subsequent increase in the number of cells or associated organelles such as nuclei.

In skeletal muscle (Goldspink, 1972), cardiac myocytes (Korecky & Rakusan, 1978) and adipocytes (Hager *et al.*, 1977) the process of replication is important to growth during gestation, although the number of cells in each of these tissues becomes relatively fixed over the course of growth, sometimes after no more than a few months of extra-uterine life.

Thereafter, the main basis for an alteration in the relative proportions of the body constituents becomes cell hypertrophy. This involves a growth in the size of existing cells – the necessary protein constituents being synthesized with the help of existing DNA templates. The process reflects, at least in part, gene activity, since the growth of muscle mass is checked if DNA-dependent RNA synthesis is inhibited by the administration of actinomycin D (Clark, 1956). The genome, therefore, has a major influence upon the development and growth of the body tissues at all stages of life.

Table 9.10. *Interclass correlations of cardiac dimensions in cultural and/or biological relatives.*

Variable	Spouse	Unrelated siblings	Parent/ adopted	Parent/ child	Biological siblings	Dizygous twins	Monozygous twins
Left ventricular wall thickness (posterior)	0.22	0.06	0.20	0.25	0.21	0.57	0.61
Interventricular septum thickness	0.22	0.46	0.38	0.34	0.32	0.34	0.52
Left ventricular volume (ml/m^2 surface area)	−0.09	0.07	0.05	0.15	0.22	0.25	0.59
Left ventricular mass (ml/m^2 surface area)	0.17	0.42	0.25	0.29	0.37	0.28	0.56

(Based on data of Bouchard, 1986.)

Enzymic defects and body composition

Possible cellular mechanisms for any inherited differences of body composition are still debated. Forbes (1987) speculated that the genes responsible for appetite and/or behaviour might be involved, but others have suggested an impact upon specific enzyme systems within the tissues immediately concerned.

The contribution of genetic factors to the genesis of obesity is well documented in animals, and indeed it is common practice to breed genetically obese variants of a species. Thus, a common model for human obesity is provided by a homozygous obese mouse. This variant has a clearly established enzymic defect, a lack of sodium-potassium ATPase (York *et al.*, 1978), an enzyme which is normally responsible for 20–50% of resting metabolism (Schull, 1986).

In humans, there is some evidence that a similar enzyme defect can develop (DeLuise *et al.*, 1980, 1982). The erythrocyte ATPase activity averages 22% less in obese individuals. The extent of this defect is correlated with deviation from the 'ideal' body mass, although Beutler *et al.* (1983) have argued that part of the relationship reflects ethnic differences between obese and thinner individuals.

10 *Environmental influences*

Habitual physical activity

We have already considered the effects of an increase of local physical activity that is restricted to a specific body part (Chapter 7). This section will examine overall interactions between habitual physical activity and body composition.

Inactivity

Prolonged bed rest, paralysis of a limb, local immobilization of a body part by suspension or application of plaster, and the loss of normal gravitational stimuli during real or simulated space missions are all well recognized as situations which have an adverse effect upon body composition, leading to increases of body fat and losses of muscle and bone.

Muscle

Inactive muscles show a progressive atrophy, with a selective loss of their contractile protein. Greenway *et al.* (1970) noted that because of paralysis below the level of the lesion, the spinally-injured had a small lean mass relative to normal individuals, with an unusually high ratio of sodium space to total body water. Ryan *et al.* (1957) had similar findings in patients paralysed by anterior poliomyelitis, while Winiarski *et al.* (1987) noted a 40% loss of muscle mass when experimental animals were immobilized by suspension.

In astronauts who undertake space missions, there is again a progressive loss of lean tissue, with deteriorations of muscle structure and function. Findings included a reduction of cross-section in the individual fibres and a decrease of tone, strength and endurance affecting particularly the slow-twitch antigravity muscles of the legs and torso (Dietrick *et al.*, 1948; Chui & Castleman, 1980; Oganov *et al.*, 1980; Kaplansky *et al.*, 1980; Szilagyi *et al.*, 1981; Grigoriev & Kozlovskaya, 1988). There is some evidence that the adverse effects of hypokinesis are exacerbated by any associated hypoxia (Szilagyi *et al.*, 1981).

Leonard *et al.* (1983) reported a 2.1 kg loss of lean mass and a 2.8 kg decrease of total body mass over an 84-day space-flight. During such missions, the astronauts show an associated decrease in body stores of potassium, calcium, nitrogen and phosphorus (Hines & Knowlton, 1937; Dietrick *et al.*, 1948; Lynch *et al.*, 1967; Whedon *et al.*, 1976; Greenleaf *et al.*, 1977). Their blood volume also decreases (Dietrick *et al.*, 1948; Leonard *et al.*, 1983), typically in association with a substantial fluid loss (Baisch & Beck, 1987).

Ushakov *et al.* (1980) noted that over an 18-day space mission, there were decreases not only in overall lean mass, but also in the mass of the skin, the musculo-skeletal system and the spleen, while the mass of both the liver and the kidneys was increased.

Changes in rat muscle were reversed within 25 days following a 22-day space-flight (Oganov *et al.*, 1980). Templeton *et al.* (1984) reported an even faster recovery from the muscular changes induced by limb suspension of the rat.

Bone

Immobilization of a limb or of the body as a whole leads to a loss of normal muscular and/or gravitational forces which normally act upon the bone. In consequence, the affected bones become lighter and thinner, with an increase in their water content and a decrease of their mineral content.

Morey-Holten & Arnaud (1985) have proposed a sequence of events that includes suppression of bone formation (Doty & Morey-Holton, 1984), possible increase of bone resorption, increase of blood calcium, changes in calcium-regulating hormones, decrease of vitamin D production, increased excretion of calcium and decreased intestinal absorption of calcium.

Claus-Walker *et al.* (1977) commented that spinally-injured individuals excreted more hydroxyproline in proportion to calcium than was the case in astronauts who were completing real or simulated space missions. It was suggested that in the spinally-injured group (who were severely handicapped by C4–C6 lesions), both skin and bone collagen underwent a progressive degradation, whereas in the astronauts (who were still mobile despite exposure to a zero gravitational field) the deterioration was limited to bone.

Donaldson *et al.* (1970) observed a 39% loss of bone mineral from the calcaneus over a six-month period of bed rest (an average of 1.5% per week), while Rambaut *et al.* (1972) noted a loss of 1.1% per week at the

Table 10.1. *Effects of a terminal 20–42 day bed rest, as examined in cadavers.*

Variable	Bed rest	Control	Percentage of control
Calcium content of ash from vertebrae (%)	34.5	36.7	93.8
Limit of strength (kN/cm^2)	462	676	64.8
Modulus of elasticity (MN/cm^2)	11.47	16.68	68.8

(Based on data of Stupakov, 1988.)

same site during a 24-week period of immobilization. Krølner & Toft (1983) found a 3.6% average decrease of vertebral density in 34 adults who were confined to bed for periods of 11–61 days on account of back pain; these changes were partially reversed within several weeks of recommencing ambulation. Stupakov (1988) carried out post-mortem examinations of patients who had been confined to bed for 20–42 days immediately prior to death. He noted not only a reduced calcium content of the residual ash, but also a decrease in the strength and modulus of elasticity of the bone relative to the control cadaver specimens (Table 10.1). The decrease in quality of the bone apparently reflected a loss of calcium ions from the surface of the hydroxyapatite crystals. The bones of the lower limbs were affected much more severely than the vertebrae, and Stupakov (1988) speculated that swelling of the intervertebral discs while the individual was recumbent may have served to maintain some pressure upon the vertebrae.

Studies of rats following an 18-day space mission (Wronski *et al.*, 1980) showed a decreased rate of periosteal bone formation, a decreased trabecular bone volume, and an increased fat content of the bone marrow. There was also a 20–25% decrease in the rate of bone resorption, secondary to a decrease in body calcium turnover (Cann *et al.*, 1980*b*).

Recent experiments have suggested that normal muscle traction helps to protect trabecular structure, and the adverse effects of weightlessness are apparently limited to bones that are normally weight-bearing (Vico *et al.*, 1987). Mack *et al.* (1967) noted that while human astronauts developed a significant bone mineral loss over a period of weightlessness, this could be checked at least partially by combining increased food intake with isometric and isotonic exercise throughout their mission. On the other hand, Greenleaf *et al.* (1977) were unable to prevent changes of body composition by providing isometric or isotonic exercises over a 14-day period of bed-rest.

Enhanced activity

Many studies purporting to examine the effects of increased physical activity upon body form have been technically unsatisfactory, and frequently anthropometric data have been collected almost as an afterthought (Malina, 1984). Available information concerns the influence of exercise upon overall body size and body composition at maturity.

Body size

At one time it was held that heavy physical labour prior to adolescence had an adverse effect upon adult stature (Wurst, 1964; Kato & Ishiko, 1966). The issue is still vigorously debated within groups such as the American Pediatric Society, although it is now generally accepted that in some early studies, the postulated adverse effects of physical loading upon growth were confounded with the influence of poor socio-economic conditions (Kotulan *et al.*, 1980).

Cross-sectional comparisons of size between athletes and sedentary adolescents may be complicated by a process of selection which is based upon body build and (at the higher levels of competition) by 'doping'. The administration of anabolic steroids and growth hormone preparations is regrettably all too prevalent among many categories of high-level performers. Such compounds promote nitrogen retention and therefore lead to an increase of lean tissue (Chapter 11), while also encouraging an early closure of epiphyses and thus a short adult stature.

Shortness of stature has been noted in gymnasts (Buckler & Brodie, 1977; Berninck *et al.*, 1983). In contrast, top-level swimmers (Åstrand *et al.*, 1963; Andrew *et al.*, 1972) and basketball players (Forbes, 1987) are taller than their peers. However, such differences reflect largely the influence of an early size-based selection of competitors (Shephard *et al.*, 1978; Osterback & Viitsalo, 1986) and (in age-class events) differences in the rate of maturation between athletes and controls. Successful male participants in certain categories of athletic competition tend to be larger, stronger and more mature than their peers, with a lower than anticipated percentage of body fat (Table 9.3). In some sports, team members are selected particularly for height, and in others for lean mass; these distinguishing characteristics becoming more marked as age increases and competitive selection becomes more intense. Female athletes are usually thinner than sedentary women of similar age, although they retain more body fat than male competitors. Women involved in a number of disciplines such as gymnastics also tend to be late maturers (Malina, 1984).

A small-scale longitudinal experiment compared vigorously exercised and control groups of male students following them from 11 through to 13 years of age; the data apparently showed a greater rate of growth in the active boys (Ekblöm, 1969), but the study did not exclude the possibility that the active group had temporarily become larger than the controls because they happened to be biologically more advanced in their development.

While it is still prudent to protect the pre-adolescent from excessive endurance activity, the general consensus thus seems that regular athletic training neither stimulates nor impairs the normal increase in body size with growth (Kotulan *et al.*, 1980; Malina, 1984; Shephard *et al.*, 1984).

Muscle

Regular physical activity encourages the development of muscle (Parizkova, 1977). Thus Boyden *et al.* (1982) noted a 1.7 kg increase of lean mass, with a corresponding reduction of body fat when female runners increased their weekly training distance from 40 to 80 km. However, the development tends to be localized to the exercised region (Malina, 1980), typically leading to an increase of upper arm and thigh circumferences (Kotulan *et al.*, 1980). Some examples of such regional development are given in Chapter 7. Von Döbeln & Eriksson (1972) noted that in 9–11-year-old boys, a period of endurance training increased body potassium by some 12 g, equivalent to the formation of 4 kg of lean tissue.

The gains of muscle strength with short-term (2–3 month) bouts of training often exceed the observed augmentation of muscle dimensions. Much of the observed increase in performance is due to an improved coordination of the contraction, rather than tissue hypertrophy (Ikai & Fukunaga, 1968; Fried & Shephard, 1970). A further possible factor is an improvement in the quality of the tissue designated as 'muscle' on a soft-tissue radiograph or CT scan of a limb. Thus Hoppeler *et al.* (1985) found that after six weeks of vigorous training, there was no increase in their estimate of total muscle area, but the proportion of the limb cross-section occupied by fat had decreased, while histological examination of biopsy specimens disclosed an increase of capillarity and mitochondrial density.

Some authors such as Alén *et al.* (1985) have found no increase of muscle cross-section even in response to power training. Other investigators have observed a progressive increase of muscle bulk in adults following appropriate forms of muscle conditioning (Mayhew & Gross, 1974; Wilmore, 1974; MacDougall *et al.*, 1977). A key variable is probably the diet that is consumed. Forbes (1987) summarized 13 training studies. In five reports where body mass had decreased, there was an average 0.7 kg increase of lean mass. In five studies with no change of total body mass, 1.4 kg of fat had been replaced by an equal mass of lean tissue, while in three studies where total body mass had increased, lean mass had also increased by an average of 2.0 kg. The upper limit of lean mass that can be realized by an effective muscle building regimen seems to be an accumulation of about 100 kg in men and 60 kg in women, some 50–60% above the expected normal values for sedentary subjects (Forbes, 1987).

Unfortunately, the advantage obtained from such muscle development is quite transient, and loss of lean tissue is obvious within a few weeks if training is discontinued.

Body fat

The impact of regular physical activity upon the extent of body fat stores probably depends upon both the total food intake of the individual and the pattern of exercise that is adopted. Parizkova (1977) found that active pre-adolescent students accumulated less body fat than their sedentary peers, and that a physically active summer camp provided an effective short-term corrective for childhood obesity; however, the obesity tended to recur when the students returned to their own homes. Other longitudinal studies of physical training in students who were living at home have found little difference in the percentage of body fat between active and inactive groups (Kotulan *et al.*, 1980; Shephard *et al.*, 1979), with no tendency to reduction of obesity in the active individuals as observations continued.

In adults, it has been suggested that individual bouts of physical activity must have a minimum duration of at least 30 minutes if any substantial mobilization of depot fat is to occur (Pollock *et al.*, 1969; Gwinup, 1975). The cumulative fat loss from a regular exercise programme is usually greater if such exercise is performed at an intensity below the anaerobic threshold, since anaerobic activity is based upon the metabolism of carbohydrate rather than fat (Pollock *et al.*, 1972).

Bone

Nilsson & Westlin (1971) documented the apparent positive influence of physical activity upon femoral bone density in a cross-sectional comparison between various classes of athletes and sedentary individuals. Average values (g/cm^3) were 0.247 in weight-lifters, 0.238 in throwers, 0.235 in runners, 0.233 in soccer players, 0.226 in swimmers, and 0.213 in active and 0.168 in inactive controls. Likewise, Dalén & Olssen (1974) reported a greater bone mineral content in athletes than in control subjects, while

Montoye (1984) showed that ballet dancers had a high mineral content in the tibia and fibula; the latter group also had an above average cortical thickness in the metatarsal bones (Mostardi *et al.*, 1982).

On the other hand, there have recently been suggestions that excessive habitual activity can lead to a demineralization of bone in young female competitors (Drinkwater *et al.*, 1984). Interpretation of such findings is in some instances complicated by a negative energy balance or a calcium poor diet (Nelson *et al.*, 1986). Snyder *et al.* (1986) found no deterioration in either radial or vertebral mineral content in elite oarswomen who had trained to the point of amenorrhoea. On the other hand, distance runners who train hard, typically show a low mineral content in the trabecular bone of the vertebrae (Cann *et al.*, 1984; Drinkwater *et al.*, 1984; Linnell *et al.*, 1984), although the cortical bone of the radius (which has a slower turnover rate) remains unaffected. If activity and/or diet is adjusted so that normal menstruation is restored, then the bone mineral is progressively replaced (Drinkwater *et al.*, 1986).

High environmental temperatures

Body form and heat production

The influence of body form and composition upon a population's ability to colonize an excessively hot or an overly cold habitat has been discussed by Harrison *et al.* (1964) and Adler (1988). However, concepts of the interactions between body surface area and heat exchange extend back much further, to Sarrus & Rameaux (1939) and Kleiber (1932, 1947).

Basal heat production

Investigators have for long accepted that basal heat production is proportional to $a(M)^n$, where M is the body mass. Usually, the term a is assumed to be a constant and the exponent n is determined experimentally, although on occasion n has been fixed and a has been determined. $(M)^n$ is sometimes described as the 'metabolic body size', and in the special case where the bodies of different individuals or species are similar in shape and density, n is equal to 0.67. Brody (1945) found an exponent of 0.73 (later rounded to 0.7) as appropriate to a wide range of species, while Kleiber (1947) argued strongly for an exponent of 0.75. A typical formulation might be:

$$\dot{V}_{O_2} = 3.8 \, (M)^{0.75}, \text{ or } \dot{V}_{O_2}/M = 3.8 \, (M)^{-0.25}$$

If the percentage of body fat is $(F/M, \%)$, then the reserve of fat that can be used to sustain body temperature in an emergency (Fr) is given by $F\%$ $(M)/a(M)^{0.75}$, or $F\% (M)^{0.25}$.

Energy cost of work

During exercise, the added energy cost of performing individual physical activities varies as $(M)^0$ to $(M)^{1.0}$, the magnitude of the exponent depending upon the cumulative vertical displacement of body mass that is required during the performance of the task, relative to the total energy cost of the required external work (Godin & Shephard, 1973). The end result for many types of activity is that the overall energy expenditure varies as $(M)^{0.75-1.0}$, with at least 75% of the consumed energy appearing in the form of heat. In dimensional terms, mass may be considered as a cube function of height (H). The body heat production of an active individual should thus be proportional to $(H)^{2.25-3.0}$.

Notice the practical implication of these calculations. Neither basal energy expenditure nor the cost of moving about are fixed quantities; both are influenced by body size and thus the state of nutrition of the individual.

Body form and heat loss

The potential for heat loss by convection and evaporation depends upon the surface area of the subject. In dimensional terms, body surface is proportional to the square of standing height. According to the classical formula of E. F. DuBois, body surface can be estimated as the product of $(M)^{0.425}$ and $(H)^{0.725}$, or (since $M = H^{3.0}$), $H^{1.275} \times H^{0.725}$.

The effectiveness of a given surface area in dissipating heat depends upon its curvature – for example, the coefficient of evaporative heat loss per unit of body surface is twice as large for a limb with a 7 cm diameter as that for a limb with a diameter of 15 cm (Harrison *et al.*, 1964). Radiant heat exchange depends also upon the height of the sun or other heat source and the exposed profile of the body.

Overall calculations of thermal balance suggest that when undertaking hard physical work in a hot environment, equilibrium is established more readily by a small than by a large person. Moreover, for any given standing height, an ectomorph has an advantage over a more spherical endomorph.

The corollary of these various statements is that the heat stress imposed by a given environment depends strongly upon the shape of the body that is exposed. The micro-environment of the naked individual is thus influenced by her or his structural and biological characteristics, as well as by gross and easily measured environmental variables, while the normal micro-environment depends also upon a person's ability to devise appropriate clothing and shelter.

Body build of successful colonists

Empirical data show that many of the populations colonizing hot environments are quite thin (Table 10.2), but this is probably a consequence of a limited food supply rather than any specific adaptation of body form to the hot climate. Since skin blood flow is greatly increased when exercising in the heat, the significance of subcutaneous fat as an impedance to heat exchange becomes much less in warm than in cooler habitats.

The concept that smallness contributes to success in populating a tropical habitat was formally expressed in the 'ecological rules' of Bergmann (1847) and Allen (1877). Somewhat more recently, Schreider (1951) commented on a gradient of body mass to surface area ratios from France ($38\,kg/m^2$) to the Andaman Islands ($32\,kg/m^2$). Roberts (1953) further suggested that the ratio of sitting height to overall stature was negatively related to the mean annual temperature of the habitat, while the ratio of arm span to height increased with average temperature.

However, the arguments in favour of a genetic adaptation of body form to a hot climate have been weakened by the observation that Europeans living in the tropics also show a lower average body mass than their peers who inhabit more temperate regions. Moreover, gradients calculated simply on the basis of latitude or mean annual temperature are hardly

Table 10.2. *Estimates of percentage body fat in selected developing nations colonizing warm or hot environments.*

Population		Men (%)	Women (%)
Masai	– tribal	6	—
	– nontribal	6	—
Samburu	– tribal	9	—
	– nontribal	12	—
Tanzanians	– urban	12	26
Nigerians	– soldiers	12	—
Upper Volta	– farmers	—	19
Colombia		15–20*	—

*Data of Barac-Nieto *et al.* (1978); estimates possibly distorted by abnormalities of fluid balance.
(Based in part on data accumulated by Norgan, 1982.)

indicative of the thermal stress imposed by a given habitat; this depends also upon summer/winter and night/day variations of temperature, absolute humidity, radiant heat load, wind speed, the intensity of physical activity necessary for survival, and any technological innovations that a given individual may have developed.

Again, traditional populations have sometimes found it convenient to live at a location where the conjunction of several climatic zones offered a variety of fauna and flora to the hunter–gatherer. Even if body form offered an advantage when colonizing one habitat, it could well prove a disadvantage to life in a second (Shephard, 1988). Finally, populations differing widely in body form and composition have apparently succeeded in colonizing the same habitat, and within some historically short populations, recent environmental changes have induced considerable increases of average body size, indicating that traditional dimensions of such samples have provided a poor indication of their genetic potential (Ghesquière & Eeckels, 1984; Ghesquière & D'Hulst, 1988).

While there remain significant differences of body form between West African negroes and Europeans, there seems little evidence that the West Africans have any greater tolerance of heat than Europeans who have become acclimatized to the same environment (Ladell, 1964; Wyndham, 1966).

Low environmental temperatures

Low environmental temperatures reduce skin blood flow almost to zero. In such circumstances, subcutaneous fat provides a very effective immediate insulation for the body, reducing overall heat loss and incidentally habituating the skin to sensations of cold. Body fat also provides a reserve of energy proportional to $F\%(M)^{0.25}$ (see above). In principle, this reserve can be utilized to sustain body heat if food is in short supply. Deep fat is particularly useful as a food store, since it can be metabolized without changing the body's insulation; however, a combination of intense cold exposure and a loss of subcutaneous fat could have serious consequences for the individual.

Because a cold-induced liberation of catecholamines mobilizes body fat, deliberate cold exposure has been suggested as a possible tactic for the treatment of obesity (see section *Basal heat production*).

Fat as an insulator

Adipose tissue provides quite an effective means of insulating the body against cold, because it has only a low rate of metabolism and thus a limited blood flow. However, it has the disadvantage that the impedance to heat flux cannot readily be adjusted to take account of varying levels of physical activity and thus internal heat production.

The problem can be illustrated by a simple calculation. The thermal conductivity of fat during immersion in cold water is such that a 1 cm layer of adipose tissue (equivalent to a skinfold thickness of about 24 mm) would allow the transfer of 1.22 kJ/min of heat per m^2 of body surface for a 1 °C temperature differential from the deeper tissues to the skin surface (Hatfield & Pugh, 1951). Thus, at rest (with a metabolic rate of some 2.1 kJ/min per m^2), the temperature gradient from deep to superficial tissue amounts to about 1.7 °C per cm of subcutaneous fat thickness, and if the metabolic rate is increased tenfold by exercise, the temperature gradient rises to 17 °C per cm of fat.

The mean weighted skinfold thickness provides a valuable index of the likely ability of an individual to swim (Keatinge *et al.*, 1986) and to survive in icy water (Nunneley *et al.*, 1985). If the body mass is M and the weighted skinfold thickness is S, then the time (T) required for a person's core temperature to drop below a critical value of 34 °C at a given water temperature of t °C can be predicted from the equation:

$$\log_n(T) = A + B \log_n[C + (D \times \text{togs}) + S]$$

where $A = -8.971 + 0.56$ (t), $B = 3.951 - 0.124$ (t), $C = 13.537 - 0.795$ (t), $D = 38.227 - (0.278\,M) + 0.897$ (t) and togs are the standard measure of clothing insulation (Hayes, 1986).

In some circumstances, inactive muscle is also poorly perfused, and this then provides an additional source of insulation.

Heat loss in women

Although women carry much more subcutaneous fat than men, they are not at any great disadvantage when attempting to lose heat in a hot climate, since in such circumstances the effective insulation of the superficial fat in both sexes is reduced by a large skin blood flow. Any residual disadvantage of impedance to heat flux in the female is compensated by a larger surface area to body mass ratio.

In a cold environment, the thick layer of subcutaneous fat gives the lightly-clothed woman a substantial advantage over a man, as evidenced by the success of females in such tasks as pearl diving (the Ama, below) and cross-channel swimming (Hatfield & Pugh, 1951).

Subcutaneous fat and exploitation of cold habitats

A substantial thickness of subcutaneous fat might be considered an important adaptive advantage when colonizing a very cold habitat. The

Table 10.3. *Influence of season and acculturation upon subcutaneous fat and muscle strength.*

Level of acculturation	Summer		Winter	
	Subcutaneous fat (average of 3 folds, mm)	Maximum quadriceps force (N)	Subcutaneous fat (average of 3 folds, mm)	Maximum quadriceps force (N)
Traditional	5.0 ± 0.8	445 ± 73	6.4 ± 1.2	457 ± 72
Intermediate	6.1 ± 2.7	447 ± 89	6.9 ± 1.8	484 ± 77
Urbanized	6.6 ± 2.8	454 ± 78	7.7 ± 4.4	482 ± 90

Note that the winter season is associated with small increases of subcutaneous fat thickness in segments of the population at all three levels of acculturation.

(Based on data of Rode & Shephard, 1973, for the Inuit community of Igloolik.)

main disadvantage of such insulation is its relative inflexibility when the individual must adapt to a substantial increase of heat flux.

A person who adopts a traditional lifestyle in an arctic habitat faces many hard physical tasks (Godin & Shephard, 1973) and a tenfold increase of metabolism is often necessary during ordinary daily activities. If the main source of insulation was subcutaneous fat rather than clothing, then the increased heat flux associated with such bouts of activity could induce an undesirably large rise of core temperature. Even if this did not cause hyperthermia, it would give rise to an unwelcome onset of sweating, drenching the person's clothing, and precipitating a disastrously rapid rate of body cooling once the immediate bout of activity had ceased. Perhaps for this reason, the skinfold thicknesses of the traditional Canadian Inuit hunter are low rather than high (Rode & Shephard, 1971), showing little tendency to increase even in the coldest winter months (Rode & Shephard, 1973b; Table 10.3). Moreover, the traditional Eskimo appears to store any reserves of fat that are needed to maintain body heat in deep rather than superficial depots (Shephard *et al.*, 1973).

The Ama (pearl divers) of Japan and Korea stand in striking contrast to the Eskimos, providing examples of populations that have made practical use of subcutaneous fat as an insulator while commercially exploiting the cold shore-waters of the Japanese and Korean coastal communities. These two groups of people encounter the most severe cold stresses during the course of their normal daily work, and plainly they face no problem of sweat accumulation while underwater. Relative to usually accepted norms, the average thickness of subcutaneous fat in the Ama is increased over the trunk, but not over the face or the extremities (Kohara, 1975). Moreover, the divers show substantial decreases in thickness of the fat blanket (2–3 cm change) in the summer months, when the water is warmer. There are nevertheless the usual differences both of average skinfold thickness and of fat distribution between men and women. In consequence, male divers work mainly in the more southerly and warmer waters, while the female divers exploit the colder northern shores.

Fat as a determinant of the metabolic response to cold

Wyndham *et al.* (1968) suggested that because of a greater thermal gradient from deep to superficial tissues, fat people became habituated to low skin temperatures, and so needed a greater cold exposure to induce shivering and other metabolic responses to cold. The ready demonstration of non-shivering thermogenesis in the traditional Ama (Hong *et al.*, 1986) is somewhat against this concept. Nevertheless, Strong *et al.* (1985) have recently confirmed that the equations relating both core and skin temperatures to resting metabolism are much steeper for thin than for fat subjects.

An insulative rather than a metabolic type of response to a cold habitat (Shephard, 1985*a*) is energy efficient, and could have adaptive value in an environment where food supplies are limited.

Fat as a metabolic resource

A combination of cold exposure and prolonged vigorous activity leads to a reduction of body fat stores that is greater than would have been encountered with performance of an equal amount of work under temperate or warm conditions.

O'Hara *et al.* (1979) carried out both field and chamber experiments on Canadian soldiers and civilians with little recent cold acclimation. They found cumulative fat losses of 2–4 kg when 1–2 week bouts of daily vigorous physical activity were performed under arctic conditions (Fig. 10.1). The subjects concerned were wearing standard arctic military

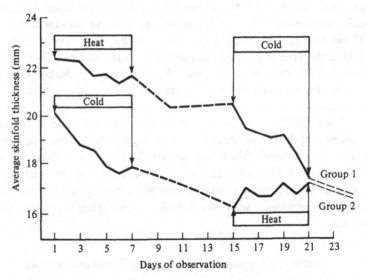

Fig. 10.1. The influence of vigorous physical activity and environment upon body composition. A crossover trial comparing the effect upon subcutaneous fat of repeated bouts of prolonged exercise in warm and cold environment. Based upon data of O'Hara *et al.* (1979).

clothing, and thus sustained little change of core temperature during the experiments. The prime stimulus to the fat loss seems to have been a local cooling of the face, since the extent of the decrease in fat mass was correlated with the changes of local skin temperatures in the facial region. Yokoyama & Iwasaki (1975) had somewhat similar findings in the Japanese Ama; they noted a cumulative 4.6 kg decrease of body mass over the diving season, a substantial component of this being attributable to the use of body fat as a heat source during diving.

Possible factors contributing to a substantial cold-induced fat loss in both the Canadian and the Japanese populations include non-shivering thermogenesis (through the initiation of 'futile' metabolic cycles in skeletal muscle and an uncoupling of metabolic reactions in brown fat, Jansky, 1973, 1976; Shephard, 1985a; Huttunen *et al.*, 1981), an incomplete combustion of fat due to the development of a 'ketone shunt' (Itoh, 1975; O'Hara *et al.*, 1979), an uncoupling of oxidative reactions due to an increased secretion of thyroxine (typically an expression of long-term acclimatization, seen over a winter of cold exposure, Itoh, 1974) and the added energy costs of physical activity (associated with the hobbling of body movement by protective clothing, the wearing of heavy boots and the need to walk through deep snow).

Cold exposure in the treatment of obesity

As cold stress induces a mobilization of depot fat which is moderate rather than extreme (O'Hara *et al.*, 1977a,b,c) – a response rather more clearly seen in obese than in thinner subjects, O'Hara *et al.* (1977c) suggested that a vigorous winter holiday might prove of clinical benefit in the treatment of moderate obesity. Murray *et al.* (1986) pointed out one immediate limitation of such therapy, in that cold-induced fat mobilization was much less marked in their experiments on young women than in the observations of O'Hara *et al.* (1977a,b,c) on young and middle-aged men. After reviewing other possible explanations of the discrepant findings between the two experiments, including the relative intensities of exercise undertaken and the availability of caffeine-containing beverages, Murray *et al.* (1986) suggested that the most likely explanation was a greater stability of body fat in women than in men. Such stability could be viewed as a biological adaptation to the potential demands of pregnancy and of lactation. Support for their hypothesis has come from recent studies of fat reserves and the techniques of cold adaptation adopted by poorly nourished women in developing countries (see below).

High altitudes

Humans who attempt to colonize such regions as the higher ranges of the Andes and the Himalayas face severe environmental challenges from both low oxygen pressure and severe cold. The plasticity of the human response is impressive, but evidence of permanent, inherited adaptations of growth patterns and adult body form in such habitats remains rather unconvincing. Many of the supposed differences of growth, adult stature and fat distribution between permanant high-altitude residents, recent sea-level migrants and continuing lowlanders (for instance, Clegg *et al.* (1972) studied children living at different altitudes in Ethiopia, and Frisancho *et al.* (1975) compared the growth patterns among Peruvian Quechua of similar genetic constitution (Table 10.4)) have been flawed by differences of socio-economic status, dietary habits and living conditions, which have tended to favour the lowlanders and recent immigrants (Habicht *et al.*, 1974; Greksa *et al.*, 1984).

On the other hand, there have been persistent reports that permanent altitude residents of the Andes, the Himalayas and of the mountainous regions of Ethiopia have large static lung volumes and large heart volumes relative to body size, with an enlarged bone marrow, and a high haemoglobin level (Barcroft, 1914; Hurtado, 1964; Burri & Weibel, 1971; Bharawaj *et al.*, 1973; Frisancho, 1978; Mueller *et al.*, 1979; Beall, 1982). In Ethiopia, the forced vital capacity of the individual increases in proportion to the altitude that has been colonized (Clegg *et al.*, 1972). Sherpas are shorter than North Americans, but have a normal height for weight relationship (Hackett *et al.*, 1980), apparently with no increase of thoracic dimensions (Pawson, 1977). Nevertheless, their vital capacity is also 4–10% greater than an age, sex and height specific Caucasian regression prediction (Hackett *et al.*, 1980). Tibetans again show a greater height-adjusted vital capacity than sea-level Asians (Beall, cited by Frisancho, 1983). Comparisons of Quechuans with Tibetans living at a similar altitude show a similar chest depth in the two populations, whereas the chest width is narrower in the Tibetans (Beall, 1982).

The large chests of the highland Amerindian seem to be acquired through an accelerated development of the thorax relative to stature during childhood and particularly during adolescence (Mueller *et al.*, 1979). Peruvians who move to altitude in early childhood develop an equal enlargement of vital capacity (Frisancho, 1983), as do Ethiopians (Harrison *et al.*, 1969). However, Greksa (1988) noted that in Europeans who became highlanders, the impact of altitude exposure depended more upon its duration than the point during growth at which it began.

Table 10.4. *Comparison between physical characteristics of Quechuans in Peruvian highlands (altitude 3700 m) and sea-level norms for the US.*

Variable	Highlanders	Lowlanders	Significance of difference
Age (yr)	21.2	22.2	
Height (cm)	161.3	178.0	$P < 0.001$
Body mass (kg)	62.5	76.4	$P < 0.001$
Residual volume (l)	1.59	1.30	$P < 0.001$

(Based on data of Frisancho, 1983.)

Several recent papers (Stinson, 1982; Schutte *et al.*, 1983; Greksa *et al.*, 1985; Greksa, 1988) have confined observations to children of European parentage and relatively high socio-economic status. On the basis of date of migration to high altitude, Greksa (1988) estimated that hypoxia had a negative effect of 4.3 cm upon growth, but positive effects on chest depth and on forced vital capacity (an increment of 1.8 cm and 384 ml, respectively, relative to recent migrants). He further commented that children who moved to high altitude at birth still apparently had some advantage of stature over families who had lived in the region for several generations, suggesting that foetal hypoxia had a negative effect upon growth. Frisancho (1983) commented that the increase of vital capacity was closely correlated with increases of lung dimensions, particularly in indigenous populations.

The large chest size of the high altitude native is retained on moving to sea-level (Beall *et al.*, 1977; Palomino *et al.*, 1979). However, given the large potential for an increase of chest dimensions when a given generation moves to altitude, it seems unlikely that inheritance plays a major role in determining adaptations of body form that are observed at altitude.

Over-nutrition

Food is required not only to balance the acute energy expenditure of physical activity, but also to sustain growth, to maintain existing tissues and to allow repair of tissue injuries. The nutrient requirements of the body thus depend upon the age, sex and size of the individual, and the activity patterns needed for survival in a given habitat.

In infancy, growth imposes a substantial added energy demand, and this need diminishes progressively as maturity is approached. On the other hand, the energy required to move the body and maintain the integrity of existing tissues depends on body size, and thus increases until a person has matured. The energy cost of daily physical activity also tends to peak in early adulthood. In women, pregnancy and lactation impose additional energy demands.

Maturation and obesity

In children, it has been argued that obesity is related in part to precocity. Obese children have high scores on indices of skeletal maturation, and their lean mass may also exceed that of their age-matched peers (Yamaoka, 1975). In some instances, the apparent problem of surplus fat may thus resolve itself as growth is completed.

However, Forbes (1987) drew attention to the fact that many obese children only become tall coincident with the onset of obesity; he thus suggested that the added height was a consequence of over-feeding rather than a mark of precocity.

Lean mass in obesity

While much of the excess body mass of an obese person is attributable to fat, there is generally some increase of lean tissue relative to a thinner individual. Thus Keys *et al.* (1955) recognized that 'obesity tissue' included muscle and water as well as fat, while Passmore *et al.* (1963) commented upon nitrogen retention in over-nourished women. Exceptions to this generalization include those forms of obesity that are secondary to Cushing's syndrome and hypothalamic disorders. In such pathological conditions, the muscle mass may actually be less than normal.

The above average lean mass of an obese person arises in part because of an associated tallness, and in part because of the additional energy cost that arises when moving a heavy body. In cross-sectional studies, Forbes (1987) found that three quarters of obese subjects had a lean body mass to height ratio that was at least one standard deviation above population norms, while Webster *et al.* (1984) estimated that in women 22–30% of any excess weight was attributable to lean tissue. Hankin *et al.* (1972) commented further on a high creatinine excretion rate in obese individuals. In 10 reported series of overweight subjects, the contribution of lean tissue to the excess body mass ranged from 19–40%, with an average of 29% (Forbes, 1987). Forbes (1987) suggested that in female subjects, the relationship between lean body mass (*LBM*) and fat mass took a semi-logarithmic form:

$$LBM \text{ (kg)} = 23.9 \log (F, \text{kg}) + 14.2$$

The implication of his equation is that lean tissue will account for a larger proportion of any increase of body mass in a person who is initially thin, whether their thinness is due to starvation, anorexia nervosa or gastrointestinal disease (Barac-Nieto *et al.*, 1979; Motil *et al.*, 1982; Forbes *et al.*, 1984). Thus Keys *et al.* (1950) found that on refeeding young men who had previously lost 23% of their body mass through a period of experimental semi-starvation, 50% of their initial weight-gain was attributable to lean tissue.

A further variable is the age of the subject. If food is abundant, children tend to accumulate lean tissue, but in adults the major part of any increase of body mass is fat (Forbes, 1964; Cheek *et al.*, 1970). One reason for this difference is that the number of both muscle and fat cells can increase fairly readily in the young person, while in the adult there is usually only potential for a hypertrophy of existing cells (Winick & Noble, 1967). Nevertheless, Faulhaber (1974) has argued that hyperplasia of adipocytes can be induced by extreme obesity, even in adults (Chapter 11).

Impact of obesity on other organs

In addition to an above-average development of muscle, the typical moderately obese person has enlarged visceral organs including the heart (Naeye & Roode, 1970), thicker than average cortical bone (Dalén *et al.*, 1975), and a small increase over the normal haemoglobin concentration (Garn & Ryan, 1982).

Wang & Pierson (1976) further reported a higher extracellular to total water ratio in obese subjects than in their normal counterparts, although the difference which they observed reflected mainly a low extracellular to total water ratio in the supposedly normal subjects. Shizgal *et al.* (1979) subsequently found no abnormalities of fluid or mineral distribution in those who were obese.

Empirical data on over-nutrition

Experimental studies of the changes induced by over-nutrition ideally require preliminary observations obtained during a period of good health and stable body mass. A substantial amount of over-feeding (>80 MJ total) is necessary to produce measurable changes of body composition, and physical activity must also be closely controlled during the period of observation, if gains of lean tissue are to be observed. In the over-feeding experiments summarized by Forbes (1987), an average of 38% of the increase in body mass was attributable to lean tissue. Likewise, Ravussin *et al.* (1985) reported that lean tissue comprised 44% of the increase in total body mass resulting from the ingestion of a 72 MJ excess of food over a 9-day period. If a very high percentage of protein is provided in the diet, more of the gain may be due to lean tissue (Müller, 1911), but if the diet is very poor in protein, weight may even be gained concomitantly with a loss of lean tissue (Miller & Mumford, 1967; Barac-Nieto *et al.*, 1979). Nevertheless, the protein needs of the average sedentary adult are quite modest. Thus, when Butterfield & Calloway (1984) provided an excess energy intake of 2.5 MJ/day, they found an increase of lean tissue, even though the protein content of the food was only 0.57 g/kg.

Studies of free-living subjects have yielded similar conclusions over longer periods of observation. A follow-up of Taiwanese over a 10-year period, during which national nutrition had improved, found an average 10.1 kg increase of body mass, 26% of which was lean tissue (Chien *et al.*, 1975a). Likewise, observation of young women during the post-partum period has suggested that on average 29% of both gains and losses of weight are attributable to lean tissue (Butte *et al.*, 1985).

Energy cost of weight gain

Calculation of the metabolic consequences of over-feeding is complicated by thermodynamic inefficiencies in the conversion of food to stored fat (Davidson *et al.* 1979; Pullar & Webster, 1977; Norgan & Durnin, 1980), the added energy cost of displacing a heavier body and increases of resting energy expenditure that tend to compensate for any increase of food intake – the 'Luxuskonsumption', discussed on p. 223. Nevertheless, given that the typical daily energy requirement of a sedentary man is around 10 MJ, a 1.0% imbalance between energy intake and expenditure can cause substantial obesity if it is sustained over many years. Forbes *et al.* (1986) found a linear relationship between any excess intake of energy and the resultant increase of body mass, the average energy cost of the added tissue amounting to some 33.7 kJ/g. A very similar relationship can be derived from the known costs of synthesizing protein (36.3 kJ/g, or 7.45 kJ/g of lean tissue) and fat (50.2 kJ/g; Spady *et al.*, 1976), assuming that 38% of the added matter is lean tissue:

$$\text{Energy cost} = 7.45\ (0.38) + 50.2\ (0.62) = 34.0\ \text{kJ/g}$$

Given that lean tissue costs less to form than fat, the energy cost of weight gain is less for a thin person who is laying down muscle than for an obese person who is increasing body fat stores (Jackson *et al.*, 1977). Thus, in anorexia nervosa, the cost of the weight gain may initially be as low as 22 kJ/g (Forbes *et al.*, 1984), while in premature infants who are thriving, figures of only 12–17 kJ/g have been observed (Forbes 1978).

Acute experiments in animals suggest a similar initial relationship between energy cost and weight gain to that observed in humans. However, if over-feeding continues, a plateau of body mass is reached (Blaxter, 1971). Given the effect of increasing body mass upon total energy expenditure, a similar plateauing of weights might be anticipated in human subjects. Inevitably, the energy cost both of maintaining body structures and of engaging in minimal activity is roughly proportional to body mass (Bessard *et al.*, 1983; Jéquier & Schutz, 1983; Van Es *et al.*, 1984; Prentice *et al.*, 1986; Forbes, 1987). Estimates based on the food intake needed to assure constancy of body weight and the 24-hour energy consumption (doubly-labelled water technique) each suggest that energy consumption is increased by about 84 kJ per kg of added body mass. Forbes (1987) has further pointed out that lean tissue (LBM) contributes proportionately more to this demand than does fat (F), the maintenance need being given by:

$$\text{Daily energy need (kJ/day)} = 149 \, (LBM, \text{kg}) + 64 \, (F, \text{kg}) + 829$$

However, the half-time of the plateauing process (an estimated 278 days, Forbes, 1987) is much longer than the period of observation in any of the deliberate over-feeding experiments that have been conducted to date.

The causes of moderate obesity

The modest accumulation of fat seen in the middle-aged adult typically represents the cumulative effect of a small imbalance between the intake of food and the energy demands of growth, maintenance and repair that has persisted for many years. The reason could be as simple as a decision to add sugar to tea or coffee, or to omit a daily walk to the local store when buying a newspaper.

Many of those who have become obese deny any indiscretion of either eating or habitual activity patterns. Often, this reflects a faulty perception of the amount of food that is being consumed or the quantity of activity that is being undertaken. Waxman & Stunkard (1980) noted that obese children usually ate more than their non-obese siblings, while Mahalko & Johnson (1980) demonstrated that the obese adult also underestimated food intake. Likewise, with respect to habitual physical activity, Chirico & Stunkard (1960), Stefanik *et al.* (1959), Bullen *et al.* (1964), Durnin (1966) and Huenemann (1972) have all documented the inactivity of the obese person relative to age-matched peers. Saris *et al.* (1980) used both activity diaries and heart rate recordings to show that the daily energy intake of obese children aged 8–12 years was essentially the same as for children of normal body build, but that levels of habitual physical activity were much lower in the obese students.

There may also be some differences of body physiology in the obese, favouring the accumulation of fat; this seems particularly likely in gross obesity (Chapter 11). Galton (1966) and Bray (1974) discussed possible enzyme defects in chronic obesity, including reductions in the activity of glycerophosphate dehydrogenase and glycerophosphate oxidase. The obese person may also have a larger number of adipocytes due to over-feeding in infancy (Angel, 1974; Björntorp *et al.*, 1973a; Björntorp, 1974). While major hormonal abnormalities are unlikely, Bray (1974) has commented on hyperinsulinaemia, a reduced release of growth hormone and an increased secretion of cortisol in obesity. A modest decrease of thyroid hormone secretion could also cause a more efficient use of food energy in the obese. Moreover, the thermal protection of a thick layer of subcutaneous fat reduces the need for body heat production in an obese person (Shephard, 1977).

The completeness of absorption of food energy from the intestines depends on the relative proportions of carbohydrate and fat in the diet (absorption of energy is more complete for carbohydrate than for fat) and the speed of passage of food through the intestines (intestinal transit is speeded by both physical activity and a high fibre diet). Normally, there is an increase of metabolism immediately after eating (specific dynamic action) and after participation in vigorous physical activity, with a particularly large metabolic response after deliberate over-eating ('Luxuskonsumption'). However, at least in animals, those with a genetic predisposition to obesity show a much smaller modification of metabolism than normal in response to these various stimuli. Again, over a long period, a reduced metabolic response could lead to the accumulation of substantial amounts of fat.

Cause of obesity and associated differences of body composition

Animal studies have looked at the impact of the cause of obesity upon body composition. An increase of lean tissue has been described in some genetically obese variants such as the yellow-obese mouse, but in other obese varieties such as the Zucker mouse there is a less than average amount of lean tissue (Pullar & Webster, 1974; Bergen *et al.*, 1975; Plocher & Powley, 1976; Deb *et al.*, 1976; Thurlby & Trayhurn, 1978).

Forced feeding gives some increase of lean mass (Rothwell & Stock, 1979), but obesity produced by hypothalamic lesions is usually associated with a loss of lean tissue (Goldman *et al.*, 1974).

Correction of obesity

The usual reason for attempting to reduce the amount of body fat is to improve physical appearance, although an excess of fat also confers the practical handicap of increasing the energy cost of a given physical task (Godin & Shephard, 1973). Moreover, obesity is associated with an increase in the risk of incurring a number of chronic diseases (Shephard, 1977), although there must be a substantial (>20 kg) accumulation of fat to produce any large change of life expectancy (Table 10.5).

Parizkova *et al.* (1971) demonstrated that in obese children, body composition was normalized by a period of controlled feeding and regular vigorous physical exercise at a summer camp; however, excess body fat was regained quite rapidly when the children returned to their own homes. The obesity was associated with a normal stature, but there was an increase of both the hydrostatically estimated lean body mass and bone breadth (bicristal diameter of pelvis, femoral condylar diameter).

In adults, dieting alone can sometimes achieve dramatic immediate reductions of body fat, but there is also a very high rate of recidivism once the desired reduction of body mass has been achieved (Sohar & Sneh, 1973; Innes *et al.*, 1974). Moreover, in many subjects, as much as 75% of the potential benefit of dietary restriction is lost through an associated reduction of physical activity (Behnke & Wilmore, 1974). Exercise alone can be quite effective in the treatment of moderate obesity (Gwinup,

Table 10.5. *Influence of excess body mass upon the incidence of various chronic diseases. All values presented as percentages of standard values for subjects of same sex, aged 15–69 years.*

Condition	Excess body mass (kg)			Excess body mass (kg)		
	+24	+33	+42	+28	+37	+46
Diabetes	179	385	629	270	242	250
Vascular diseases of brain	136	183	215	143	142	210
Heart and circulation	131	155	185	175	178	217
Pneumonia and influenza	128	103	242	148	110	—
Digestive diseases	146	230	298	93	122	—

(Based on data of Society of Actuaries, 1959.)

1975; Sidney *et al.*, 1977), although the response tends to be relatively slow (Wood *et al.*, 1983). In a review of 55 studies lasting 6–104 weeks, Wilmore (1983) noted that the average loss of body fat was only 1.6%. A better arrangement is therefore to combine a modest daily energy deficit (2–4 MJ) with a substantial increase of daily aerobic energy expenditure (American College of Sports Medicine, 1983). The latter type of recommendation helps to maintain an adequate intake of vitamins and minerals and counters the depression associated with dieting. It also causes a temporary suppression of appetite (Edholm *et al.*, 1955; Thomas & Miller, 1958; Mayer, 1960) and minimizes the protein loss induced by an energy deficit alone (Keys *et al.*, 1950; Babirak *et al.*, 1974; Walberg, 1986). Indeed, it has been claimed that in children who exercise vigorously, fat can be replaced by muscle without loss of total body mass (Parizkova, 1977; Parizkova *et al.*, 1971).

Malnutrition

Causes of malnutrition

Historically, malnutrition has often arisen from colonization of an unpromising habitat. The available food may have been inadequate in amount, quality or variety. Unfortunately, despite recent progress in agricultural technology, malnutrition still affects from 1 in 4 to 1 in 8 of the world's population (Soedjatmoko, 1981).

In developing societies, a primary energy or nutrient lack is commonly compounded by intercurrent bacterial or parasitic infection, leading to additional losses of protein and essential nutrients. Certain pathological conditions (for instance, malignancy, intestinal malabsorption, and hepatic malfunction) may also cause malnutrition because they increase the need for food and/or lead to a loss of appetite. Finally, an increasing number of girls and young women in western society deliberately reduce their food intake in order to conform with a personal perception of beauty or to meet the demands imposed by their peers in pursuits such as ballet, competitive gymnastics and cheerleading (Garfinkel & Garner, 1982).

The syndrome of 'anorexia nervosa' is an extreme example of self-imposed malnutrition. It is considered, with other pathological disturbances of body composition, in Chapter 11.

Impact on body composition

Whether the shortage of food is imposed by the harshness of the external environment, the increased demands of disease or an inner compulsion towards excessive thinness, the consequences for body composition seem the same – a loss of both fat and lean tissue (Keys *et al.*, 1950; Davies *et al.*, 1978a).

Precise determinations of the changes in body composition with malnutrition are hampered by technical factors that lead to substantial errors in many of the methods that would provide acceptable results in a person of normal body mass and composition. Dehydration and oedema both modify the compressibility of skinfolds, while also distorting all methods of analysis based upon an assumed body water content. Disturbances of mineral balance invalidate estimates based upon the assumption of fixed concentrations of potassium or nitrogen in the tissues. Body density readings may be distorted by a selective loss of muscle and a demineralization of bone. Even more direct methods of visualizing the body constituents, such as soft-tissue radiography or CT scans may be inaccurate in a malnourished individual because of changes in the composition of the gross masses already outlined.

In animal studies, precise information on current status can be obtained by sacrifice and carcass analysis, although the comparison of such data with human observations may be complicated by evisceration and/or caged inactivity prior to sacrifice. Typically, 71% of the nitrogen loss during starvation seems attributable to muscle and bone, 17% to skin, 6% to liver and 3% to the gastrointestinal tract (Uezu *et al.*, 1983).

Tissue loss

In human subjects, a loss of muscle can be demonstrated both in CT scans of the arms (Heymsfield *et al.*, 1979) and at autopsy (Arara & Rochester, 1982).

The loss of collagen seems less than the loss of other types of soft tissue (Picou *et al.*, 1966; Angeleli *et al.*, 1978). Since collagenous tissues have a low potassium content, misleading estimates of lean mass may be obtained from body potassium measurements in malnourished individuals.

Bone mass is also depleted less than muscle mass, increasing the ratio of body calcium to total body mass (Widdowson & Dickerson, 1964).

Body fluid and mineral content

Both children (Kerper-Fronius & Kovach, 1948; Flynn *et al.*, 1967) and adults (Keys *et al.*, 1950; Barac-Nieto *et al.*, 1978; LeMaho *et al.*, 1981) tend to maintain a normal plasma and extracellular fluid volume in the face of a substantial reduction of total body mass, but the total water content of the lean tissue is increased (Beddoe *et al.*, 1985).

Mineral balances are disturbed in malnourished individuals because of such factors as a poor functioning of the sodium pump, impaired renal function and any associated diarrhoea. Cells lose potassium, magnesium and phosphorus, while their sodium content is increased. In consequence, the ratio of total exchangeable sodium to total exchangeable potassium may be doubled (Shizgal, 1981). Muscle potassium accounts for a reduced proportion of total potassium (Nichols *et al.*, 1969), and the potassium to nitrogen ratio drops from the normal figure of 1.8 mE/g to about 1.6 mE/g (James *et al.*, 1984; Burkinshaw & Morgan, 1985).

The normal potassium to nitrogen ratio is restored as nutrition is improved (Brooke & Cocks, 1974), and if dietary needs are to be met by the intravenous route, provision of adequate amounts of the various minerals is important to successful restoration of body mass.

Maternal and foetal malnutrition

Maternal malnutrition leads to impaired nutrition of the foetus, with birth of an underweight child. Apte & Iyengar (1972) provided data on 41 pregnancies in malnourished women. The body content of individual elements was related to total body mass, using the equations of Shaw (1973); at birth, the infants contained only 61% of the expected quantity of iron, and 82–86% of the expected nitrogen, phosphorus and magnesium. The percentage of body fat was close to the anticipated figure, and the main impact of malnutrition was upon the development of lean tissue.

Other studies of malnourished populations have shown that the birthweight is substantially augmented when pregnant women are given dietary supplements (Adair, 1984), although maternal anthropometry remains unaltered by such treatment. Nevertheless, Venkatchalam *et al.* (1960) noted that the weight gain during pregnancy was somewhat reduced in severely malnourished Indian women, amounting to an average of about 13%, compared with an 18% gain in healthy British women (Hytten, 1980).

Presumably, in a region where food is in short supply, there is considerable evolutionary pressure for the emergence of women who can adopt energy saving tactics during pregnancy, thus accumulating sufficient tissue to allow both the birth and suckling of healthy offspring. Perhaps because of prolactin effects, Adair (1984) found that malnourished Taiwanese women showed a dramatic increase of triceps and subscapular skinfold thickness in the first 30 days after parturition, this store of fat being dissipated as the appetite of the newborn infant increased.

Child malnutrition

Nutrition during pregnancy and the first five years after birth is the primary determinant of adult size. After the age of 5 there is little possibility of 'catch-up' growth (Eveleth & Tanner, 1976). Height deficits that have not been corrected by five years of age thus persist into adolescence and adulthood (Satyanarayana *et al.*, 1980; Martorell & Habicht, 1986). Under poor socio-economic conditions, inadequate lactation is compounded by inadequate complementary feeding, and the poor condition of the child may be further worsened by repeated bouts of diarrhoea, vomiting and fever (Martorell, 1985).

Numerous studies have documented the negative influence of even a small socio-economic disadvantage upon child growth. While the average stature of the highest socio-economic stratum is remarkably constant around the world, there are major differences of adult size within genetically homogeneous populations, attributable to the impact of unfavourable economic circumstances in the early period of life (Ferro-Luzzi, 1988). Distinction must be drawn between the chronically malnourished, stunted child (with a low height and weight for age) and the acutely malnourished child (where weight is also reduced in relation to current height). Spurr (1988) noted that malnourished Colombian boys did not show the pubertal decrease of body fat that has been described in developed countries. The ages corresponding to peak height and peak weight velocity were also retarded, and muscle mass was less than anticipated. Moreover, because small children became small adults, the malnourished child was condemned to poor productivity as an adult; the cycle thus repeated itself, transmitting the disadvantage to subsequent generations (see below).

Tobias (1971) commented that adverse socio-economic conditions apparently had less effect on growth in girls than in boys – the girls showed a greater than anticipated advantage of height and muscle development over the boys in early adolescence, and there was less than the expected dimorphism in adult characteristics. Rode & Shephard (1973*a*) made similar observations in a well-nourished Inuit community. It may be that the relative enhancement of female growth in developing societies reflects differences of habitual activity patterns as well as lack of essential nutrients.

Malnutrition and adult body composition

Despite the technical difficulties of measurement, there is general agreement that moderate malnutrition is associated with very low levels of

Table 10.6. *Influence of malnutrition upon mass/height, serum albumin concentration and daily creatinine excretion.*

Degree of malnutrition	Mass/height (kg/m)	Serum albumin (g/dl)	Creatinine excretion (mg/day/m)
Mild	33.3 ± 2.1	3.8 ± 0.5	660 ± 67
Intermediate	30.8 ± 2.0	3.0 ± 0.7	559 ± 75
Severe	27.4 ± 2.1	2.1 ± 0.5	391 ± 76

(Based on data of Barac-Nieto *et al.*, 1978.)

subcutaneous fat and a reduction of lean body mass (Tables 10.2 and 10.6), the latter reflecting particularly a loss of muscle tissue (Barac-Nieto *et al.*, 1978).

Viteri (1971) concluded from observations of daily creatinine excretion in under-nourished Gautemalan agriculturalists that the deficiency of lean body mass in such populations was largely attributable to a deficiency of muscle cell mass. Likewise, Barac-Nieto *et al.* (1978) noted a correlation of weight for height with both creatinine excretion and serum albumin levels (Table 10.6).

Even if the total energy intake is adequate, a protein-poor diet may also lead to a stunting of growth (Shirai, 1975; Nagamine, 1975), with a deficiency in both the adult size and the mature mass of the long bones (Suzuki, 1975).

The body mass of a severely anorexic patient may drop as low as 38–40 kg. Keyes *et al.* (1950) did not observe any substantial change of blood volume when there was a 24% loss of total body mass due to experimentally imposed semi-starvation. However, in some of the anorexic patients studied by Fohlin (1980), the loss of body mass was associated with a substantial reduction of blood volume.

Body mass is restored progressively, typically at a rate of about 1 kg/week, as a malnourished individual undergoes hospital treatment. Much of the gain is undoubtedly lean tissue, but there is also some increase of both tissue water and body fat stores.

Manipulations of body mass and form in athletes

Several categories of athlete have been known to engage in dietary manipulation with a view to altering their body mass and/or form.

Gymnasts, ballet dancers and cheerleaders may deliberately set their energy intake substantially below daily energy expenditure (Cohen *et al.*,

1982). For instance, Van Erp-Baart *et al.* (1985) estimated the average energy intake of top 15-year-old female gymnasts at 7.28 MJ/day, whereas the corresponding energy expenditure was 8.5 MJ/day by time and motion study and 9.13 MJ/day by heart-rate recording. Such an energy imbalance is usually associated with a substantial reduction in the rate of protein synthesis (Stein *et al.*, 1983).

Wrestlers also engage in deliberate bouts of both starvation and fluid depletion in attempts to achieve a lower weight classification that is appropriate to their true body composition. Zambraski *et al.* (1974) found that in the US, even high school wrestlers decreased their body mass by as much as 9–13% at the height of a competitive season. Such attempts to obtain an unfair competitive advantage commonly lead to a loss of lean tissue and a decrease of physical performance (Tipton & Tcheng, 1970; Houston *et al.*, 1981). Given a pre-season body fat content of 8% and a competitive figure of around 5% body fat (Tipton & Tcheng, 1970), the wrestler who loses 9–13% of body weight must also have lost lean tissue and/or fluid equivalent to 6–10% of body mass. Moreover, this loss is sustained at an age when other high school students are showing a substantial month-by-month increase of lean mass (Freischlag, 1984).

Runners may also become excessively concerned about their body mass (Smith, 1980; Nelson, 1982). Indeed, the body fat content of some endurance competitors is less than 3% (Walberg, 1986).

Body composition and productivity

Severe malnutrition is a societal problem that may be perpetuated from parents to children, particularly if the body composition of the offspring is severely enough affected to restrict their physical and mental development and thus the capacity of the next generation to produce food as adults.

Martorell & Arroyave (1988) noted that many authors considering possible linkages between malnutrition, body build and productivity have looked simply at body mass. However, this approach confounds long-term nutritional influences with the effects of more immediate dietary shortages. Long-term nutritional deficiencies within a community are indicated more clearly by a shortness of average stature than by a low body mass.

Performance of children

Malina & Little (1985) indicated that in Zatopek boys, mild to moderate malnutrition did not change the normal relationship between lean mass and the performance of physical work. However, because the group was somewhat under-nourished, there was a tendency for a positive relationship between body fatness and grip strength. Likewise, Ghesquière & D'Hulst (1988) suggested that although African children had a lower working capacity than their European peers, when data were standardized for their smaller size, they could perform as well as Europeans.

On the other hand, for many of the heavy physical tasks that must be performed as an adult, the relevant criterion is the absolute working capacity rather than its size standardized derivative.

Performance of adults

The importance of adult size to productivity depends upon the total physical effort that is required for survival in a given habitat, and the proportion of the required effort that is attributable to the raising and lowering of body mass. If a very large total power output is needed to carry out essential daily tasks, then a small person is at a grave disadvantage, but if light physical work is to be undertaken in a hilly environment then a small person with a low body mass may have an advantage over a larger individual (Martorell & Arroyave, 1988).

Immink *et al.* (1981) observed no relationship of the productivity of sugar-cane cutters to either the upper arm muscle area to height ratio or the percentage weight to height ratio, but nevertheless commented that tall workers were more productive, particularly if account was taken of the fact that their health permitted them to work a greater number of days per week. Likewise, Spurr (1988) found a positive relationship between standing height and sugar production. Heywood (1974), equally, commented that sugar-cane cutters who weighed less than 85% of the standard mass for their height had a poor productivity, while Brook *et al.* (1979) observed a significant relationship between weight for height and the productivity of manual workers engaged in road construction. Deolalikar (1984) also noted the impact of weight for height upon the productivity of farm labourers, while Satyanarayana *et al.* (1977) commented that there was a positive relationship between weight for height and productivity in those engaged in the relatively light work of detonator fuse manufacture.

On the other hand, Davies (1973) found no relationship between the ability to cut sugar cane and height, body mass, summed skinfolds, lean mass, leg volume or limb circumference in an East African population. Possibly, poor motivation and chronic health problems such as parasitic infections were more important determinants of low productivity in this group.

Spurr *et al.* (1977) indicated that the productivity of Colombian sugar-cane cutters was correlated with maximum oxygen intake, which in turn depended upon mass for height (M/H), the log sum of triceps and subscapular skinfold thicknesses (ΣS), total haemoglobin (Hb) and creatinine excretion (Cr), according to the equation:

$$\dot{V}_{O_2} \max = 0.095 \, M/H - 0.152 \, \Sigma \, S + 0.087 \, \text{Hb} \\ + 0.031 \, Cr - 2.550$$

Some 80% of the difference in aerobic power between mild and severely malnourished subjects was due to the difference of muscle cell mass (Spurr, 1988), the residual deficiency reflecting a low haemoglobin concentration and/or nutritional deterioration of cardiac muscle (Araujo *et al.*, 1970). Inadequate nutrition may also lead to qualitative changes in skeletal muscle composition, including a reduction in glycogen stores (Heymsfield *et al.*, 1982*a*), a decrease of ATP and phosphocreatine content (Lopes *et al.*, 1982), and reduced activity of oxidative enzymes (Spurr, 1988).

Starvation

Starvation may arise through accidents affecting the crew of ocean yachts and wilderness explorers, deliberate attempts to treat the grossly obese, hunger strikes, and occasional famines. The observed pattern of tissue loss is substantially influenced by initial body mass and composition.

Body composition and food reserves

Initial body composition, particularly the extent of non-essential fat reserves, influences an individual's tolerance of total starvation (Van Itallie & Yang, 1984). An obese person can withstand as much as 120 days of total starvation (Drenick *et al.*, 1964; Drenick, 1967), decreasing their body mass by as much as 50 kg over this period, whereas initially thinner hunger strikers have died within 60–70 days of commencing their fast (Forbes, 1987).

In a typical, relatively thin 65 kg man, fat stores can be reduced from 9.0 to 2.5 kg, and protein stores from 11.5 to 8.5 kg. Initial glycogen stores provide at most a reserve of 500 g of carbohydrate. Some tissues such as brain and red cells can only metabolize carbohydrate, creating a minimum demand of 80–90 g/day for this type of fuel. If the diet does not provide the essential minimum of carbohydrate, glucose is formed by a roughly equivalent breakdown of protein (at least 60–70 g/day). Early studies where young men were placed on diets of 7.95 MJ/day (Benedict *et al.*, 1919) or 6.70 MJ/day (Keys *et al.*, 1950) showed that lean tissue contributed 46 and 57% of the observed weight loss, respectively.

Even if little or no physical activity is required, energy expenditure is unlikely to be less than 6.5–7.0 MJ/day. As continuing metabolism exhausts reserves of non-essential fat, the usage of protein is increased, accelerating the deterioration of physical condition.

Mathematics of tissue nitrogen and mineral loss

The rate of nitrogen loss is less than would be predicted from the decrease of body mass, averaging about 10 g/kg of weight loss in the obese, and 20 g/kg in thin subjects (Forbes & Drenick, 1979). One formulation (Forbes, 1987) expresses the relationship between lean tissue loss and body fat mass (F, kg) as:

$$d(LBM)/d(M) = 10.4/(F + 10.4)$$

However, the constants to this relationship depend upon the extent of the energy deficit.

There is often a 5 kg decrease of body mass over the first few days of total starvation, but this represents mainly fluid loss, associated with metabolic acidosis, the obligatory water loss that accompanies solute excretion, and water bound to protein and glycogen. Thereafter, the average rate of weight loss of an initially obese individual settles down to about 0.5 kg/day, this figure being sustained for a period of up to two months (Drenick, 1967).

As much as 1.5 kg of protein can be metabolized over two months of starvation (Drenick, 1967). A two-component model can be calculated for the course of nitrogen loss during prolonged food restriction, although the generality of the model is limited by the substantial impact of initial obesity upon the rate constants (Durrant *et al.*, 1980). In a fat subject, 6% of the total nitrogen loss occurs with a half-time of 10 days and the remainder with a half-time of 433 days, while in a thin subject the corresponding half-times are 2.4 and 116 days (Forbes, 1987). Observations by Wynn *et al.* (1985) have confirmed a continuing loss of nitrogen after 150 days of food deprivation.

There is a substantial sodium loss over the first five days of starvation, but over the next 10 days sodium stores are replenished. Potassium losses are more persistent, and substantial losses of magnesium, calcium and phosphorus are also observed (Drenick, 1967).

Lean tissue loss and protein-sparing diets

Protein breakdown does not seem to be avoided completely by the adoption of a 'protein-sparing' regimen when dieting. However, the extent of nitrogen loss may be reduced in this manner (Fisler *et al.*, 1982).

Forbes (1987) suggests that those claiming complete protein sparing have ignored subtle losses of nitrogen through desquamation and excretion of sweat (about 0.5 g/day), or have been misled by the small size of the overall negative nitrogen balance (sometimes only 1 g/day). Brown *et al.* (1983) found that even with a protein-sparing diet, 36% of the weight loss over five weeks and 26% of the loss over 20 weeks was attributable to the metabolism of lean tissue.

Some authors have held that tissue protein can be spared if an energy deficit is created by vigorous physical activity rather than food restriction. The data cited earlier in this chapter, where fat was replaced by lean tissue during endurance training would seem to support this view, and Rigotti *et al.* (1984) have suggested that physical activity helps to conserve lean tissue even in patients with anorexia nervosa. On the other hand, the encouragement of vigorous exercise among study participants did not prevent protein loss in the semi-starvation experiments of Keys *et al.* (1950), while Göranzon & Forsum (1985) suggested that the nitrogen loss for a given energy deficit was similar for active and sedentary individuals. Harvey *et al.* (1979) reported that lean tissue accounted for two thirds of the 3.3 kg average loss of body mass observed during a mountaineering expedition, while Hagan *et al.* (1986) found that those who exercised while adhering to a 5000 kJ diet actually lost more lean tissue than those who remained sedentary (although, of course, they also had a larger negative energy balance).

Bortz (1969) had a 304 kg patient lose 213 kg over two years. Changes in urinary creatinine excretion suggested that lean tissue accounted for 18% of the loss. If body protein is to be metabolized, the prime target is skeletal muscle (Uezu *et al.*, 1983), but both radiographs (Keys *et al.*, 1950) and echocardiography (MacMahon *et al.*, 1986) show that there is also a loss of heart muscle, while CT scans suggest that in patients with anorexia nervosa there may even be a subtle reduction of brain size (Datlof *et al.*, 1986). Because protein losses from muscle and bone are greater than from skin and hair, the body potassium to nitrogen ratio decreases during starvation (Vaswani *et al.*, 1983).

Overall weight loss and associated water loss

The utilization of stored glycogen during weight reduction is associated with an early body water loss of some 1.5 kg, mainly from the intracellular compartment.

There is then an overall weight-loss ranging from 0.32% per day in the obese to 0.55% per day in thin subjects. Because energy consumption and thus the rate of weight loss is proportional to body mass, the absolute rate of weight loss decreases in an exponential fashion. The half-time of this process ranges from 220 days in an obese person to 127 days in a thin individual (Forbes, 1987).

As starvation continues, the pattern of weight loss is commonly complicated by the development of oedema. While this is sometimes associated with a loss of serum proteins and thus a disturbance of osmotic balance, this is not always the explanation (Beattie *et al.*, 1948). Another possible factor is a fall of tissue tension as fat is lost from within the fascia (Youmans, 1936). The interstitial pressure is normally subatmospheric, but in a well-nourished person it must drop to -15 mmHg before oedema develops (Guyton *et al.*, 1971).

As death approaches, there is an increase in the rate of weight loss and a higher proportion of the total loss is due to nitrogen.

Adaptation to malnutrition?

Nutritionists agree that a limited intake of food during growth restricts adult size, but some have maintained that a slowing of growth is a healthy adaptation to a habitat with limited food resources (Seckler, 1980; Calloway, 1982).

It has been argued that further substantial adaptation to a shortage of food is possible through a reduction of resting energy expenditure, and occupational and leisure activity (James & Shetty, 1982; Sukhatme & Margen, 1982; Ferro-Luzzi, 1988; but not Beaton, 1985). In essence, a new and lower plateau of body mass is supposedly established through a reduction in basal metabolism, a reduction in the energy cost of work, and the avoidance of 'inessential' activities.

However, Beaton (1984) has argued that this last tactic represents a substantial social cost in terms of the resultant reduction in the quality of life; the situation may be a rationalization of the unwillingness of developed societies to share resources, rather than a positive adaptation on the part of those who lack food.

Basal metabolism

A decrease in basal metabolism is observed over the first few weeks of semi-starvation; the maximum decrease is about 16% (James & Shetty, 1982). Such a change cannot be attributed to tissue loss, since the amount of lean tissue is not usually reduced appreciably at this stage (Keys *et al.*, 1950; Garrow & Warwick, 1978).

One possible explanation is that there has been a reduction in non-shivering thermogenesis. James & Trayhurn (1976) argued that if food was in short supply, children who were genetically thermo-responsive would be inclined to die, giving an evolutionary pressure towards a lack of thermal sensitivity in a cold habitat with limited food resources. In support of this hypothesis, groups such as the Kalahari bushmen (Wyndham & Morrison, 1958) and Australian aborigines (Scholander *et al.*, 1958*a*) have a reduced response to cold, and are also very prone to obesity on moving to an urban environment where the food supply is greater. However, it is less certain that insensitivity to cold is an inherited adaptation. Pregnant women also develop a lesser metabolic sensitivity to cold, possibly related to their attempts at fat conservation, with associated increases in blood levels of prolactin. Prentice (1984) noted that prolactin levels were also high in under-nourished Gambian women, although Darby *et al.* (1983*a*) did not observe any increase of prolactin concentrations in patients with anorexia nervosa.

A second possible factor is an elimination of the increase of resting metabolism that accompanies feeding, the so-called 'Luxuskonsumption' (Garrow, 1978). Garrow has suggested that when body fat stores fall below a critical level, the 'Luxuskonsumption' may be eliminated. The evolutionary pressures of a limited food supply may well favour the emergence of an isolate with a reduced 'Luxuskonsumption', since there is good evidence that the metabolic response to feeding has a genetic component (Pittet *et al.*, 1976; Shetty *et al.*, 1979).

Energy costs of work

If there is a decrease in total body mass of, for instance 14%, then the cost of body displacement and thus of performing most types of physical work should also be reduced. In practice, the correlation between body mass and the corresponding daily energy expenditure is quite variable (Ferro-Luzzi, 1982; Table 10.7). Nevertheless, the combined effect of the change in basal metabolism discussed above and the inevitable impact of body size upon the cost of performing standard physical tasks leads to a decrease in daily energy expenditure of about 23%, without any change

Table 10.7. *Coefficients of correlation between body mass and daily intake of energy (kJ) in farming populations.*

Population	Coefficient of correlation
Italian	0.63
Finns	0.29
US 'Americans'	0.45
Papua New Guinea	
Coastal – men	0.43
– women	0.04
Highland – men	0.16
– women	0.16

(Based on data of Ferro-Luzzi, 1982.)

of overall daily activity patterns (Ferro-Luzzi, 1988). Further savings of energy usage are achieved by either cutting out specific daily tasks and pastimes entirely, or slowing their pace and reducing their intensity. In some instances, the decrease of pace is obvious, but in other situations there may be quite subtle changes in the economy of movement.

It remains difficult to disentangle long-standing cultural adaptations from the more immediate effects of food shortages (Durnin & Drummond, 1988), or indeed to decide whether a population is as active as it wishes to be. Those who believe that food supply is not an important determinant of activity patterns note that studies of subsistence horticulturists from Papua New Guinea (Spencer & Heywood, 1983) have shown seasonal variations in both body mass and activity patterns despite a relatively constant food supply. Likewise, provision of food supplements has not increased body mass or skinfold thicknesses in apparently malnourished Taiwanese women (Adair, 1984), nor has it eliminated the typical decrease of body mass during the 'farming/hungry' season in Gambia (Prentice, 1984). On the other hand, any gain of body mass when food supplies are increased remains much less than the amount of food energy that is provided. Apparently, some of the additional energy is allocated to increased physical activity – perhaps because the people now feel more inclined to take voluntary exercise.

Comparisons of habitual activity patterns between developing and developed countries support the view that deliberate restriction of physical activity is not a common method of meeting a limitation of food supply (Ferro-Luzzi, 1988). If there were any substantial adaptation of this sort, it would be difficult to explain why body mass rises with over-nourishment in western society, or why body mass fluctuates with the seasons in developing countries. Moreover, even if it is accepted that activity can be restricted to meet the challenge of immediate food shortages, the long-term effects are counter-productive. The resultant reduction of food production and lessened quality of leisure lead to a slow but progressive deterioration in the physical and mental condition of the individual (Durnin & Drummond, 1988). A population where most people are malnourished will therefore have difficulty in making productive use of their chosen habitat.

Any burden of malnourishment in a primitive community typically falls most heavily upon women. They must operate a household under primitive conditions, contribute to agricultural production and bear a growing family. A common price of food shortages is thus an inability to bear and suckle healthy children. This can hardly be considered as a positive adaptation to a given habitat.

Changes during re-feeding

In the Minnesota experiment (Keys *et al.*, 1950), 36 young men underwent severe food restriction until their body mass had decreased by 20%. Re-feeding was then permitted, and their average body mass was increased to 10% above its initial value, perhaps as a consequence of persisting metabolic adaptions of the type discussed above. During re-feeding, the body fat content rose to 40% above initial, pre-starvation values.

11 Pathological disturbances of body composition

Gross obesity

The National Institutes of Health (1986) suggested that an increase of body mass 20% or more above the actuarial 'ideal' value constituted an established health hazard. The equivalent body mass index, around 27.8 in males and 27.3 in females, is not uncommon in older North American adults. More occasionally, some individuals show gross obesity, with weights 100% or more above the ideal body mass. Potential causes include genetic abnormalities, hypothalamic lesions, endocrine disorders and possibly gross over-feeding. Each of these potential aetiologies will be considered briefly.

The adverse health consequences of moderate obesity become disproportionately exaggerated as a person's body mass rises further above the 'ideal' value. Nevertheless, if statistical allowance is made for other risk factors that are commonly associated with obesity (particularly hypercholesterolaemia and hypertension), the accumulation of body fat has little independent influence upon either cardiac or overall mortality (Stallones, 1969). It thus becomes a matter of semantics whether obesity is itself regarded as a major public health problem or whether the adverse prognosis is attributed to the associated risk factors.

Genetic abnormalities

A number of forms of inherited obesity are well recognized in experimental animals (Bray, 1974). In the yellow mouse, obesity is an autosomal dominant trait, controlled by a single gene locus. Three separate pairs of genes can each give rise to a recessive type of obesity, as cross-breeding of non-obese, heterozygotic carriers of the three strains never gives rise to the birth of any fat offspring. Finally, there are polygenic forms of obesity, illustrated by New Zealand obese and Japanese KK mice.

Spontaneous degeneration of the glucostat in the ventero-medial area of the hypothalamus may also be attributable to a specific gene combination. Mayer (1972) suggested that the fundamental metabolic error in certain types of obesity was the absence of a repressor substance. While all cells contained the genetic code for the formation of the enzyme glycerokinase, this was only expressed in the white adipose tissue of obese individuals. The resultant increase in the concentration of glycerophosphate can lead to such problems as hyperglycaemia, hypertrophy of the Islets of Langerhans, and hyperinsulinaemia (discussed in Chapter 10).

A familial clustering of obesity in humans has been recognized for many years (Mayer, 1972). Up to three quarters of obese children boast one if not two obese parents. In contrast, less than 10% of children of average weight parents are obese. However, when interpreting such observations, it is difficult to disentangle formal genetic inheritance from the cultural transmission of certain eating and activity habits (Chapter 9). Angel (1949) commented on a high incidence of obesity in first and second generation Americans. Mayer (1972) indicated that in the early 1970s, obesity was common in white men and negro women from the southern part of the US; however, he was inclined to attribute this to the active occupations of negro men and the absence of social pressures for weight control among negro women of that era. More recently, Pérusse *et al.* (1988a) pointed out a strong familial resemblance of energy intake among French Canadians; this would appear to be the result of a common environment rather than a genetic transmission of eating patterns. Likewise, exercise and sport participation was characterized by a cultural inheritance of 12%, without the transmission of any significant genetic influence (Pérusse *et al.*, 1988b).

Early studies showed a high concordance of body mass among identical twins ($r = 0.973$, Mayer, 1972). The variance in body mass was 1.4% when the twins were living together, and 3.6% when they were reared independently, suggesting that there were influences from both inheritance and environment. There also seems to be some segregation in the transmission of body build from one generation to the next. Variability in the body build of the offspring is greatest when matings occur between stout and slim individuals. The variability is less for stout × stout, and least for slim × slim matings. It has thus been postulated that stout subjects carry gametes for slenderness, but that slender subjects rarely carry gametes for stoutness. Withers (1964) found that while the weight of natural children was well correlated with that of their parents, the weight of adopted children showed no such correlation. A further puzzling feature is that there are fewer males yet fewer larger families among the progeny of slim × slim or fat male × slim female matings than among fat × fat or slim male × fat female matings (Mayer, 1972). It has thus been suggested that obesity is not only sex-linked, but is also associated with a fatal recessive characteristic.

Other data on the inheritance of body composition are discussed in Chapter 9.

Hypothalamic lesions

Animal experiments have shown that electrical or chemical stimulation of certain parts of the ventro-medial hypothalamus can modify the intake of food energy (Lepkovsky, 1973; Panksepp, 1974). Likewise, damage to the ventro-medial part of the hypothalamus leads to over-eating and obesity (Hetherington & Ranson, 1942), while damage to the lateral part of the hypothalamus is associated with a progressive loss of appetite and a reduction of body mass (Anand, 1967). It has thus been hypothesized there there is a 'satiety centre' in the ventromedial hypothalamus which inhibits eating, and a feeding centre in the lateral hypothalamus which stimulates the ingestion of food (Tepperman, 1980).

Probably through the influence of such local regulating centres, eating is increased in response to cold, dilution of food or food deprivation, but is decreased in experimental obesity. The sensory input includes the taste and smell of food, a feeling of short-term satiety linked to fullness of the stomach and an adequate blood sugar level, and a long-term error signal that is apparently related to body nutrient depletion and repletion.

The regulating centres have a characteristic set-point in any given individual. This set-point shows a circadian variation, but is also modified by hormonal factors (blood levels of insulin, glucagon, oestrogens and adrenal hormones) and mood state. Abnormalities in the set-point of one or both of these regulatory centres may be an important factor in both gross obesity and anorexia nervosa.

Endocrine disorders

Gross obesity is usually accompanied by a series of endocrine disorders, including hyperinsulinaemia, a reduced release of growth hormone in response to normal stimuli, and an increased rate of corticosteroid secretion (Sims *et al.*, 1974). In some cases, the endocrine defects may be primary, but Mayer (1972) has suggested that altered endocrine responses are normally secondary to an enzymic defect in the white adipose tissue. Glucose tolerance is impaired despite the increased secretion of insulin.

Obese patients also have a less than normal metabolic response to both feeding and cold exposure (Kaplan & Leveille, 1974; Pittet *et al.*, 1976; Schutz *et al.*, 1984). Although the diminution of response is only around 10–15%, the cumulative consequences for body composition can be substantial (see Chapter 10).

Gross over-feeding

The metabolic and endocrine changes associated with the onset of spontaneous obesity can be reproduced in direction, if not in degree, by deliberate over-eating (Sims *et al.*, 1974).

Nutritionists have argued that one factor encouraging recidivism after the apparently successful treatment of gross obesity is that the individual concerned has an above average metabolic set-point. This could reflect abnormal 'feedback' from adipocytes, due to the normal filling of an increased number of fat cells (hyperplastic obesity) rather than the overloading of a normal number of fat cells (hypertrophic obesity). If each of an excessive number of adipose cells were to regain a normal fat content after dieting has ceased, the individual concerned would inevitably become obese once again. Most authors maintain that the fat cell count is determined in early life (Björntorp, 1974). Faulhaber (1974) has argued that the number of adipose cells can be increased by gross over-feeding even when an adult, but his evidence for this hypothesis (a lack of correlation between the amount of excess weight and fat cell diameter, plus a leftward skewing of fat-cell dimensions towards cells with a small diameter in those who are grossly obese) could probably be explained by hyperplasia of the adipose tissue at an earlier point in life. Irrespective of the age of onset, the practical consequence seems that a combination of an *ad libitum* diet with bouts of vigorous exercise produces a slower than normal fat loss, no loss, or even a gain of weight in subjects with hyperplastic obesity (Björntorp *et al.*, 1970, 1973a, 1973b).

Grossly obese subjects have difficulty in developing and/or sustaining any substantial increase of energy expenditure, and for this reason exercise may not be a particularly effective method of reducing their body fat stores, unless the normalization of body composition is viewed as a very long-term process. One quite popular alternative is complete or nearly complete starvation (Chapter 10).

Anabolic steroids and growth hormone

The administration of testosterone and its synthetic analogues encourages the development of a positive nitrogen balance by stimulating the incorporation of amino acids into muscle; the anabolic steroids also increase red cell mass by stimulating the renal production of erythropoietin.

Legitimate use

Legitimate clinical use of anabolic hormones is limited, although there have been reports of the successful administration of such compounds in older adults as a means of encouraging protein synthesis following injury and extensive burns, in wasting diseases such as tuberculosis, muscular dystrophy and hemiplegia, and in gonadal dysfunction (Kochakian, 1976).

Doping

The abuse of anabolic steroids by various classes of power athlete has attracted widespread publicity over the past decade, and elaborate techniques have now been developed to detect this particular form of doping. Commonly, it has seemed that dishonest athletes who are intent upon increasing their mass of lean tissue have remained one step ahead of the detecting laboratories, administering anabolic compounds (including human growth hormone and chorionic gonadotropin) that could not be distinguished from natural products during the most sophisticated types of urinalysis (Di Pasquale, 1984).

For a long period, sports physicians attempted to discourage the illegal use of anabolic steroids by maintaining that such compounds had no beneficial effect upon muscle mass – they argued that if steroids induced an increase of body weight, this was water rather than muscle, and if muscle was indeed increased in size it still did not become stronger. Finally, it had to be admitted that gains of strength were occurring, in part because the doses of steroid used by many athletes far exceeded those that could ethically be tested in sport-research laboratories, although it was still argued by some physicians that the athletes concerned could have won an equivalent response by a vigorous training programme, or that the steroids had produced their beneficial effect by improving appetite, making the competitor more aggressive or stimulating vigorous training, rather than by directly stimulating protein synthesis.

Steroids and body composition

Forbes (1987) has suggested a semi-logarithmic relationship between the total dose of steroid that is administered and the resultant increment of lean body mass. Little gain of lean tissue is observed with a total dose of under 2000 mg; however, in some of the studies in which Forbes himself has been involved, the cumulative dose of steroid has amounted to 4500–9760 mg, with a 8.5–19.1 kg increase of lean mass measured by ^{40}K counting, and at the same time a substantial decrease of body fat. Alén *et al.* (1984) likewise reported that a variety of steroids, when self-administered in doses of 4300–5200 mg, increased body mass by an average of 6%, muscle fibre area by 16% and muscle force by 15%, while skinfold readings decreased.

Growth hormone and body composition

Growth hormone essentially serves as a linear amplifier of the changes induced by anabolic hormones, further increasing the uptake of amino acids by muscle. When growth hormone is given to treat either cases of severe burns or under-sized children, there is a background of circulating androgens; an increase of lean mass and extracellular fluid volume is thus seen, accompanied by a decrease of subcutaneous fat (Tanner & Whitehouse, 1967; Collipp *et al.*, 1973; Wilmore *et al.*, 1974; Parra *et al.*, 1979).

Muscle wasting

Some examples of the changes in body composition induced by spinal injury and poliomyelitis have been given in Chapter 10. Loss of lean tissue may arise in many other clinical conditions, including infections, trauma, various forms of cancer, other visceral and hormonal disorders, muscular dystrophies, and the prolonged therapeutic administration of cortisol. In many of these problems, it is quite difficult to disentangle the primary effect of the disease process from the secondary consequences of bed rest and a limited intake of nutritious foods.

Other clinical conditions such as hypertension, cardiac failure and renal disease may alter the fluid content of the body tissues.

Infections

The metabolic demands of the body are increased by a 'fever'. In accordance with the Law of Arrhenius, a 10 °C increase of body temperature doubles the rate of metabolism. At the same time, most bacterial and viral illnesses lead to a loss of appetite, while vomiting and diarrhoea may further deplete body stores of food, minerals and fluid.

A negative nitrogen balance has been demonstrated in such conditions as malaria, typhoid fever, tularaemia, and sand-fly fever (Shaffer &

Coleman, 1909; Beisel *et al.*, 1967). Beisel *et al.* (1967) further demonstrated that in tularaemia, the nitrogen loss was about three times as great as that observed in control subjects who were provided with an identical amount of food.

Trauma

Trauma inevitably leads to some loss of blood, fluid and tissue. Often, the effects of the immediate injury are compounded by a substantial period of bed rest. In addition to these problems, the tissue damage seems to initiate a more sustained increase of metabolism and an increased production of cortisol (Forbes, 1987), with protein catabolism. Neutron activation techniques suggest that there is a nitrogen loss of 39–42 g/kg (Hill *et al.*, 1978; Bogle *et al.*, 1985), substantially greater than the loss earlier calculated by the somewhat fallible nitrogen balance technique. In addition to losses of nitrogen and potassium, there is an internal redistribution of fluids, with an increase of muscle sodium content (Bergström, 1962).

A comparable protein breakdown occurs in healthy subjects if their levels of serum cortisol are increased to the figures observed after trauma (Bessey *et al.*, 1984). Moreover, the nitrogen loss of the injured patients can be minimized or avoided if a high energy high nitrogen content diet is administered intravenously.

Cancer

In most forms of cancer, the overall metabolic rate is increased by the vigorous growth of the tumour cells. Gastrointestinal tumours further interfere with the normal intake of nutrients, and in the late stages of disease, metabolism may be further distorted by infiltration of the liver and/or the absorption of toxic products from degenerating tissue.

Formal studies of body composition in patients with tumours (Shizgal, 1985; Heymsfield & McManus, 1985) generally show a progressive loss of both lean tissue and fat, with muscle accounting for a high proportion of the lean tissue breakdown. Partly because of a decrease in serum protein levels, extracellular fluid levels are increased.

Other visceral diseases

The body content of lean tissue is decreased in a variety of other visceral disorders that disturb the normal processes of digestion, absorption and metabolism of food, for example cystic disease of the pancreas, Crohn's disease and cirrhosis of the liver; the metabolic problems inherent in the last condition may be exacerbated by a poorly balanced diet, particularly if there is associated alcoholism.

Muscular dystrophies

Both progressive muscular dystrophy of the Duchenne type (Edmonds *et al.*, 1985) and adult myotonic dystrophy (Griggs *et al.*, 1983) are associated with a substantial loss of lean tissue and a large increase in the sodium to potassium ratio for the body. As subjects lose their ability to walk, there is usually an associated increase in the proportion of body fat.

Hormonal disorders

An excess of glucocorticoids are produced in Cushing's syndrome. The proportion of body fat is increased, the amount of lean tissue is decreased, and there is an associated loss of both potassium and calcium from the body (Aloia *et al.*, 1974; Ellis & Cohn, 1975).

Complications arising from the prolonged therapeutic use of corticoids are attracting increasing interest in connection with transplant operations. Horber *et al.* (1985) noted that there was a 16% reduction of muscle cross-section with an increase in body fat in CT scans of the thigh following renal transplants; moreover, the mottled appearance of the 'muscle' in such patients suggested a deterioration in quality of the residual tissue. Loss of skeletal musle is particularly important following cardiac transplantation, since this leads to an associated increase of cardiac after-loading (Kavanagh *et al.*, 1988), with adverse effects upon endurance performance. Moreover, the data of Horber *et al.* (1985) suggests that the tendency to muscle breakdown can be at least partially reversed by an appropriate regimen of isokinetic exercise.

Another group of patients who sometimes require prolonged corticoid therapy are those suffering from severe asthma. Adinoff & Hollister (1983) noted that there was a 15% loss of trabecular bone density following a prolonged course of corticoids, although cortical bone density remained normal.

Over-production of aldosterone leads to an increase of extracellular sodium ion concentrations with expansion of the extracellular fluid volume (Chobanian *et al.*, 1961). Conversely, a deficiency of mineralocorticoids leads to a depletion of body sodium reserves, with an associated retention of potassium (Skrabal *et al.*, 1972).

Insulin has an anabolic effect. Patients suffering from diabetes tend to show muscle wasting, with reductions in body potassium and bone density. Although the prime cause of the tissue loss is a lack of insulin, a metabolic acidosis probably contributes to these findings in many patients (Ellis & Cohn, 1975; Levin *et al.*, 1976).

Thyroid hormone stimulates the release of growth hormone by the anterior pituitary gland, and a thyroid deficiency during childhood therefore leads to dwarfism. The thyroid hormone 'uncouples' carbohydrate metabolism, thus leading to increases in both overall metabolic rate and protein turnover, with a tendency to a loss of lean tissue (Lamki *et al.*, 1973).

There is still a lack of agreement concerning the influence of the female growth hormones upon body composition. Oestrogens stimulate the release of the growth hormone, but from the viewpoint of body composition they seem important mainly as determinants of fat deposition in the breasts and around the hips. Nevertheless, a deficiency of oestrogen can also contribute to the development of osteoporosis (below). Synthetic oestrogens also stimulate the growth of lean tissue in cattle, but apparently this effect does not extend to other species. Small doses of progesterone increase the excretion of both nitrogen and sodium, in part by blocking the action of aldosterone (Landau, 1973), but if larger doses of progesterone are given, they may be converted to androgens, with a resultant anabolic effect.

Hypertension

An excessive intake of salt has been suggested as a cause of hypertension in some racial groups, although the analysis of data on mildly hypertensive patients is complicated by currently prescribed modifications of diet and the administration of diuretic drugs that reduce serum potassium levels.

Beretta-Piccoli *et al.* (1982) and Williams *et al.* (1982) noted modest correlations between extracellular sodium ion concentrations and the level of both systolic and diastolic pressures in hypertensive patients. The increase of sodium and the decrease of potassium levels are more marked in patients with malignant hypertension and primary aldosteronism (Hollander *et al.*, 1961; Chobanian *et al.*, 1961).

Cardiac failure

Right ventricular failure leads to oedema, and thus an increase of tissue fluid volumes. High sodium and low potassium levels are commonly seen if the patient has received no treatment, but as in hypertension the observed results for patients with cardiac failure are usually modified by the adoption of a low salt diet and the prescription of various diuretics.

Thomas *et al.* (1979) commented on a low total body nitrogen in cases of cardiac failure. This could reflect either inactivity, secondary to poor clinical condition, or an effect of the cardiac disorder upon hepatic function and thus protein synthesis.

Renal disease

Renal disease is typically associated with an increase of extracellular fluid and disturbances of mineral balance. The later stages of renal failure are now treated by dialysis and/or a kidney transplant. Favourable changes in body composition have been described in subjects who undertake regular bouts of endurance exercise while they are receiving the dialysis treatment.

Anorexia nervosa

Clinical picture

Anorexia nervosa was initially confused with other wasting diseases, including tuberculosis and primary insufficiency of the pituitary hypophysis. It is now recognized as a distinct clinical syndrome, associated with a distorted body image and a relentless pursuit of thinness (Garfinkel & Garner, 1982). The individual (commonly an adolescent girl) is obsessed with a perceived fatness when in reality she is substantially below the 'ideal' body mass for her height. Typically, the person concerned is only 50–85% of their ideal weight, and 25% (American Psychiatric Association, 1980) below their highest personal weight; nevertheless, they persistently deny their thinness, hunger and fatigue (Mayer, 1972).

Prevalence

The majority of females between the ages of 14 and 18 years have some concerns about their weight (Nylander, 1971), and an undesirable reduction of body mass develops in as many as 1% of this age group (Crisp *et al.*, 1976; Jones *et al.*, 1980; Schleimer, 1983). Unfortunately, cultural factors over recent years have been pushing young women of western society to adopt an even thinner body frame (Garfinkel & Garner, 1982), with an apparent increase in the prevalance of the syndrome (Nylander, 1971; Kendall *et al.*, 1973; Willi & Grossman, 1983; Hall, 1987).

Body composition changes

The changes of body composition that are observed in the anorexic patient parallel some of the features of severe malnutrition, but greater amounts of protein and vitamins are available to the anorexic individual, and such patients may also show differences attributable to the deliberate induction of vomiting (bulimia) or compulsive exercising (Oscai & Holloszy, 1969; Darby *et al.*, 1983). The interest of the anorexic person in exercise (Essén *et al.*, 1981) stands in marked contrast with the behaviour of malnourished people in underdeveloped countries; the latter typically show apathy and socio-cultural deprivation. In other forms of eating disorder, bulimia may occur in individuals who have a normal body mass, or who are even obese (Fairburn & Garner, 1986). Garfinkel *et al.* (1980) noted that 48% of bulimic patients and 28% of those with a pure restriction of food intake had obese mothers. Many of the bulimic group also had a personal history of obesity, suggesting that this may have been a factor in the genesis of the disorder.

In the more usual type of anorexia, lean tissue is lost not only from muscle but also from other viscera, including the liver, spleen and kidneys, with a decrease of total body water (Fohlin, 1977; Davies *et al.*, 1978*b*). Lean tissue accounts for 15–45% of the total decrease in body mass, the extent of muscle breakdown depending in part upon the amount of exercise that the patient is undertaking. In severe cases, losses of cerebral tissue have been described. Haemoglobin and serum protein levels usually remain normal, and oedema is uncommon. Mineral levels are commonly within normal limits, although loss of potassium and chloride ions may occur if there has been much vomiting (Pertschuk *et al.*, 1983). Extracellular fluid may also be increased relative to lean mass (Ljunggren *et al.*, 1961).

Hospital treatment

When the anorexic patient is first admitted to hospital, the thickness of the triceps skinfold may be as little as 2–3 mm (Pertschuk *et al.*, 1983). Muscle circumferences are also much below their anticipated values, the total body potassium content averages only about 40–50 g, and the haematocrit averages no more than 33%. Because of the loss of lean tissue, the ratio of maximum oxygen intake to heart volume is substantially decreased (Fohlin *et al.*, 1978). One consequence of the loss of subcutaneous fat is that normal mechanisms of thermoregulation are distorted; in one experiment, the heat loss by radiation and convection rose to 65%, compared with 21% for normal subjects exposed to the same environment (Davies *et al.*, 1978*a*).

The course of recovery of the anorexic individual depends in part on whether there is a previous history of obesity (Stordy *et al.*, 1977) and in part on the success of the physician in controlling obsessional hyperactivity. The patient typically shows an early weight gain, associated with a restoration of tissue fluids and a replenishment of glycogen stores. This complicates the accurate assessment of food needs. As treatment continues, there is a slower increase of both skinfolds and muscle girths. Martin *et al.* (1985) described the changes of body composition occurring over the first five weeks of rehabilitation. They observed increases of both skinfold thickness and muscle bulk, particularly over the proximal part of the limbs. In one individual, there was a 40% increase of lean mass, but surprisingly no increase of grip strength. Hill *et al.* (1979) noted a strong correlation between gains of body mass and increments of body water ($r = 0.90$).

Increases of lean tissue during the first few weeks of re-feeding (Russell *et al.*, 1983) have been demonstrated by both ^{40}K and total body nitrogen determinations. One problem in data interpretation is that a selective resynthesis of muscle occurs. The added tissue has a high potassium content, so that gains of potassium and water exceed gains of nitrogen. Russell & Mezey (1962) found that on re-feeding, 77% of the weight gain was attributable to fat, 7% to protein and 16% to water. Forbes *et al.* (1984) attributed a much higher proportion of the increased mass to lean tissue: 70% if subjects were given a high protein diet, and 61% if they received a 10% protein diet. Blom *et al.* (1962) also indicated that lean tissue accounted for 70% of the increase in body mass. Contrary to earlier medical opinions, it is now agreed that moderate exercise during the latter part of rehabilitation speeds the restoration of lean tissue (Goldberg *et al.*, 1977; Halmi *et al.*, 1979; Falk *et al.*, 1985).

The energy cost of tissue resynthesis is somewhat lower than that associated with weight gain in heavier individuals (Chapter 10), because lean tissue accounts for a substantial fraction of the added mass in an anorexic individual. Despite the need to replenish body protein stores, the rate of weight gain seems no different between a moderate (10% protein) and a high protein diet (Forbes *et al.*, 1984); however, the latter regimen does increase the rate of lean tissue formation.

Inter-individual differences

Perhaps because such individuals have a greater need for rehydration and protein regeneration, the rate of increase in body mass is greatest in those patients who have the lowest body mass on admission to hospital (Walker *et al.*, 1979; Crisp *et al.*, 1986). However, the rate of weight gain is also greater if patients have a history of pre-morbid obesity (Stordy *et al.*, 1977), possibly because such individuals have a higher 'set-point' for their body mass. Our data (Shinder, Garner, Garfinkel & Shephard, unpublished) show that patients with a high pre-morbid body mass not only gain weight faster, but do so with a smaller total input of energy. Unfortunately, many patients continue to have difficulty in appreciating their body size appropriately; they thus become uncomfortable as body mass is restored, and there is a strong tendency to recidivism (Garfinkel & Garner, 1982).

Target weights

One important issue is the target weight that should be set for patients during their treatment. Possibly, bulimic individuals have a higher 'set-point' than pure food restrictors (reflecting corresponding differences in the body complement of adipocytes, Faust *et al.*, 1977). If so, bulimic individuals should be encouraged to stabilize their body mass at a figure somewhat above the population-based norms for height (Garner, 1985; Pirke *et al.*, 1985). Garner (1985) has suggested aiming for a value that is 10% below the patient's highest stable pre-morbid body mass.

Pregnancy

Given the age of the average patient with anorexia nervosa, a number of cases become complicated by pregnancy. The proportion of pregnant patients would be larger but for the effect of starvation in causing amenorrhoea. Treasure & Russell (1988) described seven instances of pregnancy in anorexic individuals, noting that the average weight gain during pregnancy was only 8 kg, compared with a recommended minimum of 11 kg. Growth of the foetus was subnormal during the third trimester, and at birth the abdominal circumferences of the infant were below the third percentile. However, most of the children studied showed 'catch-up' growth over the first year of their life.

Osteoporosis

Forms of bone disorder

A distinction is sometimes drawn between osteopenia (a reduction in bone mass), osteoporosis (the combination of osteopenia and resultant changes of gross structure such as fractures) and osteomalacia (where there is a deterioration in the quality of the bone matrix, most easily identified by iliac crest biopsy). However, there is increasing evidence (Hansson *et al.*, 1975; Patterson-Buckendahl *et al.*, 1987) that at least two forms of osteoporosis (due to prolonged bed rest and weightlessness) are associated with a deterioration in bone quality.

Causation

In osteopenia, there is a deficit of bone mass that is due to an increase of osteoclastic activity and/or a decrease of osteoblastic function (Heaney *et al.*, 1978; Wronski *et al.*, 1987). The trabecular bone becomes more porous, while cortical bone undergoes a progressive thinning. However, residual tissue retains a normal chemical composition (Glunky, 1983).

The problem of osteopenia can arise in relatively young individuals, but it is more common with advancing years. The two main types are 'post-menopausal' and 'age-related' (Riggs *et al.*, 1986). Contributing factors are a lack of weight-bearing physical activity and (in older women) a decline of oestrogen levels (Aloia, 1981; Martin *et al.*, 1981; Smith *et al.*, 1981*b*; Black-Sandler *et al.*, 1982; Heaney *et al.*, 1982; Krølner *et al.*, 1983; Jacobsen *et al.*, 1984; Chow *et al.*, 1988), an inadequate calcium intake (the need of a post-menopausal, oestrogen-deficient woman may be 1500 mg/day, compared with a likely daily intake of 400–500 mg; Heaney *et al.*, 1977), smoking (perhaps an indirect effect upon the age of menopause or the body mass, Daniell, 1976) and an excessive intake of protein (with a resultant excretion of calcium, Margen *et al.*, 1974, but not Suzuki *et al.*, 1969, the latter author using rats rather than human subjects).

Other more specific potential causes of osteoporosis are oophorectomy (Lindsay *et al.*, 1976; Cann *et al.*, 1980*a*), alcoholism (even in the absence of liver damage, Feitelberg *et al.*, 1987), Addison's disease (where patients may be affected by a lack of androgenic steroids, Devogelaer *et al.*, 1987), Cushing's disease (where there is an over-production of cortisone and, therefore, excessive catabolism), hyperprolactinaemia (Klibanski *et al.*, 1980; Schlechte *et al.*, 1983), anorexia nervosa (Treasure *et al.*, 1987), muscle paralysis (Abramson & Delagi, 1961), prolonged bed rest (Dietrick *et al.*, 1948; Donaldson *et al.*, 1970; Hulley *et al.*, 1971) and the loss of gravitational stimulation during space exploration (Rambaut *et al.*, 1975; Spengler *et al.*, 1979), see further Chapter 10. In the last three instances, the deterioration of the bone can be at least partially corrected by an appropriate daily physical activity programme.

Diagnosis and course

The diagnosis of osteoporosis is based on the radiographic appearance of the femoral trabeculae and the vertebrae, various photon absorptiometric measurements taken from the metacarpals, the lumbar vertebrae and the ankle, and total body calcium determinations. Unfortunately, the correlation between simpler indices and total body calcium determinations is rather poor ($r = 0.5$–0.7) to allow confident diagnoses to be made in an individual patient (Aloia *et al.*, 1977).

A weakening of trabecular structure can sometimes be detected as early as 20 years of age, but on the other hand, cortical bone density does not usually peak until 30–35 years of age. Thereafter, there is a progressive loss of bone tissue, the rate of change apparently being fastest in those who have the greatest initial bone density (Smith *et al.*, 1976*a,b*). In older women, losses typically average about 8% per decade, with some acceleration for a few years around the time of the menopause (Horsman & Simpson, 1975; Cohn *et al.*, 1976; Smith, 1982). In men, the rate of loss is slower (about 3% per year over a broad age span, but accelerating to 0.7% per year after the age of 50, Cohn *et al.*, 1976). Smith *et al.* (1976*a,b*) pointed out that such changes remain close to the precision of duplicate measurements, and the rate of loss thus cannot be determined for an individual unless data are collected for a period of several years.

By the age of 70 years, Smith (1982) estimated the average person showed a 45% loss of bone density from the vertebrae, a 30% loss from the radius as a whole and a 39% loss from the distal part of the radius. The main source of information on the status of the general population remains the changes observed in standard radiographs, rather than the specialized measurements of dual photon absorptiometry and calcium activation. Unfortunately, a substantial calcium deficit (30–35%) is necessary to diagnose abnormal bone loss from routine films, although even on this crude criterion it has been estimated that between the ages of 45 and 79 years, some 29% of women and 18% of men have developed significant osteoporosis (Alvioli, 1976; Singer, 1981). Some authors have suggested that as many as 80% of women over the age of 65 years have clinically significant vertebral atrophy.

Osteoporosis of young athletes

In young women, a decrease of bone density has been described following repeated bouts of prolonged endurance running (Cann *et al.*, 1984; Drinkwater *et al.*, 1984; Nelson *et al.*, 1986). A partial explanation of this finding may lie in the reduction of serum oestrogen levels which is induced by strenuous physical activity, since oestrogen lack is thought to be a factor in senile osteoporosis. However, a further important consideration in many young female athletes is the search for a slim body form, with the development of a negative energy balance and an inadequate intake of calcium (Nelson *et al.*, 1986). Mechanical stimulation, both from muscular contraction and from gravitational forces, apparently has a beneficial influence upon both the development and the maintenance of bone structure (Aloia, 1981; Smith, 1982; Black-Sandler *et al.*, 1982; Bauer & Griminger, 1983), and interestingly female oarswomen are able to exercise to the point of amenorrhoea without developing osteoporosis, at least if their training programme also includes some weight-lifting (Evers *et al.*, 1985). Drinkwater *et al.* (1986) have further shown that the bone mineral loss is made good quite rapidly if normal menstruation is restored by a moderation of training and/or greater energy intake.

Exercise and prevention

The stress required to maintain the bone in healthy condition apparently lies well within physiological limits (Lanyon & Rubin, 1983). Cross-sectional studies have shown a positive relationship between the maximum oxygen intake of the individual and bone mass (Chow *et al.*, 1986; Pocock *et al.*, 1986). Lumbar vertebral density is also correlated positively with trunk extensor strength (Sinaki *et al.*, 1986) and with psoas muscle mass (Doyle *et al.*, 1970). Longitudinal studies have yet to be completed in humans, but in mice, regular training increases the non-collagenous component of the organic matrix (particularly the concentration of glycosaminoglycans), an indication of a biologically 'young' ground substance (Kiiskinen & Heikkinen, 1978).

Effects of bed rest

During bed rest, the decrease of total body calcium becomes more rapid after the first two weeks. Calcium balance studies suggest a decrease of 0.5–0.7% per month, but the loss from the os calcis can be 10 times as rapid. Losses over a 12.6-day lunar flight have amounted to 0.2% of total body calcium (Chapter 10). In addition to lack of weight-bearing, Rambaut *et al.* (1975) suggested that poor calcium absorption may have been involved.

12 *Human adaptability and body composition*

General considerations and terminology

Dejours (1987) has pointed out that physico-chemical conditions vary quite widely in the water or hydrosphere, in the air or atmosphere and in the soil or lithosphere (both from place to place, and at any given location within a day and across a season). Nevertheless, the range of environmental variation which is compatible with human life remains much narrower, and indeed even a brief approach to the upper or lower tolerated limits of temperature, pressure, gravitational forces or nutrient supply can be quite stressful. Problems of adaptation are exacerbated because the stressors do not always act independently of each other. For example, a hot environment imposes greater stress if fluids are in short supply, while a cold environment is more stressful if it is encountered at high altitude, or if subcutaneous fat and/or glycogen stores have been depleted by a period of malnutrition. Humankind has thus faced severe challenges in developing a functional relationship with and in colonizing many extreme habitats.

Processes occurring within a lifetime (phenotypic change) must be distinguished from genetic mutations that favour the adaptation of succeeding generations to specific stresses. The terminology remains confusing, and it is worth summarizing a few of the terms commonly used in the context of human adaptability (see Glossary). Other information will be found in Bligh & Johnson (1973), Folk (1974), Yousef (1985) and Dejours (1987).

Stress and strain

Particular confusion has arisen with respect to the terms stress and strain. While these two words evoke clearly defined concepts of applied force and reaction for the physicist, Selye (1950) has described the response of the human to a stressor as a stress reaction rather than as a strain; some environmental physiologists speak of the stress imposed by a given habitat, but others regard it as a strain!

Local versus general adaptations

Adaptation can be either a local or a general process. Both types of reaction can contribute to success in colonizing a particular habitat. For example, local adaptation of the hands to cold, by increasing regional blood flow and manual dexterity under arctic conditions, probably played a key role in the evolutionary success of the Eskimo when exploiting the circumpolar regions.

Acclimatization, conditioning and habituation

Acclimatization may be considered the process of adapting to an unfamiliar natural environment. It is sometimes distinguished from acclimation, the adaptation that occurs on exposure to a simulated environment, for instance the conditions encountered in a climatic chamber. Usually, the stress is greater, and any adaptive response more complete with acclimatization to a natural environment.

Conditioning is essentially the transfer of an existing response to a new stimulus. Habituation is a form of negative conditioning, whereby the responses to a hostile environment become progressively less marked as exposure to the adverse situation is repeated. For example, the increase of heart rate and blood pressure diminish progressively if the hand is plunged repeatedly into a bucket of cold water. While there may be more subtle adaptations involved, a large part of the reduced response represents the habituation to a cold stimulus.

Many of the useful adaptations to hostile environments are behavioural, but others depend upon structural changes affecting such variables as metabolism, insulation and heat storage. An increase of metabolism, for example, might seem a simple and appropriate expedient in a cold climate, but it becomes an inappropriate method of increasing the chances of survival in a habitat where food is in short supply.

Historical views on human adaptability

Lamarck (1809) first expounded the doctrine of transformism, suggesting that progressive changes in the environment had forced all living beings to adapt and evolve. In his view, the adaptations were then transmitted to subsequent generations of the species.

Darwin (1859) stressed the role of natural selection in the evolution of species. According to his reasoning, the survival and propagation of an individual was favoured by small but advantageous pre-existing differences of body form.

Prosser (1964) added to these concepts the idea of an inherited potential for adaptation. The ability to adapt decreased the adverse effects that would otherwise have been exerted by an unfavourable environment; moreover, it contributed to the evolutionary process by enhancing the survival prospects of the individual concerned.

However, the detailed genetic mechanisms of successful evolution have remained obscure. We now recognize that any adaptation, for instance to an extreme of climate, persists only during exposure to the stressor, and disappears rapidly when the individual returns to a normal environment. Moreover, successful genetic adaptation has been to an average environment rather than the extreme conditions encountered in a particular habitat at a particular season, and the changes of environmental tolerance offered by spontaneous mutations have generally been small. Finally, with the notable exception of the current acculturation of traditional societies to western technology, the pace of environmental change has been slow; moreover there have been wide superimposed variations about the average conditions encountered in any newly colonized habitat. These tendencies, together with the absence of any directionality among spontaneous mutations should have forced subsequent generations back towards the pre-existing mean of adaptability, rather than leading to emergence of a particularly well-adapted subspecies.

International programmes for the study of human adaptability

The Human Adaptability Project of the International Biological Programme and successor organizations such as *Man and Biosphere* and the *Decade of the Tropics* have been concerned not only to provide very practical world-wide baseline data for health and nutritional programmes, but also to explore the detailed ecology of inter-relationships between the human colonist and challenging habitats at a period in human history marked by rapid changes in the distribution, density and lifestyle of many of the world's populations.

Until recently, cultural factors such as the skillful design of clothing and houses have played a major role in human adaptation to high and low environmental temperatures. Now, modern air-conditioning has all but eliminated exposure to the historic extremes of heat and cold. On the other hand, technological advances are exposing humans to other stressful situations such as prolonged space voyages, while less privileged groups in the tropics must still battle against malnutrition and disease. Occasional explorers must now adapt to environments with highly abnormal gravititional fields and extremes of ambient pressure. The average city dweller of western society must also adapt, albeit to a smaller extent as the body has to adjust to a level of habitual activity that is far below the anthropological ideal, with potentially adverse implications for fitness, health and genetic constitution.

The response to a hostile environment, whether an immediate acclimatization or a long-term constitutional adaptation, combines a substantial cultural inheritance (behavioural, social and structural, including particularly such items as an appropriate choice of housing and clothing, Samueloff, 1987) with alterations in physiological control mechanisms (Werner, 1987) and morphological changes. In this final chapter, particular consideration is given to the morphological component, that is to say the potential contribution of changes in body composition to the adaptive process, both in ancient times and at the present day.

Account is taken of both phenotypic gradients and their underlying determinants on local, regional and world-wide scales. Unfortunately, much early research in physical anthropology was concerned with the definition of racial characteristics through detailed comparisons of cranial and facial dimensions, and it is only in more recent years that attention has turned to those features of human morphology that have more obvious adaptive value.

Body composition and cold

Among possible biological tactics for protection against a cold habitat, we must distinguish regulatory changes such as an increase of metabolic rate (metabolic acclimation, reflecting a combination of voluntary activity, shivering and non-shivering thermogenesis) and a decrease of peripheral temperature (insulative acclimation, reflecting a decrease of peripheral blood flow) from morphological changes such as increase in the amount or type of body fat.

Metabolic acclimation

Body movement
The simplest method of adaptation to a cold climate might seem a voluntary increase of physical activity, but because body movement increases heat loss (Shephard, 1982*b*), this is a very costly tactic even

when food supplies are abundant. It is not a practical option when attempting to colonize an arctic wilderness without external logistic support.

Shivering

Shivering is a second method of increasing heat production. It is a possible solution for an emergency, but is again impractical as a long-term solution. The involuntary and unpredictable body movements of shivering limit the performance of any skilled task, and the intense isometric contractions which are induced depend upon intramuscular glycogen reserves, which become depleted quite quickly, but are only refilled over 1–2 days of re-feeding.

Non-shivering thermogenesis

Small mammals are further able to sustain their core temperature by a process of non-shivering thermogenesis, associated with the hypertrophy of brown adipose tissue (Heldmaier, 1974). Young children also have a capacity for non-shivering thermogenesis, and there is some evidence that well-nourished adults who are exposed to prolonged bouts of cold, either in arctic air (Scholander *et al.*, 1958*b*; Huttunen *et al.*, 1981) or in diving (Hong *et al.*, 1986) develop a similar increase of metabolism. Typically, deposits of brown fat are found in the inguinal, axillary and subscapular regions. The heat producing capacity of this tissue is such that as little as 35 g of brown fat could account for the total added heat production of an infant during exposure to the cold.

Unlike the single fat vacuole of the normal white adipose tissue, brown fat contains many vacuoles and many mitochondria. A cell that is well filled with lipid has a typical brown colour, but it is less readily distinguished from white adipose tissue when its fat content is depleted (Mount, 1979).

There may also be increases of metabolism in muscle and white adipose tissue, through the development of 'futile' metabolic cycles. Inevitably, the potential for adaptation by any form of increased metabolism depends in part on the extent and the availability of food reserves within a given habitat. A common cause of hypothermic death is the failure of both shivering and non-shivering thermogenesis when the food supplies of an injured person are exhausted, and in many parts of the arctic food shortages were episodic before the arrival of modern technology, thus limiting any potential metabolic response to cold.

Insulative reactions

Perhaps because human exposure to cold is more intermittent than that of other mammals, a more usual tactic of adaptation to cold is a restriction of blood flow to the extremities, allowing a substantial drop of limb temperatures (Skreslet & Aarefjord, 1968; Shephard, 1985a). This approach has been described in such groups as the Australian aborigines, Kalahari bushmen and Korean diving women (Hammel, 1964). As blood flow becomes restricted, not only fat but also muscle can serve as an insulator, increasing the temperature gradient from core to skin surface (Veicsteinas, 1987).

In fat subjects, the effective insulation offered by the superficial tissues continues to increase down to a water temperature of 12 °C, but in thin subjects maximum insulation may be reached at a water temperature as high as 33 °C (Cannon & Keatinge, 1960; Jéquier *et al.*, 1978).

Insulative acclimation is an effective tactic for an emergency, and it could also enhance survival prospects in a hibernating species. However, any limitation of peripheral blood flow rapidly reduces manual dexterity, and is thus a counter-productive long-term adaptation for a society that must fish or hunt game under arctic conditions. Moreover, the insulation provided by the muscles disappears if limb blood flow is increased by even mild exercise, while the thermal protection of subcutaneous fat is also greatly decreased by moderate exercise. Again, an insulative reaction could hardly be a long-term solution to successful colonization of the arctic.

Morphological changes

In a cold environment the weight of fur in animals is substantially increased (Héroux, 1963; Werner, 1987) and the pig, for example, responds to growth by a shortening of its adult limb and ear length (Ingram & Weaver, 1969).

In animals, at least, the chemical composition of stored fat also changes with exposure to cold. In particular, there is an increase in the proportion of unsaturated fat, so that the temperature of solidification is reduced (Mount, 1979). In humans, the main possible morphological adaptation to cold would seem to be an increase in the thickness of the subcutaneous fat layer. From the metabolic point of view, this would be a more economical adaptation than an increase of metabolism through shivering or non-shivering thermogenesis, and an evolutionary advantage might be gained at periods when the food supply was limited. However, the problem of gaining insulation from a thickening of subcutaneous fat is that there is then difficulty in increasing heat exchange during bouts of vigorous work.

Samueloff (1987) did note a slight and statistically significant increase in the body weight of Jewish immigrants during the colder winter months, but this could reflect no more than a seasonal difference in activity patterns. At one time, the Inuit were also said to be a fat people, but this idea seems to have arisen mainly from a misinterpretation of substantial mass to height ratios. There is no evidence that the thickness of subcutaneous fat is increased in traditional circumpolar residents relative to those dwelling in more temperate climes, nor is there any indication that skinfold readings are increased among arctic residents at cold seasons of the year (Chapter 10).

Technological adaptations

There seem good reasons why technology provides a better adaptation to cold than any morphological change. From the earliest neolithic times, animal skins have provided effective and adjustable amounts of insulation. A double layer of caribou skin, for example, offers 11 CLO units of thermal protection, compared with the 4 CLO units provided by current arctic military clothing. Studies of both Eskimos and Antarctic explorers have suggested that good winter clothing can sustain a very warm micro-climate even at ambient temperatures of −40 °C and below, and this immediate protection is commonly reinforced by other social and behavioral adaptations such as the building of snow shelters. Moreover, the flexible external insulation of heavy clothing provides a more appropriate solution to the problems of a cold climate than the deposition of a thick layer of fat. The latter tactic could lead to excessive sweating when it was necessary to increase the work-rate, with subsequent deaths from severe frost-bite and hypothermia as sweat degraded the insulation of clothing.

Given the effectiveness of technological solutions, and the disadvantages of morphological changes, it is not surprising that it is difficult to demonstrate any constitutional adaptation to an arctic habitat.

Body composition and heat adaptation

In general, adaptations to heat tend to be the opposite of those providing successful adaptation to a cold climate. Adaptation to a warm habitat is helped by efficiency in the conversion of food energy to useful external

work rather than heat, together with tolerance of a rise in core temperature, tolerance of dehydration and efficiency in dissipating heat.

An obese person moves in a much less efficient manner than someone
who is slim, thus producing more heat. Tolerance of accumulating heat
depends in part upon body mass and thus the rise of temperature per unit
of heat storage, while the ability to withstand dehydration depends in part
upon initial body water volumes. Successful heat dissipation depends
upon the efficiency of sweating and therefore the relationship of body
mass to body surface area.

In adapting to a hot climate, the human again relies more on technology and behavioural adaptations than upon specific changes in body
composition. Housing is designed to maximize ventilation and in some
environments to provide an effective barrier to radiant heat, clothing is
minimized, and work is avoided at the hottest hours of the day.

Attempts to demonstrate morphological adjustments to heat remain
unconvincing. The Allen and Bergmann rules have already been criti-
cized (Chapter 10). Many tropical peoples are quite thin, but this is a
common feature of under-developed societies, and reflects a combination of vigorous physical activity and limited food supplies rather than a
specific adaptation to heat. Regular heat exposure undoubtedly places a
substantial additional stress upon the circulation, and a 5–40% increase
of blood volume has been described over the first few weeks in a tropical
environment (Leithead & Lind, 1964). Sweat production is also
enhanced. However, a telling argument against a genetic contribution to
any of these adaptations is the fact that well acclimatized Europeans have
a very similar heat tolerance to the indigenous people of the tropics.
Moreover, most of the heat acclimatization which can be observed is lost
within a few days of return to a cooler climate.

As with a cold climate, the main explanation of the lack of genetic
change is probably the success of human innovations. While heat stress
can kill an inexperienced marathon runner in North America, it is not a
common cause of selective mortality in young African hunters.

High and low pressure environments

Exposure of humans to high pressure environments is quite a recent
phenomenon, and the availability of modern technology has eliminated
any evolutionary pressures towards adaptation, although there is some
evidence that vulnerability to decompression sickness is greater in obese
subjects than in their thinner peers (Shephard, 1982*b*).

Small and isolated populations have colonized high altitude regions for many years, and there is evidence that they are much more tolerant of low oxygen pressures than lowlanders (Chapter 10). Experiments on migrants to high altitude have not fully clarified whether the adaptations are inherited, or are acquired through a lifetime of physical activity in the stressful environment. However, there is some evidence of morphological change, particularly the development of barrel-chests and a large lung volume.

Space travel

Space travel is again a recently encountered stress, and humans meeting this challenge are provided with a high level of technological support. Nevertheless, certain features of body composition – for instance dense and stable bones – may favour the conduct of long space missions, and in the very distant future when the natural resources of the earth have become depleted, the ability to undertake such missions may have survival value.

Malnutrition and disease

Malnutrition and disease are important in that these are readily modifiable features of the environment. It would thus be very useful to establish just how far the individual can adapt to limitations in the quality and quantity of the diet.

Tanner (1966) has noted that the body of a growing child typically shows some adaptation to short-term periods of malnutrition or starvation by a homeorrhetic response. Energy demands are reduced by a slowing of growth, and if the malnutrition is not too prolonged there is a compensating phase of 'catch-up' growth once supplies of food become more plentiful. Moreover, natural selection tends to act against those children in whom the homeorrhetic response is poorly developed. Death may occur immediately, maturation may be delayed, or there may be factors hampering reproduction (a narrow pelvis in the female, a weak adult form in the male).

In theory, malnutrition could affect the rate of growth, ultimate body size, ultimate body shape and ultimate body composition (Tanner, 1966; Lasker & Womack, 1975). The immediate response to poor nutrition is a slowing of the growth rate but after a certain time, probably shorter for the young child and the adolescent than for a child of intermediate age, permanent effects upon adult size occur. However, an adverse nutritional environment does not usually affect body shape unless rate and size changes have already become severe. The main potential change is a relative enlargement of the head, brought about by starvation after the head has completed most of its growth. With regard to body composition, there is plainly scope to utilize muscle protein as an energy reserve if food is short, but it is unclear whether an unusual capacity to exploit this resource explains the slender muscles of some East African tribes. In contrast to the environmental stability of body shape and composition, genetic factors have a marked influence on these two variables.

Adult adaptations to a habitat where food shortages are common may include a small body size (since energy needs are proportional to body mass), the absence of any metabolic response to cold, a good mechanical efficiency and a low level of habitual activity (Chapters 10 and 11).

It is finally worth commenting that when examining the physical characteristics of poorly nourished populations, a well-nourished sub-set must be isolated in order to establish the growth potential of the ethnic group concerned. Recent studies along these lines have suggested that many of the supposed differences of stature between ethnic groups have a nutritional rather than a genetic basis. For instance, in India (Malhotra, 1966), the average height of the male declines from 168.4 cm in the Punjab, where the daily food intake is 13.9 MJ to 163.7 cm in Madras, where the diet provies only 8.7 MJ. However, Indian Air Force Personnel from the two states are taller and show much less difference of stature (170.0, 167.6 cm, respectively). Nevertheless, the issue is complex, and Tanner (1966) has pointed out the potential roles of assortative mating, together with the ability to gain economic advantage from a genetically determined tall stature.

The effects of disease are generally similar to those of malnutrition, and indeed may act in part by inducing a disturbance of energy balance. Short periods of disease are associated with a homeorrhetic checking of growth, but there is a later 'catch-up phase'. On the other hand, chronic diseases such as tuberculosis (Rode & Shephard, 1971), asthma and cystic fibrosis (Shephard, 1982b), cyanotic heart disease (de Knecht & Binkhorst, 1980) and poorly controlled diabetes can all lead to a permanent stunting of the adult. In general, resistance to disease is increased by substantial body reserves of protein, fat and fluid.

Currently, medical services are such that the selective effects of disease are small. However, in the recent past, large fractions of some indigenous populations such as the Inuit were destroyed by imported diseases to which there was no natural immunity. The magnitude of the selective pressures exerted by such epidemics relative to that of other selective forces such as skill in hunting and an ability to adapt to extreme climatic conditions remains unclear.

Adaptations to physical inactivity

The physical inactivity faced by the typical city dweller is a new challenge to human adaptability which has emerged with the energy-saving technology of the twentieth century. It is typically accompanied by substantial changes in body composition, including a progressive accumulation of subcutaneous fat, a loss of lean tissue from the muscles, a decrease of plasma volume (Fortney, 1987), and a loss of bone mineral. An excess of food and a lack of physical activity seem particularly stressful for indigenous populations that have until recently worked hard and had limited food supplies. In such groups, gross obesity and metabolic diseases such as diabetes are widely prevalent.

In theory, physical inactivity and over-feeding would have an adverse impact upon human adaptability in all ethnic groups, but in practice their selective pressure has been weakened in most modern societies by an onset of associated diseases in middle or later adult life, a shortening of the effective fertile period (so that children are born before the onset of the disease) and a highly effective medical technology that counters the decline in health with aggressive secondary and tertiary medical care. Although physical inactivity and obesity are linked to an increased mortality from ischaemic heart disease and other chronic medical conditions (Shephard, 1987), a poor tolerance of hot environments (Fortney, 1987) and a high mortality from falls in the elderly, it is not clearly established that these factors have had any evolutionary impact. Indeed, the current association of a poor lifestyle with a low socio-economic status and an above-average family size may be tending to perpetuate rather than to eliminate these problems.

Final comments

As the investigators for the International Biological Programme envisaged, acculturation to a highly technological western society has now removed most of the traditional environmental pressures, thereby limiting the possibility of further study of human adaptability. The skills of modern technology have now established a rather uniform milieu extérieur in most parts of the world, and (perhaps in response to this equalization of opportunity, perhaps as a result of interbreeding) many of the long-established inter-racial differences of body form also seem to be disappearing.

The final flurry of research, stimulated by the International Biological Programme, adduced little evidence that evolutionary pressures had modified body form or function. While some investigators were surprised by this, mature reflection suggests that a longer search with more sophisticated tools would be unlikely to yield a different conclusion. There are several major arguments against such environmental selection.

The first is the diversity of habitat. Many colonists choose to live at the conjunction of two or more distinct ecological zones. But even when such environmental variation is not deliberately sought, there is rarely a distinctive 'climate' to which a person could adapt. In the arctic, for example, different seasons call for the trapping of fox and bear by sled, the shooting of wild duck, the hauling of seals and walrus onto ice-floes, fishing by kayak, the hunting of whales in the umiak, and the land pursuit of caribou (Shephard, 1978a). Plainly, an adaptation favouring survival in one season could well hamper performance in a second.

A second factor is the cooperative nature of human society. Evolutionary theory presupposes a competition for survival, but in many indigenous populations the emphasis has been on cooperation rather than upon competition in meeting environmental challenges. The dominant pressure in many habitats has been a shortage of food, but the societal norm has been to share available resources, with the elderly deliberately sacrificing themselves at periods of extreme deprivation.

An even more important issue is the dominance of the cerebrum in any competition that may exist. Continuing study of body form in various races may eventually delineate stable and transmissible differences, but these will be small, and of marginal significance for survival in a hostile environment. From an anthropological perspective, the big difference both between humans and lower primates, and within members of the human race is not in overall body composition, but rather in the development of intelligence. In our studies of the Inuit hunters (Godin & Shephard, 1973), we were much impressed with the disparity between the energy expended and the rewards accruing to the individual. One man set 48 traps, and by an 11 km sled journey was able to capture 21 foxes in the space of 24 hours. Another man set 24 traps on each of two occasions, journeying 24 and 136 km, yet managed to capture nothing! Clearly, physical characteristics were not helping the latter individual substantially in any 'battle' for survival.

While technology has always been the dominant force in human adaptability, a final concern for future generations must be the failure to transmit such technology at the level of the individual family. In the arctic, skills of building a snow-shelter and of fashioning weather-proof clothing are being lost, while in the urban society of the south air-conditioning and power appliances are continually narrowing the necessary range of personal adaptability. If indeed there is a selective pressure at the present time, it is directed to the evolution of a people who are fertile, attractive to the opposite sex, and yet have a very limited capacity to respond to environmental challenge through personal lifestyle, physiological regulation or adaptations of body form. This variant may well face serious difficulties in a world where non-renewable energy reserves are quickly becoming exhausted.

Glossary

Some terms used in human adaptability, based in part upon Yousef & Samueloff (1987) and Dejours (1987).

acclimation A physiological change which occurs during the lifetime of a subject and minimizes the stress imposed by an *experimental* stressor (for example, residence in a climatic chamber).

acclimatization A physiological change which may have a phenotypic or a genetic basis and minimizes the strain caused by a severe *natural* environment.

acculturation Control of the micro-environment by behavioural means (clothing, heating, cooling, etc). Also used to describe the process of change from a traditional culture to an urbanized westernized lifestyle.

adaptation A change minimizing the physiological strain resulting from a stressful environment.

Allen's rule In warm-blooded species there tends to be an increase of body surface area by an enlargement of the relative size of protruding body structures, such as ears, legs and tail, with increasing temperature of the habitat, insofar as they play a role in temperature regulation.

Bergmann's rule Within a polytypic warm-blooded species, or closely related species, the body size of the sub-species usually increases as the temperature of its habitat decreases.

biosphere The region of the earth's surface, including air and water, in which life can exist.

biotope A region that is uniform with respect to environmental and soil conditions and the population of plants and animals for which it is the habitat.

conditioning The transfer of an existing response to a new stimulus.

ecology The branch of biology that deals with the complex network of mutual interactions between living organisms and their environment, both physical and biological, or the study of living organisms interacting among themselves and their environment.

habituation   A decrease in the response to a given environmental stimulus as it is presented repeatedly.

microclimate   The immediate environment in the habitat of plants and animals, or the environment under the clothing worn by humans.

resistance (to a stressor)   Morpho-physiological (inherited or acquired) ability of the organism to tolerate cold, heat, hypoxia, hyperoxia, etc, connected with the development of some defence response to such factors.

strain   Condition of the animal body resulting from exposure to stress, with reversible or irreversible changes in performance, physiological or biochemical characteristics.

stress   Environmental conditions that can elicit strain, for instance, extremes of temperature or ambient pressure. Also, for Selye (1950), the reaction of the human organism to a stressor.

tolerance   Ability to endure more than minimal displacement of normal physiological levels in response to a stressor.

References

Works by two authors are listed alphabetically by second author and those by three or more authors are listed chronologically.

Abramson, A. A. & Delagi, E. F. (1961). Influence of weight-bearing and muscle contraction on disuse osteoporosis. *Arch. Phys. Med. Rehab.*, **42**, 147–51.

Adair, L. S. (1984). Marginal intake and maternal adaptation: the case of rural Taiwan. In *Energy Intake and Activity*, ed. E. Pollitt & P. Amante, pp. 33–55. New York: Liss.

Adams, J., Mottola, M., Bagnall, K. M. & McFadds, K. D. (1982). Total body fat content in a group of professional football players. *Can. J. Appl. Spt. Sci.*, **7**, 36–40.

Adams, J. E. (1968). Bone injuries in very young athletes. *Clin. Orthop.*, **58**, 129–40.

Adams, J. M., Best, T. W. & Edholm, O. G. (1961). Weight changes in young men. *J. Physiol.*, **156**, 38P.

Adams, P., Davies, G. T. & Sweetnam, P. (1970). Osteoporosis and the effects of aging on bone mass in elderly men and women. *Quart. J. Med.*, **39**, 601–15.

Adinoff, A. D. & Hollister, J. R. (1983). Steroid-induced fractures and bone loss in patients with asthma. *New Engl. J. Med.*, **309**, 265–8.

Adler, J. H. (1988). An exercise in the evaluation of body fat content as illustrated by the sand rat (*Psammomys obesus*). In *Adaptive Physiology to Stressful Environmemts*, ed. S. Samueloff & M. K. Yousef, pp. 181–6. Boca Raton: Chemical Rubber Company Press.

Afting, E. G., Bernhardt, W., Janzen, R. W. C. & Röthig, H. J. (1981). Quantitative importance of non-skeletal muscle N-methyl histidine and creatinine in human urine. *Biochem. J.*, **200**, 449–52.

Agostini, E., Gurnter, G., Torri, G. & Rahn, H. (1966). Respiratory mechanics during submersion and negative pressure breathing. *J. Appl. Physiol.*, **21**, 251–8.

Akers, R. & Buskirk, E. R. (1969). An underwater weighing system utilizing 'force cube' transducers. *J. Appl. Physiol.*, **26**, 649–52.

Albrink, M. J. & Meigs, J. W. (1964). Interrelationship between skinfold thickness, serum lipids and blood sugar in normal men. *Am. J. Clin. Nutr.*, **15**, 255–61.

Alén, M., Häkkinen, K. & Komi, P. V. (1984). Changes in neuromuscular performance and muscle fiber characteristics of elite power athletes self-administering androgenic and anabolic steroids. *Acta Physiol. Scand.*, **122**, 535–44.

Alexander, M. K. (1964). The post-mortem estimation of total body fat, muscle and bone. *Clin. Sci.*, **26**, 193–202.

Allen, J. A. (1877). The influence of physical conditions in the genesis of the species. *Radical Rev.*, **1**, 108–40.

Allen, T. H., Peng, M. T., Chen, K. P., Huang, T. F., Chang, C. & Fang, H. S. (1956). Prediction of blood volume and adiposity in man from body weight and cube of height. *Metabolism*, **5**, 328–45.

Allen, T. H., Anderson, E. C. & Langham, W. H. (1960). Total body potassium and gross body composition in relation to age. *J. Gerontol.*, **15**, 348–57.

Almond, D. J., King, R. F. G. J., Burkinshaw, L., Oxby, C. B. & McMahon, M. J. (1984). Measurement of short-term changes in the fat content of the body: a comparison of three methods in patients receiving intravenous nutrition. *Br. J. Nutr.*, **52**, 215–25.

Aloia, J. F. (1981). Exercise and skeletal health. *J. Am. Geriatr. Soc.*, **15**, 104–107.

Aloia, J. F., Roginsky, M., Ellis, K., Shukla, K. & Cohn, S. (1974). Skeletal metabolism and body composition in Cushing's syndrome. *J. Clin. Endocrinol. Metab.*, **39**, 981–5.

Aloia, J. F., Vaswani, A., Atkins, H., Zanzi, I., Ellis, K. & Cohn, S. H. (1977). Radiographic, morphometry and osteopenia in spinal osteoporosis. *J. Nucl. Med.*, **18**, 425–31.

Aloia, J. F., Cohn, S. H., Babu, T., Abesamis, C., Kalici, N. & Ellis, K. (1978a). Skeletal mass and body composition in marathon runners. *Metabolism*, **27**, 1793–6.

Aloia, J. F., Cohn, S. H., Ross, P., Vaswani, A., Abesamis, C., Ellis, K. & Zanzi, I. (1978b). Skeletal mass in post-menopausal women. *Am. J. Physiol.*, **235**, E82–7.

Alsmeyer, R. H., Hiner, R. L. & Thornton, J. W. (1963). Ultrasonic measurements of fat and muscle thickness of cattle and swine. *Ann. N.Y. Acad. Sci.*, **110**, 23–30.

Alvioli, L. (1976). Senile and post-menopausal osteoporosis. *Adv. Intern. Med.*, **21**, 391–415.

American College of Sports Medicine (1983). Position statement on proper and improper weight loss program. *Med. Sci. Sports Exerc.*, **15**, ix–xii.

American Psychiatric Association (1980). *Diagnostic and Statistical Manual.* Washington, D.C.: American Psychiatric Association.

Amor, A. F. (1978). A survey of physical fitness in the British army. In *Proceedings of the first RSG4 Physical Fitness Symposium with Special Reference to Military Forces*, ed. C. Allen. Downsview, Ont.: Defence and Civil Institute of Environmental Medicine.

Anand, B. K. (1967). Central chemosensitive mechanisms related to feeding. In *Handbook of Physiology*, Section 6. *Alimentary Canal*, vol. I, ed. C. F. Code & W. Heidel, p. 249. Washington, D.C.: American Physiology Society.

Andersen, K. L. (1967). Ethnic group differences in fitness for sustained and strenuous muscular exercise. *Can. Med. Assoc. J.*, **96**, 832–5.

Andersen, K. L. (1969). Racial and inter-racial differences in work capacity. *J. Biosoc. Sci. Suppl.*, **1**, 69–80.

Anderson, E. C. (1963). Three component body composition analysis based on potassium and water determinations. *Ann. N.Y. Acad. Sci.,* **110**, 189–212.

Anderson, E. C. (1965). Determination of body potassium by 4D gamma counting. In *Radioactivity in Man,* II, ed. G. R. Meneely & S. M. Linde, pp. 211–31. Springfield, Ill: C. C. Thomas.

Anderson, E. C. & Langham, W. H. (1959). Average potassium concentration of the human body as a function of age. *Science,* **130**, 713–14.

Anderson, P. & Saltin, B. (1985). Maximal perfusion of skeletal muscle in man. *J. Appl. Physiol.,* **366**, 233–50.

Anderson, T. L., Muttart, C. R., Bieber, M. A., Nicholson, J. F. & Heird, W. C. (1979). A controlled trial of glucose versus glucose and amino acids in premature infants. *J. Pediatr.* **94**, 947–51.

Anderson, T. W., Brown, J. R., Hall, J. W. & Shephard, R. J. (1968). The limitations of linear regressions for the prediction of vital capacity and forced expiratory volume. *Respiration,* **25**, 140–58.

Andres, R. (1985). Mortality and obesity: the rationale for age-specific height–weight tables. In *Principles of Geriatric Medicine,* ed. R. Andres, E. L. Bierman & W. R. Hazzard, pp. 311–18. New York: McGraw-Hill.

Andrew, G. M., Becklake, M. R., Guleria, J. S. & Bates, D. V. (1972). Heart and lung functions in swimmers and nonathletes during growth. *J. Appl. Physiol.,* **32**, 245–51.

Angel, A. (1974). Patho-physiology of obesity. *Can. Med. Assoc. J.,* **110**, 540–8.

Angel, J. L. (1949). Constitution in female obesity. *Am. J. Phys. Anthrop.,* **7**, 433–71.

Angeleli, A. Y. O., Burini, R. G. & Campana, A. O. (1978). Body collagen in protein-deficient rats. *J. Nutr.,* **108**, 1147–54.

Apte, S. V. & Iyengar, L. (1972). Composition of the human foetus. *Br. J. Nutr.,* **27**, 305–12.

Arara, N. S. & Rochester, D. F. (1982). Effect of body weight and muscularity on human diaphragm muscle mass, thickness and area. *J. Appl. Physiol.,* **52**, 64–70.

Araujo, J., Sanchez, G., Gutierrez, J. & Perez, F. (1970). Cardiomyopathies of obscure origin in Cali, Colombia: clinical, etiologic and laboratory aspects. *Am. Heart J.,* **80**, 162–70.

Arborelius, M., Balldin, U. I., Lila, B. & Lungren, C. E. G. (1972). Regional lung function in man during immersion with the head above water. *Aerospace Med. Biol.,* **43**, 701–7.

Archibald, E. H., Harrison, J. E. & Pencharz, P. B. (1983). Effect of a weight reduction high-protein diet on the body composition of obese adolescents. *Am. J. Dis. Childr.,* **137**, 658–62.

Asahina, K. (1975). Growth acceleration in Japan. In *JIBP Synthesis: Physiological Adaptability and Nutritional States of the Japanese. B. Growth, Work Capacity and Nutrition of Japanese,* ed. K. Asahina & R. Shigiya, pp. 23–6. Tokyo: Japanese Committee for the International Biological Programme, Science Council of Japan.

Asatoor, A. M. & Armstrong, M. D. (1967). 3-methyl histidine, a component of actin. *Biochem. Biophys. Res. Comm.,* **26**, 168–74.

Ashwell, M., Chin, S., Stalley, S. & Garrow, J. S. (1982). Female fat distribution

– a simple classification based on two circumference measurements. *Int. J. Obesity*, **6**, 143–52.

Ashwell, M., Cole, T. J. & Dixon, A. K. (1985). Obesity: new insight into the anthropometric classification of fat distribution by computed tomography. *Br. Med. J.*, **290**, 1691–4.

Asmussen, E. (1964). Muscular exercise. In *Handbook of Physiology. Respiration*, part 2, ed. W. Fenn & H. Rahn, pp. 939–78. Washington, D.C.: American Physiological Society.

Asmussen, E. (1973). Growth in muscular strength and power. In *Physical Activity, Human Growth and Development*, ed. G. L. Rarick, pp. 60–79. New York: Academic Press.

Asmussen, E. & Christensen, E. H. (1967). *Kompendium: Legemsovelsernes Specielle Teori*. Copenhagen: Kobenhavns Universitets Fond til Tilvejebringelse af Laremidler.

Åstrand, P. O. & Rodahl, K. (1985). *Textbook of Work Physiology*, 3rd edn. New York: McGraw-Hill.

Åstrand, P. O., Engström, L., Eriksson, B., Karlberg, P., Nylander, I., Saltin, B. & Thorén, C. (1963). Girl swimmers. *Acta Paediatr. Scand.* Suppl. **147**, 1–75.

Atkinson, P. J., Buckler, J. M. H., Burkinshaw, L. & Brodie, D. A. (1978). Longitudinal measurements of the changes in bone mineral density during growth. *Proceedings of the 4th International Conference on Bone Mineral Measurement, University of Toronto*, pp. 113–120.

Austin, D. M., Ghesquiere, J. & Azma, M. (1979). Work capacity and body morphology of Bantu and Pygmoid groups of Western Zaire. *Hum. Biol.*, **50**, 79–89.

Babirak, S. P., Dowell, R. T. & Oscai, L. B. (1974). Total fasting and total fasting plus exercise: effects on body composition of the rat. *J. Nutr.*, **104**, 452–7.

Babu, D. S. & Chuttani, C. S. (1979). Anthropometric indices independent of age for nutritional assessment in school children. *J. Epidemiol. Comm. Health*, **33**, 177–9.

Baecke, J. A., Burema, J. & Deurenberg, P. (1982). Body fatness, relative weight and frame size in young adults. *Br. J. Nutr.*, **48**, 1–6.

Bailey, D. A., Bell, R. D. & Howarth, R. E. (1973). The effect of exercise on DNA and protein synthesis in skeletal muscle of growing rats. *Growth*, **37**, 323–31.

Baisch, F. & Beck, L. (1987). Body impedance measurement during spacelab mission D1. *Physiologist* **30** (1). Suppl., s47–8.

Bakker, H. K. & Struidenkamp, R. S. (1977). Biological variability and lean body mass estimates. *Hum. Biol.*, **49**, 187–202.

Balazs, E. A. (1977). Intercellular matrix of connective tissue. In *Handbook of the Biology of Aging*, ed. C. Finch & L. Hayflick, pp. 222–40. New York: Van Nostrand, Reinhold.

Baldwin, B. T. (1925). Weight–height–age standards in metric units for American born children. *Am. J. Phys. Anthropol.*, **8**, 1–10.

Barac-Nieto, M., Spurr, G. B., Lotero, G. B. & Maksud, M. G. (1978). Body composition in chronic undernutrition. *Am. J. Clin. Nutr.*, **31**, 23–40.

Barac-Nieto, M., Spurr, G. B., Lotero, H., Maksud, M. G. & Dahners, H. W.

(1979). Body composition during nutritional repletion of severely under-nourished men. *Am. J. Clin. Nutr.*, **32**, 981–91.

Barcroft, J. (1914). *The Respiratory Function of the Blood.* Cambridge: Cambridge University Press.

Bar-Or, O., Zwiren, L. D. & Ruskin, H. (1974). Anthropometric and developmental measurements of 11 to 12 year old boys, as predictors of performance two years later. *Acta Paediatr. Belg.*, **28**, Suppl., 214–20.

Bauer, K. D. & Griminger, P. (1983). Long-term effects of activity and of calcium and phosphorus intake on bones and kidneys of female rats. *J. Nutr.*, **113**, 2011–21.

Baumgartner, R. N., Chumlea, W. C. & Roche, A. F. (1987). Associations between bioelectric impedance and anthropometric variables. *Hum. Biol.*, **59**, 235–44.

Baumrind, S. (1986). Integrated surface and deep structure mapping of the human anatomy. In *Perspectives in Kinanthropometry,* ed. J. A. P. Day, pp. 269–84. Champaign, Ill.: Human Kinetics Publishers.

Bayer, L. M. & Bayley, N. (1959). *Growth Diagnosis.* Chicago: University of Chicago Press.

Beall, C. M. (1982). A comparison of chest morphology in high altitude Asian and Andean populations. *Hum. Biol.*, **54**, 145–63.

Beall, C. M., Baker, P. T., Baker, T. S. & Haas, J. D. (1977). The effects of high altitude on adolescent growth in southern Peruvian Amerindians. *Hum. Biol.*, **49**, 109–24.

Beaton, G. H. (1984). Adaptation to and accommodation of long term low energy intake: A commentary on the conference on energy intake and activity. In *Energy Intake and Activity,* ed. E. Pollitt & P. Amante, pp. 395–403. New York: Liss.

Beaton, G. H. (1985). The significance of adaptation in the definition of nutrient requirements and for nutrition policy. In *International Symposium on Nutritional Adaptation in Man,* ed. J. C. Waterlow, pp. 219–32. London: John Libbey.

Beattie, J., Herbert, P. & Bell, D. J. (1948). Famine oedema. *Br. J. Nutr.*, **2**, 47–65.

Becque, M. D., Katch, V. L. & Moffatt, R. J. (1986). Time course of skin–pulus–fat compression in males and females. *Hum. Biol.*, **58**, 33–42.

Beddoe, A. H. & Hill, G. L. (1985). Clinical measurement of body composition using in vivo activation analysis. *J. Parent Ent. Nutr.*, **9**, 504–20.

Beddoe, A. H., Streat, S. J. & Hill, G. L. (1984). Evaluation of an in vivo prompt gamma neutron activation facility for body composition studies in critically ill intensive care patients: results in 41 normals. *Metabolism,* **33**, 270–80.

Beddoe, A. H., Streat, S. J. & Hill, G. L. (1985). Hydration of fat-free body in protein-depleted patients. *Am. J. Physiol.*, **249**, E227–33.

Bedell, G. N., Marshall, R., DuBois, A. B. & Harris, J. H. (1956). Measurement of the volume of gas in the gastro-intestinal tract. Values in normal subjects and ambulatory patients. *J. Clin. Invest.*, **35**, 336–45.

Beeston, J. W. U. (1965). Determination of specific gravity of live sheep and its correlation with fat percentage. In *Human Body Composition. Approaches and Applications,* ed. J. Brozek, pp. 49–56. Oxford: Pergamon Press.

Behnke, A. R. (1942). Physiologic studies pertaining to deep-sea diving and aviation, especially in relation to the fat content and composition of the body. *Harvey Lecture Series*, **37**, 198–226.

Behnke, A. R. (1961). Quantitative assessment of body build. *J. Appl. Physiol.*, **16**, 960–8.

Behnke, A. R. & Wilmore, J. (1974). *Evaluation and Regulation of Body Build and Composition*. Englewood Cliffs, N.J.: Prentice-Hall.

Behnke, A. R., Feen, B. G. & Welham, W. C. (1942). The specific gravity of healthy men: body weight/volume as an index of obesity. *J. Amer. Med. Assoc.*, **118**, 495–501.

Behnke, A. R., Guttentag, O. E. & Brodsky, C. (1959). Quantification of body weight and configuration from anthropometric measurements. *Hum. Biol.*, **31**, 213–34.

Beisel, W. R., Sawyer, W. D., Ryll, E. D. & Crozier, D. (1967). Metabolic effects of intracellular infections in man. *Ann. Int. Med.*, **67**, 744–79.

Benedict, F. G., Miles, W. R., Roth, P. & Smith, H. M. (1919). *Human Vitality and Efficiency Under Prolonged Restricted Diet*. Washington: Carnegie Institute.

Benedict, J. D., Forsham, P. H., Roche, M., Soloway, S. & Stettin, D. (1950). The effect of salicylates and adrenocorticotropic hormone upon the miscible pool of uric acid in gout. *J. Clin. Invest.*, **29**, 1104–11.

Benn, R. T. (1971). Some mathematical properties of weight-for-height indices used as measures of adiposity. *Br. J. Prev. Soc. Med.*, **25**, 42–50.

Beretta-Piccoli, C., Davies, D. L., Boddy, K., Brown, J. J., Cumming, A. M. M., East, B. W., Fraser, R., Lever, A. F., Padfield, P. L., Semple, P. F., Robertson, J. I. S., Weidmann, P. & Williams, E. D. (1982). Relation of arterial pressure with body sodium, body potassium and plasma potassium in essential hypertension. *Clin. Sci.*, **63**, 257–70.

Berg, K. & Bjure, J. (1974). Preliminary results of long-term physical training of adolescent boys with respect to body composition, maximal oxygen uptake and lung volume. *Acta Paediatr. Belg.*, **28**, Suppl., 183–90.

Bergen, W. G., Kaplan, M. L., Merkel, R. A. & Leveille, G. A. (1975). Growth of adipose and lean tissue mass in hind limbs of genetically obese mice during preobese and obese phases of development. *Am. J. Clin. Nutr.*, **28**, 157–61.

Bergmann, C. (1847). Uber die Verhältnisse der Wärmeökonomie des Thiere zu ihrer Grosse. *Gottinger Studien*, **3**, 595–708.

Bergström, J. (1962). Muscle electrolytes in man. *Scand. J. Clin. Lab. Invest.*, **14**, Suppl., 68.

Berninck, M. J. E., Erich, W. B. M., Peltenburg, A. L., Zonderland, M. L. & Huisveld, I. A. (1983). Height, body composition, biological maturation and training in relation to socio-economic status in girl gymnasts, swimmers and controls. *Growth*, **47**, 1–12.

Bertrand, H. A., Stacey, C., Masoro, E. J., Yu, B. P., Murata, I. & Maeda, H. (1984). Plasticity of fat cell number. *J. Nutr.*, **114**, 127–31.

Bessard, T., Schutz, Y. & Jéquier, E. (1983). Energy expenditure and postprandial thermogenesis in obese women before and after weight loss. *Am. J. Clin. Nutr.*, **38**, 680–93.

Bessey, P. Q., Walters, J. M., Aoki, T. T. & Wilmore, D. W. (1984). Combined hormone infusion simulates the metabolic response to injury. *Ann. Surg.*, **200**, 264–81.

Best, W. R., Kuhl, W. J. & Consolazio, L. F. (1953). Relation of creatinine co-efficients to leanness–fatness in man. *J. Lab. Clin Med.*, **42**, 784.

Bettmann, O. L. (1956). *A Pictorial History of Medicine*, p. 143. Springfield, Ill.: C. C. Thomas.

Beunen, G., Ostyn, M., Renson, R., Simons, J., Swalus, P., Van Gerven, D. & Willems, E. J. (1972). Skeletleeftijd en fysische ontwikkeling bij twaalfjarige jongens. *Arch. Belg. Med Soc.*, **30**, 102–19.

Beunen, G., de Beul, G., Ostyn, M. *et al.* (1978*a*). Age of menarche and motor performance in girls aged 11 through 18. In *Pediatric Work Physiology*, ed. J. Borms & M. Hebbelinck, pp. 118–23. Basel: S. Karger, A.G.

Beunen, G., Ostyn, M., Renson, R. *et al.* (1978*b*). Motor performance as related to chronological age and maturation. In *Physical Fitness Assessment*, ed. R. J. Shephard & H. Lavallée, pp. 229–37. Springfield, Ill.: C. C. Thomas.

Beutler, E., Kuhl, W. & Sacks, P. (1983). Sodium-potassium-ATPase is influenced by ethnic origin and not by obesity. *N. Engl. J. Med.*, **309**, 756–60.

Bharawadj, H., Singh, A. P. & Malhotra, M. S. (1973). Body composition of the high altitude natives of Ladakh. A comparison with sea-level residents. *Hum. Biol.*, **45**, 423–34.

Billewicz, W. Z., Kemsley, W. F. & Thomson, A. M. (1962). Indices of obesity. *Br. J. Prev. Soc. Med.*, **16**, 183–8.

Billewicz, W. Z., Thompson, A. M. & Fellowes, H. M. (1983). Weight for height in adolescence. *Ann. Hum. Biol.*, **10**, 119–24.

Bishop, C. W., Bowen, P. E. & Ritchey, S. J. (1981). Norms for nutritional assessment of American adults by upper arm anthropometry. *Am. J. Clin. Nutr.*, **34**, 2530–9.

Bistrian, B. R. & Blackburn, G. L. (1983). Assessment of protein-calorie malnutrition in the hospitalized patients. In *Nutrition Support of Medical Practice*, 2nd edn, ed. H. A. Schneider, C. E. Anderson & D. B. Coursin. Philadelphia: Harper & Row.

Bistrian, B. R., Blackburn, G. L. & Sherman, M. (1975). Therapeutic index of nutritional depletion in the hospitalized patient. *Surg. Gyn. Obstetr.*, **141**, 512–16.

Björntorp, P. (1974). Size, number and function of adipose tissue cells in human obesity. In *Lipid Metabolism, Obesity and Diabetes Mellitus: Impact upon Atherosclerosis*, ed. H. Greten, R. Levine, E. F. Pfeiffer & A. E. Renold. Stuttgart: Georg Thieme.

Björntorp, P. (1983*a*). Physiological and clinical aspects of exercise in obese persons. *Ex. Spt. Sci. Rev.*, **11**, 159–80.

Björntorp, P. (1983*b*). Role of adipose tissue in human obesity. In *Contemporary Issues in Clinical Nutrition: Obesity*, ed. M. R. C. Greenwood. Edinburgh: Churchill-Livingstone.

Björntorp, P., De Jounge, K., Sjöstrom, L. & Sullivan, L. (1970). The effect of physical training on insulin production in obesity. *Metabolism*, **19**, 631–8.

Björntorp, P., de Jounge, K., Sjöstrom, L. & Sullivan, L. (1973*a*). Physical

training in human obesity. II. Effects on plasma insulin in glucose intoler-
ant subjects without marked hyperinsulinaemia. *Scand. J. Clin. Lab. Invest.*, **32**, 41–5.

Björntorp, P., de Jounge, K., Krotkiewski, M., Sullivan, L., Sjöstrom, L. & Steinberg, J. (1973*b*). Physical training in human obesity: 3. Effects of long term physical training on body composition. *Metabolism*, **22**, 1467–75.

Black-Sandler, R., Laporte, R. E., Sashin, D., Kuller, L. H., Sternglass, E., Cauley, J. A. & Link, M. M. (1982). Determinants of bone mass in menopause. *Prev. Med.*, **11**, 269–80.

Blair, D., Habicht, J-P., Sims, E. A. H., Sylvester, D. & Abraham, S. (1984). Evidence for an increased risk for hypertension with centrally located body fat and the effects of race and sex on this risk. *Am. J. Epidemiol.*, **119**, 526–40.

Blair, S. N., Ludwig, D. A. & Goodyear, N. N. (1988). A canonical analysis of central and peripheral subcutaneous fat distribution and coronary heart disease risk factors in men and women aged 18–65 years. *Hum. Biol.*, **60**, 111–22.

Blaxter, K. L. (1971). Methods of measuring the energy metabolism of animals and interpretation of results obtained. *Fed. Proc.*, **30**, 1436–43.

Bligh, J. & Johnson, K. G. (1973). Glossary of terms for thermal physiology. *J. Appl. Physiol.*, **35**, 941–61.

Bliznak, J. & Staple, T. W. (1975). Roentgenographic measurement of skin thickness in normal individuals. *Radiology*, **118**, 55–60.

Blom, P. S., de Graeff, J., Kassenaar, A. A. H. & Sonneveldt, H. A. (1962). Nitrogen balance studies: the relation of protein metabolism to excessive shifts of sodium and potassium. In *Protein Metabolism*, ed. F. Gross, pp. 137–49. Berlin: Springer Verlag.

Boddy, K., King, P. C., Tothill, P. & Strong, J. A. (1971). Measurement of total body potassium with a shadow shield whole body counter. Calibration and errors. *Phys. Med. Biol.*, **16**, 275–82.

Boddy, K., King, P. C., Hume, R. & Weyers, E. (1972). The relationship of total body potassium to height, weight and age in normal adults. *J. Clin. Path.*, **25**, 512–17.

Bogin, B. & MacVean, R. B. (1981). Nutritional and biological determinants of body fat patterning in urban Guatemalan children. *Hum. Biol.*, **53**, 259–68.

Bogle, S., Burkinshaw, L. & Kent, J. T. (1985). Estimating the composition of tissue gained or lost from measurements of elementary body composition. *Phys. Med. Biol.*, **30**, 369–84.

Boileau, R. A., Buskirk, E. R., Horstman, H., Mendez, J. & Nicholas, W. C. (1971). Body composition changes in obese and lean men during physical conditioning. *Med. Sci. Sports*, **3**, 183–9.

Boileau, R. A., Horstman, D. H., Buskirk, E. R. & Mendez, J. (1972). The usefulness of urinary creatinine excretion in estimating body composition. *Med. Sci. Sports*, **4**, 85–90.

Boileau, R. A., Wilmore, J. H., Lohman, T. G., Slaughter, M. H. & Riner, W. F. (1981). Estimation of body density from skinfold thicknesses, body circumferences and skeletal widths in boys aged 8 to 11 years; comparison of two samples. *Hum. Biol.*, **53**, 575–92.

Boling, E. A. (1963). Determination of K^{42}, Na^{24}, Br^{82} and tritiated water concentration in man. *Ann. N.Y. Acad. Sci.*, **110**, 246–54.

Bondi, K. R., Young, J. M., Bennett, R. M. & Bradley, M. E. (1976). Closing volumes in man immersed to the neck in water. *J. Appl. Physiol.*, **40**, 736–40.

Booth, F. W. & Seider, M. J. (1979). Early change in skeletal muscle protein synthesis after limb immobilization of rats. *J. Appl. Physiol. Resp. Environ. Ex. Physiol.*, **47**, 974–7.

Booth, R. A. D., Goddard, B. A. & Paton, A. (1966). Measurement of fat thickness in man: a comparison of ultrasound, calipers and electrical conductivity. *Br. J. Nutr.*, **20**, 719–27.

Borisov, B. K. & Marei, A. N. (1974). Weight parameters of adult human skeleton. *Health Phys.*, **27**, 224–9.

Borkan G. A. & Norris, A. H. (1977). Fat redistribution and the changing body dimensions of the adult male. *Hum. Biol.*, **49**, 495–514.

Borkan, G. A., Gerzof, S. G., Robbins, A. H., Hults, D. E., Silbert, C. K. & Silbert, J. E. (1982*a*). Assessment of abdominal body fat by computed tomography. *Am. J. Clin. Nutr.*, **36**, 172–7.

Borkan, G. A., Hults, D. E., Cardarelli, J. & Burrows, B. A. (1982*b*). Comparison of ultrasound and skinfold measurements in assessment of subcutaneous and total fatness. *Am. J. Phys. Anthrop.*, **58**, 307–13.

Borkan, G. A., Hults, D. E., Gerzof, S. G., Robbins, A. H. & Silbert, S. K. (1983*a*). Age changes in body composition revealed by computed tomography. *J. Gerontol.*, **38**, 673–7.

Borkan, G. A., Hults, D. E., Gerzof, S. G. & Robbins, A. H. (1983*b*). Relationships between computed tomography tissue areas, thicknesses and total body composition. *Ann. Hum. Biol.*, **10**, 537–46.

Borkan, G. A., Hults, D. E., Gerzof, S. G. & Robbins, A. H. (1985). Comparison of body composition in middle-aged and elderly males using computed tomography. *Am. J. Phys. Anthrop.*, **66**, 289–95.

Bortz, W. M. (1969). A 500 pound weight loss. *Am. J. Med.*, **47**, 325–31.

Bottemley, P. A. (1983). Nuclear magnetic resonance: beyond physical imaging. *Adv. Technol.*, **20**, 30–8.

Bouchard, C. (1978). Family resemblance in selected biological traits: a preliminary report. In *Croissance et développement de l'enfant*, ed. H. Lavallée & R. J. Shephard, pp. 122–30. Trois Rivières: Université de Québec à Trois Rivières.

Bouchard, C. (1985). Inheritance of fat distribution and adipose tissue metabolism. In *Metabolic Complications of Human Obesities*, ed. J. Vague *et al.*, pp. 87–96. Amsterdam: Elsevier.

Bouchard, C. (1986*a*). Genetics of aerobic power and capacity. In *Sport and Human Genetics*, ed. R. M. Malina & C. Bouchard, pp. 59–88. Champaign, Ill.: Human Kinetics Publishers.

Bouchard, C. (1986*b*). Heritability and trainability. In *Sports Medicine for the Mature Athlete*, ed. J. R. Sutton & R. M. Brock, pp. 81–90. Indianapolis: Benchmark Press.

Bouchard, C. & Després, J.-P. (1989). Variation in fat distribution with age and health implications. In *Physical Activity and Aging. The Academy Papers*

266   *Body composition in biological anthropology*

22, ed. W. W. Spirduso & H. Ekert, pp. 78–106. Champaign Ill.: Human
Kinetics.
Bouchard, C., Hollmann, W. & Herkenrath, G. (1968). Relations entre le niveau
de maturité biologique, la participation à l'activité physique et certaines
structures morphologiques et organiques chez les garçons de huit à dix-huit
ans. *Biométrie Humaine*, **3**, 101.
Bouchard, C., Thibault, M-C. & Jobin, J. (1981). Advances in selected areas of
human work physiology. *Yearbook of Physical Anthropology*, **24**, 1–36.
Bouchard, C., Savard, R., Deprés, J-P., Tremblay, A. & Leblanc, C. (1985).
Body composition in adopted and biological siblings. *Hum. Biol.*, **57**, 61–
75.
Boyden, T. W., Pamenter, R. W., Gross, D., Stanforth, P., Rotkis, T. &
Wilmore, J. H. (1982). Prolactin responses, menstrual cycles and body
composition of women runners. *J. Clin. Endocrinol. Metab.*, **54**, 711–14.
Bracco, E. F., Yang, M. U., Segal, K., Hashim, S. A. & Van Itallie, T. B. (1984).
A new method for estimation of body composition in the live rat. *Proc. Soc.
Exp. Biol. Med.*, **174**, 143–6.
Bradbury, M. W. B. (1961). Urea and deuterium oxide spaces in man. *Br. J.
Nutr.*, **15**, 177–82.
Bradfield, R. B., Schutz, Y. & Lechtig, A. (1979). Skinfold changes with weight
loss. *Am. J. Clin. Nutr.*, **32**, 1756 (letter).
Brandon, L. J., Boileau, R. A. & Lohman, T. G. (1981). Comparison of two
procedures employed in measuring body density in children. *Med. Sci.
Sports Exerc.*, **13**, 122 (Abstract).
Brans, Y. W., Summers, J. E., Dweck, H. S. & Cassady, G. (1974). A non-
invasive approach to body composition in the neonate: dynamic skinfold
measurements. *Ped. Res.*, **8**, 215–22.
Bray, G. (1974). Genetic and regulatory aspects of obesity of animals and man. In
*Lipid Metabolism, Obesity and Diabetes Mellitus: Impact Upon Athero-
sclerosis*, ed. H. Greten, R. Levine, E. Pfeiffer & A. Renold, pp. 63–9.
Stuttgart: Georg Thieme.
Bray, G. (1975). *Obesity in Perspective*. Washington, D.C.: National Institutes of
Health (DHEW Publication NIH 75–708).
Bray, G. A., Greenway, F. L., Molitch, M. E., Dahms, W. T., Atkinson, R. L. &
Hamilton, K. (1978). Use of anthropometric measures to assess weight
loss. *Am. J. Clin. Nutr.*, **31**, 769–73.
Breymeyer, A. I. & Van Dyne, G. M. (1979). *Grassland Systems Analysis and
Man*. Cambridge: Cambridge University Press.
Brody, S. (1945). *Bioenergetics and Growth*. New York: Reinhold.
Brook, C. G. D., Huntley, R. M. C. & Slack, J. (1975). Influence of heredity and
environment in determination of skinfold thickness in children. *Br. Med.
J.*, **2**, 719–21.
Brooke, O. G. & Cocks, T. (1974). Resting metabolic rate in malnourished
babies in relation to total body potassium. *Acta Paediatr. Scand.*, **63**, 817–
25.
Brooks, R. M., Latham, M. C. & Crompton, D. W. T. (1979). The relationship of
nutrition and health to worker productivity in Kenya. *East Afr. Med. J.*, **56**,
413–21.

Brown, M. (1973). Role of activity in the differentiation of slow and fast muscles. *Nature,* **244**, 178–9.

Brown, M. R., Klish, W. J., Hollander, J., Campbell, M. A. & Forbes, G. B. (1983). A high protein, low calorie liquid diet in the treatment of very obese adolescents: long term effect on body mass. *Am. J. Clin. Nutr.,* **38**, 20–31.

Brozek, J. (1952). Changes of body composition in man during maturity and their nutritional implications. *Fed. Proc.* **11**, 784–93.

Brozek, J. (1965). *Human Body Composition. Approaches and Applications.* Oxford: Pergamon Press.

Brozek, J. & Kinsey, W. (1960). Age changes in skinfold compressibility. *J. Gerontol.,* **15**, 45–51.

Brozek, J. & Mori, H. (1958). Some interrelations between somatic, roentgeno-graphic and densitometric criteria of fatness. *Hum. Biol.* **30**, 322–36.

Brozek, J., Grande, F., Anderson, T. & Keys, A. (1963). Densitometric analysis of body composition: revisions of some quantitative assumptions. *Ann. N.Y. Acad. Sci.,* **110**, 113–40.

Bruce, A., Andersson, M., Arvidsson, B. & Isaksson, B. (1980). Body composition. Prediction of normal body potassium, body water and body fat in adults with basis of body height, body weight and age. *Scand. J. Clin. Lab. Invest.,* **40**, 461–74.

Buchanan, T. A. S. & Pritchard, J. J. (1970). DNA content of tibialis anterior of male and female white rats measured from birth to 50 weeks. *J. Anat.,* **107**, 185 (Note).

Buckler, J. M. H. & Brodie, D. A. (1977). Growth and maturity characteristics of schoolboy gymnasts. *Ann. Hum. Biol.,* **4**, 455–63.

Bugyi, B. (1972). Lean body weight estimation in 6–16-year-old children based on wrist breadth and body height. *J. Sports Med. Phys. Fitness,* **12**, 171–3.

Bulcke, J. A., Termote, J-L., Palmers, Y. & Crolla, D. (1979). Computed tomography of the human skeletal muscular system. *Neuroradiology,* **17**, 127–36.

Bullen, B. A., Reed, R. B. & Mayer, J. (1964). Physical activity of obese and non-obese adolescent girls. Appraised by motion picture samples. *Am. J. Clin. Nutr.,* **14**, 211–23.

Bullen, B. A., Quaade, F., Olesen, E. & Lund, S. A. (1965). Ultrasonic reflections used for measuring subcutaneous fat in humans. *Hum. Biol.,* **37**, 377–84.

Buller, A. J., Eccles, J. C. & Eccles, R. M. (1960). Interactions between motoneurons and muscles in respect of the characteristic speed of their responses. *J. Physiol.* (Lond.), **150**, 417–39.

Burgert, S. L. & Anderson, C. F. (1979). A comparison of triceps skinfold values as measured by the plastic McGaw caliper and the Lange caliper. *Am. J. Clin. Nutr.,* **32**, 1531–3.

Burkinshaw, L. & Morgan, D. B. (1985). Mass and composition of the fat-free tissues of patients with weight-loss. *Clin. Sci.,* **68**, 455–62.

Burkinskaw, L. & Spiers, F. W. (1967). *Whole-body Radiation Counters.* Leeds: Medical Research Council Environmental Radiation Research Unit, General Infirmary.

268   *Body composition in biological anthropology*

Burkinshaw, L., Jones, P. R. M. & Krupowicz, D. W. (1973). Observer error in skinfold thickness measurements. *Hum. Biol.*, **45**, 273–9.

Burkinshaw, L., Hill, G. L. & Morgan, D. B. (1978). Assessment of the distribution of protein in the human by in vivo neutron activation analysis. In *International Symposium on Nuclear Activation Techniques. Life Sciences.* Report SM 227/39, pp. 787–96. Vienna: International Atomic Energy Authority.

Burkinshaw, L., Morgan, D. B., Silverton, N. P. & Thomas, R. D. (1981). Total body nitrogen and its relation to body potassium and fat-free mass in healthy subjects. *Clin. Sci.*, **61**, 457–62.

Burmeister, W. & Bingert, A. (1967). Die quantitativen Veranderungen der menschlichen Zellmasse zwischen dem 8. und 90. Lebensjahr. *Klin. Wschr.*, **45**, 409–16.

Burri, P. H. & Weibel, E. R. (1971). Morphometric evaluation of changes in lung structure due to high altitude. In *High Altitude Physiology; Cardiac and Respiratory Aspects,* ed. R. Porter & J. Knight, p. 15–30. Edinburgh: Churchill-Livingstone.

Buskirk, E. R. (1961). Underwater weighing and body density: a review of procedures. In *Techniques for Measuring Body Composition,* ed. J. Brozek, pp. 90–107. Washington, D.C.: National Academy of Science.

Buskirk, E. R. & Mendez, J. (1984). Sports science and body composition analysis: emphasis on cell and muscle mass. *Med. Sci. Sports Exerc.*, **16**, 584–93.

Buskirk, E. R., Andersen, K. L. & Brozek, J. (1956). Unilateral activity and bone and muscle development in the forearm. *Res. Quart.*, **27**, 127–31.

Butte, N. F., Wills, C., Smith, E. O. & Garza, C. (1985). Prediction of body density from skinfold measurements in lactating women. *Br. J. Nutr.*, **53**, 485–9.

Butterfield, G. E. & Calloway, D. H. (1984). Physical activity improves protein utilization in young men. *Br. J. Nutr.* **51**, 171–84.

Byard, P. J., Poosha, D. V. R. & Satyanarayana, M. (1985). Genetic and environmental determinants of height and weight of families from Andra Pradesh, India. *Hum. Biol.,* **57**, 621–33.

Byard, P. J., Sharma, K., Russell, J. M. & Rao, D. C. (1984). A family study of anthropometric traits in a Punjabi community. II. An investigation of familial transmission. *Am. J. Phys. Anthrop.* **64**, 97–104.

Byrd, P. J. & Thomas, T. R. (1983). Hydrostatic weighing during different stages of the menstrual cycle. *Res. Quart.*, **54**, 296–8.

Calloway, D. H. (1982). Functional consequences of malnutrition. *Rev. Inf. Dis.,* **4**, 736–45.

Calloway, D. H. & Kurzer, M. S. (1982). Menstrual cycle and protein requirements of women. *J. Nutr.,* **112**, 356–66.

Cameron, J. R. & Sorenson, J. A. (1963). Measurement of bone mineral in vivo. An improved method. *Science,* **140**, 230–2.

Cameron, J. R., Mazess, R. B. & Sorenson, J. A. (1968). Precision and accuracy of bone mineral by direct photon absorptiometry. *Invest. Radiol.,* **3**, 141–50.

Campbell, J. & Rastogi, K. S. (1969). Action of growth hormone: enhancement of insulin utilization with inhibition of insulin effect on blood glucose in dogs. *Metab. Clin. Exp.*, **18**, 930–44.

Campbell, M. J., McComas, A. J. & Petito, F. (1973). Physiological changes in ageing muscles. *J. Neurol. Neurosurg. Psychiatry*, **36**, 174–82.

Cann, C. E., Genant, H. K., Ettinger, B. & Gordon, G. S. (1980*a*). Spinal mineral loss in oophorectomized women. *J. Am. Med. Assoc.*, **244**, 2056–9.

Cann, C. E., Adachi, R. R. & Holton, E. M. (1980*b*). Bone resorption and calcium absorption in rats during spaceflight. *Physiologist*, **23** (6), S83–6.

Cann, C. E., Martin, M. C., Genant, H. K. & Jaffe, R. B. (1984). Decreased spinal mineral content in amenorrheic women. *J. Am. Med. Assoc.*, **251**, 626–9.

Cannon, P. & Keatinge, W. R. (1960). The metabolic rate and heat loss of fat and thin men in cold and warm water. *J. Physiol.*, **154**, 329–44.

Carlberg, K. A., Buckman, M. T., Peake, G. T. & Riedesel, M. L. (1983). Body composition of oligo/amenorrheic athletes. *Med. Sci. Sports Exerc.*, **15**, 215–17.

Carr, L. S., Davies, S. & Dressendorfer, R. H. (1985). Estimation of percentage body fat by four methods before and after weight reduction in obese women. *Med. Sci. Sports Exerc.*, **17**, 243 (Abstract).

Carter, J. L. (1980). *The Heath-Carter Somatotype Method*. San Diego: San Diego State University Syllabus Service.

Carter, J. L. (1984). *Physical structure of Olympic athletes*. Part 1. *The Montreal Olympic Games Anthropological Project*. Part 2. *The Kinanthropometry of Olympic athletes*. Basel: Karger.

Carter, J. E. L. (1980). The contribution of somatotyping to kinanthropometry. In *Kinanthropometry* II, ed. M. Ostyn, G. Beunen & J. Simons. Baltimore: University Park Press.

Carter, J. E. L. & Heath, E. H. (1988). *Somatotyping – Development and Applications*. Cambridge: Cambridge University Press.

Castenfors, J., Mossfeldt, F. & Piscator, M. (1967). Effect of prolonged heavy exercise on renal function and urinary protein excretion. *Acta Physiol. Scand.*, **70**, 194–206.

Cavalli-Sforza, L. L. & Bodmer, W. C. (1971). *The Genetics of Human Populations*. San Francisco: Freeman.

Cerny, L. (1969). The results of an evaluation of skeletal age of boys 11–15 years old with different regime of physical activity. In. *Physical Fitness Assessment*, ed. V. Seliger, pp. 56–9. Prague: Charles University.

Cerovaska, J., Petrasek, R., Hajino, K. & Kaucka, J. (1977). Indices of obesity and body composition in four groups of Czech population. *Z. Morph. Anthrop.*, **68**, 213–9.

Chalmers, J. & Ho, K. C. (1976). Geographical variations in senile osteoporosis. *J. Bone Joint Surg.*, **52** (B), 667–75.

Cheek, D. B. (1968). *Human Growth*. Philadelphia: Lea & Febiger.

Cheek, D. B. & West, C. D. (1955). An appraisal of methods of tissue chloride analysis: the total carcass chloride, exchangeable chloride, potassium and water of the rat. *J. Clin. Invest.*, **34**, 1744–55.

Cheek, D. B., Schultz, R. B., Parra, A. & Reba, C. (1970). Overgrowth of lean and adipose tissue in adolescent obesity. *Pediatr. Res.*, **4**, 268–79.

Cheek, D. B., Wishart, J. MacLennan, A. H. & Haslam, R. (1984). Cell hydration in the normally grown, the premature and low weight for gestational age infant. *Early Hum. Develop.*, **10**, 75–84.

Chen, K. P. (1953). Report on measurement of total body fat in American women estimated on the basis of specific gravity as an evaluation of individual fatness and leanness. *J. Formosan Med. Assoc.*, **52**, 271–6.

Chestnut, C. H., Nelp, W. B., Denney, D. & Sherrard, D. J. (1973). Measurement of total body calcium (bone mass) by neutron activation analysis: applicability to bone-wasting disease. In *Clinical Aspects of Metabolic Bone Disease*, ed. B. Frame, A. M. Parfitt & H. Duncan, pp. 50–4. Amsterdam: Excerpta Medica.

Chien, S., Peng, M., Chen, K., Huang, T., Chang, C. & Fang, H. S. (1975a). Longitudinal studies on adipose tissue and its distribution in human subjects. *J. Appl. Physiol.*, **39**, 825–30.

Chien, S., Peng, M. T., Chen, K. P., Huang, T. F., Chang, C. & Fang, H. S. (1975b). Longitudinal measurements of blood volume and essential body mass in human subjects. *J. Appl. Physiol.*, **39**, 818–24.

Chinn, K. S. K. (1967). Prediction of muscle and remaining tissue protein in man. *J. Appl. Physiol.*, **23**, 713–15.

Chirico, A. M. & Stunkard, A. J. (1960). Physical activity and human obesity. *New Engl. J. Med.*, **263**, 935–40.

Chobanian, A. V., Burrows, B. A. & Hollander, W. (1961). Body fluid and electrolyte composition in arterial hypertension. II. Studies in mineralocorticoid hypertension. *J. Clin. Invest.*, **40**, 416–22.

Chow, R. C., Harrison, J. E., Brown, C. F. & Hojek, V. (1986). Physical fitness effect on bone mass in post-menopausal women. *Arch. Phys. Med. Rehab.*, **67**, 231–4.

Chow, R. C., Harrison, J. E., Sturtridge, W., Josse, R., Murray, T. M., Bayley, A., Dornan, J. & Hammond, T. (1988). The effect of exercise on bone mass of osteoporotic patients on fluoride treatment. *Clin. Invest. Med.*, **10** (2), 59–63.

Christiansen, C. & Rödbro, P. (1975). Bone mineral content and estimated body calcium in normal adults. *Scand. J. Clin. Lab. Invest.*, **35**, 433–9.

Christiansen, C., Rödbro, P. & Jensen, H. (1975). Bone mineral content in the forearm measured by photon absorptiometry. *Scand. J. Clin. Lab. Invest.*, **35**, 323–30.

Chui, L. A. & Castleman, K. R. (1980). Morphometric analysis of rat muscle fibers following space flight and hypogravity. *Physiologist*, **23** (6), S76–8.

Chumlea, W. C., Roche, A. F & Rogers, E. (1984). Replicability for anthropometry in the elderly. *Hum. Biol.*, **56**, 329–37.

Chumlea, W. C., Roche, A. F., Guo, S. & Woynarowska, B. (1987). The influence of physiologic variables and oral contraceptives on bioelectric impedance. *Hum. Biol.*, **59**, 257–69.

Clark, B. A. & Mayhew, J. L. (1980). An inexpensive method of determining body composition by underwater weighing. *Br. J. Spts. Med.*, **14**, 225–8.

Clark, D. A., Kay, T. D., Tatsch, R. F. & Theis, C. F. (1977). Estimation of body composition by various methods. *Aviat. Space Environ. Med.*, **48**, 701–4.

Clark, L. C., Thompson, H. L., Beck, E. I. & Jacobson, W. (1951). Excretion of creatine and creatinine by children. *Am. J. Dis Childr.*, **81**, 774–83.

Clark, P. J. (1956). Heritability of certain anthropometric characters as ascertained from measurements of twins. *Am. J. Hum. Genet.*, **8**, 49–54.

Clarkson, P. M., Katch, F. I., Kroll, W., Lane, R. & Kanen, G. (1980). Regional adipose cellularity and reliability of adipose cell size and determination. *Am. J. Clin. Nutr.*, **33**, 2245–52.

Clarys, J. P. & Marfell-Jones, M. J. (1986). Anatomical segmentation in humans and the prediction of segmental masses from intra-segmental anthropometry. *Hum. Biol.*, **58**, 771–82.

Clarys, J. P., Drinkwater, D. T., Martin, A. D. & Marfell-Jones, M. J. (1987). The skinfold: myth or reality? *J. Spt. Sci.*, **5**, 3–33.

Clarys, J. P., Martin, A. D. & Drinkwater, D. T. (1984). Gross tissue weights in the human body by cadaver dissection. *Hum. Biol.*, **56**, 459–73.

Claus-Walker, J., Singh, J., Leach, C. S., Hatton, D. V., Hubert, C. W. & DiFerrante, N. (1977). The urinary excretion of collagen degradation products by quadriplegic patients and during weightlessness. *J. Bone Joint Surg.*, **59** (A), 209–12.

Clauser, C. E., McConville, J. T. & Young, J. W. (1969). *Weight, Volume and Center of Mass of Segments of the Human Body*, pp. 69–70. Dayton, Ohio: Wright-Patterson Air Force Base Report, AMRL TR.

Clegg, E. J. & Kent, C. (1967). Skinfold compressibility in young adults. *Hum. Biol.*, **39**, 418–29.

Clegg, E. J., Pawson, I. G., Ashton, E. H. & Flinn, R. M. (1972). The growth of children at different altitudes in Ethiopia. *Phil. Trans. R. Soc. Lond. B*, **264**, 403–37.

Cloninger, C. R., Rao, D. C., Rice, J., Reich, T. & Morton, N. E. (1983). A defence of path analysis in genetic epidemiology. *Am. J. Hum. Genet.*, **35**, 733–56.

Cohen, J. B., Gemmell, H., Hayes, P. A., Smith, F. W. & Sowood, P. J. (1986). A comparison of magnetic resonance imaging, ultrasound and skinfold caliper techniques to measure subcutaneous fat thickness in humans. *J. Physiol.*, **380**, 61P.

Cohen, J. L., Chung, S. K., May, P. B. & Ertel., N. H. (1982). Exercise, body weight and professional ballet dancers. *Phys. Sports Med.*, **10** (4), 92–101.

Cohn, C. (1963). Feeding frequency and body composition. *Ann. N.Y. Acad. Sci.*, **110**, 395–409.

Cohn, S. H. & Dombrowski, C. S. (1970). Absolute measurements of whole body potassium by gamma-ray spectrometry. *J. Nucl. Med.*, **11**, 239–46.

Cohn, S. H. & Ellis, N. F. (1976). Validity of the absorptiometric measurement of bone mineral content of the radius. In *Third International Conference on Bone Mineral Measurement*, ed. R. B. Mazess. *Am. J. Roentegenol.*, **126**, 1286–7.

Cohn, S. H. & Parr, R. M. (1985). Nuclear-based techniques for the in vivo study of human body composition. *Clin. Physics Physiol. Meas.*, **6**, 275–301.

Cohn, S. H., Shukla, K. K. & Fairchild, R. G. (1972). Design and calibration of a broad beam ^{238}Pu,Be source for total body neutron activation analysis. *J. Nucl. Med.*, **13**, 487–92.

Cohn, S. H., Ellis, K. J. & Wallach, S. (1974). In vivo neutron activation analysis. Clinical potential in body composition studies. *Am. J. Med.*, **57**, 683–6.

Cohn, S. H., Vaswani, A., Zanzi, I., Aloia, J. F., Roginsky, M. S. & Ellis, K. J.

(1976a). Changes in body chemical composition with age measured by total body neutron activation. *Metabolism,* **25**, 85–95.

Cohn, S. H., Vaswani, A., Zanzi, I. & Ellis, U. J. (1976b). Effect of aging on bone mass in adult women. *Am. J. Physiol.,* **230**, 143–8.

Cohn, S. H., Abesamis, C., Yasumura, S., Aloia, J. F., Zanzi, J. & Ellis, K. J. (1977). Comparative skeletal mass and radial bone content in black and white women. *Metabolism,* **26**, 171–8.

Cohn, S. H., Vartsky, D., Yasumura, S., Sawitsky, A., Zanzi, I., Vaswani, A. N. & Ellis, K. J. (1980). Compartmental body composition based on total body nitrogen, potassium and calcium. *Am. J. Physiol.,* **239**, E524–30.

Cohn, S. H., Vaswani, A. N., Yasumura, S., Yuen, K. & Ellis, K. J. (1984). Improved methods for determination of body fat by in vivo neutron activation. *Am. J. Clin. Nutr.,* **40**, 255–59.

Cole, D. F. (1952). The effects of oestradiol on the skeletal muscle and liver of the rat. *J. Endocrinol.,* **8**, 179–86.

Cole, T. J. (1985). A critique of the NCHS weight for height standard. *Hum. Biol.,* **57**, 183–96.

Colling-Saltin, A. S. (1980). Skeletal muscle development in the human foetus and during childhood. In *Children and Exercise IX,* ed. K. Berg & B. O. Eriksson, pp. 193–207. Baltimore: University Park Press.

Collipp, P. J., Curti, V., Thomas, J., Sharma, R. K., Maddaiah, V. T. & Cohn, S. H. (1973). Body composition changes in children receiving human growth hormone. *Metabolism,* **22**, 589–95.

Colliver, J. A., Stuart, F. & Arthur, F. (1983). Similarity of obesity indices in clinical studies of obese adults: a factor analysis. *Am. J. Clin. Nutr.,* **38**, 640–7.

Comroe, J. H., Forster, R. E., Dubois, A. B., Briscoe, W. A. & Carlsen, E. (1962). *The Lung: Clinical Physiology and Pulmonary Function Tests.* Chicago: Yearbook Medical Publishers.

Comstock, G. W. & Livesay, V. T. (1963). Subcutaneous fat determinations from a community-wide chest X-ray survey in Muscogee County, Georgia. *Ann. N.Y. Acad. Sci.,* **110**, 475–91.

Consolazio, C. F., Johnson, R. E. & Pecora, L. T. (1963). *Physiological Measurements of Metabolic Functions in Man* . New York: McGraw-Hill.

Conway, J. M., Norris, K. H. & Bodwell, C. E. (1984). A new approach for the estimation of body composition: infrared interactance. *Am. J. Clin. Nutr.,* **40**, 1123–30.

Coombs, G. F., DeWys, J., Iaconi, J. & Hubbard, V. S. (1986). Report on the second conference for federally supported human nutrition research units and centres. *Am. J. Clin. Nutr.,* **43**, 325–9.

Corbin, C. B. & Zuti, W. B. (1982). Body density and skinfold thicknesses of children. *Am. Corr. Therap. J.,* **36**, 50–5.

Correnti, V. & Zauli, B. (1964). *Olimpionici 1960.* Rome: Tipolitografia Marves.

Corsa, L., Olney, J. M., Steenberg, A. W., Bell, M. R. & Moore, F. D. (1950). The measurement of exchangeable potassium in man by isotope dilution. *J. Clin. Invest.,* **29**, 1280–95.

Craig, A. B. & Ware, D. E. (1967). Effect of immersion in water on vital capacity and residual volume of the lungs. *J. Appl. Physiol.,* **23**, 329–43.

Crenier, E. J. (1966). La prédiction du poids corporel 'normal'. *Biométr. Hum.,* **1**, 10–24.

Crisp, A. H., Palmer, R. L. & Kalucy, R. S. (1976). How common is anorexia nervosa? A prevalence study. *Br. J. Psychiatr.,* **128**, 549–54.

Crisp, A. H., Mayer, C. N. & Bhat, A. V. (1986). Pattern of weight gain in a group of patients treated for anorexia nervosa. *Int. J. Eating Disorders,* **5**, 1007–24.

Cronk, C. E. & Roche, A. F. (1982). Race- and sex-specific reference data for triceps and subscapular skinfolds and weight/stature^2. *Am. J. Clin. Nutr.,* **35**, 351–4.

Cronk, C. E., Roche, A. F., Kent, R., Berkey, C., Reed, R. B., Valadian, I., Eichorn, D. & McCammon, R. (1982). Longitudinal trends and continuity in weight/stature^2 from 3 months to 18 years. *Hum. Biol.,* **54**, 729–49.

Crook, G. H., Bennet, G. A., Norwood, W. D. & Mahaffey, J. A. (1966). Skinfold measurements and weight chart to measure body fat. *J. Amer. Med. Assoc.,* **198**, 39–44.

Cruz-Coke, R., Donoso, H. & Barrera, R. (1973). Genetic ecology of hypertension. *Clin. Sci. Mol. Med.,* **45**, 55s–65s.

Culebras, J. M. & Moore, F. D. (1977). Total body water and the exchangeable hydrogen. Theoretical calculations of nonaqueous exchangeable hydrogen in man. *Am. J. Physiol.,* **232**, R54–9.

Cumming, G. R., Garand, T. & Borysk, L. (1972). Correlation of performance in track and field events with bone age. *J. Pediatr.,* **80**, 970–3.

Cunningham, J., Calles, J., Bell, L., Jacob, R., Snyder, P., Loke, J. & Felig, P. (1983). Prediction of maximal oxygen consumption from muscle mass in healthy men and women. *Am. J. Clin. Nutr.,* **37**, 706 (Abstract).

Cureton, K. J. (1984). A reaction to the manuscript of Jackson. *Med. Sci. Sports Exerc.,* **16**, 621–2.

Cureton, K. J., Boileau, R. A. & Lohman, T. G. (1975). A comparison of densitometric, potassium-40 and skinfold estimates of body composition in prepubescent boys. *Hum. Biol.,* **47**, 321–36.

Dahlback, G. O. & Lundgren, C. E. (1972). Pulmonary air-trapping induced by water immersion. *Aerospace Med.,* **43**, 768–74.

Dalén, N. & Olsson, K. E. (1974). Bone mineral content and physical activity. *Acta Orthop. Scand.,* **45**, 170–4.

Dalén, N., Hallberg, D. & Lamke, B. (1975). Bone mass in obese subjects. *Acta Med. Scand.,* **197**, 353–5.

Dallaire, L. (1978). Facteurs héréditaires et environnementaux en rapport avec la croissance et le développement des enfants. In *Croissance et développement de l'enfant,* ed. H. Lavallée & R. J. Shephard, pp. 131–40. Trois Rivières: Université de Québec à Trois Rivières.

Daniell, H. W. (1976). Osteoporosis of the slender smoker. *Arch. Int. Med.,* **136**, 298–304.

Darby, P. L., Brown, G. M. & Garfinkel, P. (1983*a*). Patterns of prolactin secretion in Anorexia Nervosa. In *Anorexia Nervosa: Recent Developments in Research,* ed. P. L. Darby, P. E. Garfinkel, D. M. Garner & D. V. Coscina, pp. 279–84. New York: A. Liss.

Darby, P. L., Garfinkel, P. E., Garner, D. M. & Coscina, D. V. (1983*b*). *Anorexia Nervosa: Recent Developments in Research.* New York: A. R. Liss.

Darwin, C. (1859). *On the Origin of Species by Means of Natural Selection or the Preservation of Favoured Races in the Struggle for Life.* London: Watts & Co.

Datlof, S., Coleman, P. D., Forbes, G. B. & Kreipe, R. E. (1986). Ventricular dilatation on CAT scans of patients with anorexia nervosa. *Am. J. Psychiatr.,* **143,** 96–8.

Dauncey, M. J., Gandy, G. & Gairdner, D. (1977). Assessment of total body fat in infancy from skinfold thickness measurements. *Arch. Dis. Childh.,* **52,** 223–7.

Davidson, S., Passmore, R. Brock, J. F. & Truswell, A. S. (1979). *Human Nutrition and Dietetics,* 7th edn. Edinburgh: Churchill-Livingstone.

Davies, B. N. (1982). A method of deciding which formulae to use for the estimation of body fat from anthropometric measurements. *J. Physiol.,* **325,** 11P.

Davies, C. T. M. (1971). Body composition in children: a reference standard for maximum aerobic power output on a stationary bicycle ergometer. *Acta Paediatr. Scand.,* **217,** 136–7.

Davies, C. T. M. (1973). Relationship of maximum aerobic power output to productivity and absenteeism of East African sugar cane workers. *Br. J. Industr. Med.,* **30,** 146–54.

Davies, C. T. M., Fohlin, L. & Thorén, C. (1978*a*). Thermoregulation in anorexia patients. In *Pediatric Work Physiology,* ed. J. Borms & M. Hebbelinck, pp. 96–101. Basel: Karger.

Davies, C. T. M., Von Döbeln, W., Fohlin, L., Freyschuss, U. & Thorén, C. (1978*b*). Total body potassium, fat-free weight and maximal aerobic power in children with anorexia nervosa. *Acta Paediatr. Scand.,* **67,** 229–34.

Dean, R. F. A. (1965). Some effects of malnutrition on body composition. In *Human Body Composition. Approaches and Applications,* ed. J. Brozek, pp. 267–72. Oxford: Pergamon Press.

Deb, S., Martin, R. J. & Hershberger, T. V. (1976). Maintenance requirement and energetic efficiency of lean and obese Zucker rats. *J. Nutr.,* **106,** 191–7.

Debry, G. (1982). Nutrition: de la carence à la surcharge. In *Gérontologie: Biologie et clinique,* ed. F. Bourlière, pp. 191–212. Paris: Flammarion.

Dejours, P. (1987). What is a stressful environment? In *Adaptive Physiology to Stressful Environments,* ed. S. Samueloff & M. K. Yousef, pp. 11–16. Boca Raton: Chemical Rubber Company.

De Knecht, S. & Binkhorst, R. A. (1980). Physical characteristics of children with congenital heart disease: body characteristics and physical working capacity. In *Children and Exercise IX,* ed. K. Berg. & B. O. Eriksson, pp. 333–46. Baltimore: University Park Press.

DeLuise, M., Blackburn, G. L. & Flier, G. S. (1980). Reduced activity of the red-cell sodium-potassium pump in human obesity. *New Engl. J. Med.,* **303,** 1017–22.

DeLuise, M., Rappoport, E. & Flier, J. S. (1982). Altered erythrocyte Na^+-K^+ pump in adolescent obesity. *Metabolism,* **31,** 1153–8.

Delwaide, P. A. & Crenier, E. J. (1973). Body potassium as related to lean body mass measured by total water determination and by anthropometric method. *Hum. Biol.,* **45**, 509–26.

Dequecker, J. (1976). Quantitative radiology: radiogrammetry of cortical bone. *Br. J. Radiol.,* **49**, 912–20.

Deolalikar, A. B. (1984). *Are there Pecuniary Returns to Health in Agricultural Work? An Econometric Analysis of Agricultural Wages and Farm Productivity in Rural South India.* Economics Progress Report 66. Andra Pradesh, India: ICRISAT. Cited by Martorell & Arroyave (1988).

Deskins, W. J., Winter, D. C., Sheng, H. P. & Garza, C. (1985). Use of a resonating cavity to measure body volume. *J. Acoust. Soc. Am.,* **17**, 756–8.

Després, J. P., Bouchard, C., Tremblay, A., Savard, R. & Marcotte, M. (1985). Effects of aerobic training on fat distribution in male subjects. *Med. Sci. Sports Exerc.,* **17**, 113–18.

Devogelaer, J. P., Crabbé, J. & Deuxchaisnes, C. N. de. (1987). Bone mineral density in Addison's disease: evidence for an effect of adrenal androgens on bone mass. *Br. Med. J.,* **294**, 798–800.

Devor, E. J. & Crawford, M. H. (1984). Family resemblance for neuromuscular performance in a Kansas Mennonite community. *Am. J. Phys. Anthropol.,* **64**, 289–96.

Devor, E. J., McGue, M., Crawford, M. H. & Lin, P. M. (1986). Transmissible and non-transmissible components of anthropometric variation in the Alexanderwohl Mennonites. II. Resolution by path analysis. *Am. J. Phys. Anthropol.,* **69**, 83–92.

Dietrick, J. E., Whedon, G. D. & Shorr, E. (1948). Effects of immobilization upon various metabolic and physiologic functions of normal man. *Am. J. Med.,* **45**, 3–36.

DiGirolamo, M. (1986). Body composition. *Phys. Sports Med.,* **14** (3), 144–6.

DiGirolamo, M., Mendlinger, S. & Fertig, J. W. (1971). A simple method to determine fat cell size and number in four mammalian species. *Am. J. Physiol.,* **221**, 850–8.

Dill, D. B., Robinson, S. & Ross, J. C. (1967). A longitudinal study of 16 champion runners. *J. Sports Med.,* **7**, 4–27.

Di Pasquale, M. G. (1984). *Drug Use and Detection in Amateur Sports*, pp. 1–123. Warkworth, Ont.: M.G.D. Press.

Dixon, A. K. (1983). Abdominal fat assessed by computed tomography: sex differences in distribution. *Clin. Radiol.,* **34**, 189–191.

Dixon, T. (1984). Simple proton spectroscopic imaging. *Radiology,* **153**, 189–94.

Domeruth, W., Veum, T. L., Alexander, M. A., Hendrick, H. M., Clark, J. & Ekland. D. (1976). Prediction of lean body composition of live market weight swine by indirect methods. *J. Anim. Sci.,* **43**, 967–76.

Donaldson, C., Halley, S. B., Voge, J. M., Hattner, R. S., Bayers, J. H. & MacMillan, D. E. (1970). Effect of prolonged bedrest on bone mineral. *Metabolism,* **19**, 1071–84.

Donnelly, J. E. & Sintek, S. S. (1986). Hydrostatic weighing without head submersion. In *Perspectives in Kinanthropometry*, ed. J. A. P. Day, pp. 251–56. Champaign, Ill.: Human Kinetics Publishers.

Donnelly, J. E., Sintek, S. S., Anderson, J. T. & Pellegrino, L. (1984).

Hydrostatic weighing without head submersion. *Proceedings of 94th Nebraska Academy of Sciences,* p. 20. Lincoln: University of Nebraska.

Dossetor, J. B., Haystead, J., Howson, W. T., Lockwood, B., McConnachie, P. R., Schaefer, O., Smith, L. & Wilson, J. (1971). *The HL-A Antigens in Eskimos.* Annual Report 3, HA Project, Igloolik, NWT. Toronto, Ont.: Dept. Anthropology, University of Toronto.

Doty, S. B. & Morey-Holton, E. (1984). Alterations in bone forming cells due to reduced weight bearing. *Physiologist,* **27** (6), S80–1.

Douglas, J. W. B. & Simpson, H. R. (1964). Height in relation to puberty, family size and social class. *Millbank Memorial Fund Quarterly,* **40**, 20–35.

Doyle, F., Brown, J. & Lachance, C. (1970). Relation between bone mass and muscle weight. *Lancet,* (i), 391–3.

Drenick, E. J. (1967). Weight reduction with low-calorie diets. *J. Am. Med. Assoc.,* **202**, 136–38.

Drenick, E. J., Swendseid, M. E., Blahd, W. H. & Tuttle, S. G. (1964). Prolonged starvation as treatment for severe obesity. *J. Am. Med. Assoc.,* **187**, 100–5.

Dreyer, G., Ray, W. & Walker, E. W. A. (1912). The size of the trachea in warm blooded animals and its relationship to the weight, the surface area, the blood volume and the size of the aorta. *Proc. R. Soc. B.,* **86**, 56–65.

Drinkwater, D. T. & Ross, W. D. (1980). Anthropometric fractionation of body mass. In *Kinanthropometry* II, ed. M. Ostyn, G. Beunen & J. Simons, pp. 178–89. Baltimore: University Park Press.

Drinkwater, B. L., Nilson, K. L. & Chestnut, C. S. (1984). Bone mineral content of amenorrheic and eumenorrheic athletes. *New Engl. Med. J.,* **311**, 277–81.

Drinkwater, B. L., Nilson, K., Ott, S. & Chestnut, C. H. (1986a). Bone mineral density after resumption of menses in amenorrheic athletes. *J. Amer. Med. Assoc.,* **256**, 380–2.

Drinkwater, D. T., Martin, A. D., Ross, W. D. & Clarys, J. P. (1986b). Validation by cadaver dissection of Matiegka's equations for the anthropometric estimation of anatomical body composition in adult humans. In *Perspectives in Kinanthropometry,* ed. J. A. P. Day, pp. 221–7. Champaign, Ill.: Human Kinetics Publishers.

Dubois, J. (1972). Water and electrolyte content of human skeletal muscle. Variations with age. *Eur. J. Clin. Biol. Res.,* **17**, 505–15.

Dugdale, A. E., May, G. M. S. & O'Hara, V. M. (1980). Ethnic differences in the distribution of subcutaneous fat. *Ecol. Food Nutr.,* **10**, 19–23.

Dunn, W. L., Wahner, H. W. & Riggs, B. L. (1980). Measurement of bone mineral content in human vertebrae and hip by dual photon absorptiometry. *Radiology,* **136**, 485–7.

Durnin, J. V. G. A. (1961). Basic physiological factors affecting calorie balance. *Proc. Nutr. Soc.,* **20**, 52–8.

Durnin, J. V. G. A. (1965). Somatic standards of reference. In *Human Body Composition. Approaches and Applications,* ed. J. Brozek, pp. 73–84. Oxford: Pergamon Press.

Durnin, J. V. G. A. (1966). Age, physical activity and energy expenditure. *Proc. Nutr. Soc.,* **25**, 107–13.

Durnin, J. V. G. A. & Drummond, S. (1988). The role of working women in a rural environment when nutrition is marginally inadequate: problems of assessment. In *Capacity for Work in the Tropics,* ed. K. J. Collins & D. F. Roberts, pp. 77–84. Cambridge: Cambridge University Press.

Durnin, J. V. G. A. & Satwanti, K. (1982). Variations in the assessment of the fat content of the human body due to experimental technique in measuring body density. *Ann. Hum. Biol.,* **9,** 221–5.

Durnin, J. V. G. A. & Taylor, A. (1960). Replicability of measurements of density of the human body as determined by underwater weighing. *J. Appl. Physiol.,* **15,** 142–4.

Durnin, J. V. G. A. & Womersley, J. (1974). Body fat assessed from total body density and its estimation from skinfold thickness: measurements on 481 men and women aged from 16 to 72 years. *Br. J. Nutr.,* **32,** 77–97.

Durrant, M. L., Garrow, J. S., Royston, P. Stalley, S. F., Sunkin, S. & Warwick, P. M. (1980). Factors influencing the composition of the weight loss by obese patients on a reducing diet. *Br. J. Nutr.,* **44,** 275–86.

East, B. W., Boddy, K. & Price, W. H. (1976). Total body potassium content in males with X and Y chromosome abnormalities. *Clin. Endocrinol.,* **5,** 43–52.

Edelman, I. S., Olney, J. M., James, A. H., Brooks, L. & Moore, F. D. (1952). Body composition: studies in the human being by the dilution principle. *Science,* **115,** 447–54.

Edgerton, V. R. (1973). Exercise and growth of muscle tissue. In *Physical Activity, Human Growth and Development,* ed. G. L. Rarick, pp. 1–31. New York: Academic Press.

Edholm, O. G., Fletcher, J. G., Widdowson, E. M. & McCance, R. A. (1955). Energy expenditure and food intake of individual men. *Br. J. Nutr.,* **9,** 286–300.

Edholm, O. G., Humphrey, S., Lourie, J. A., Tredre, B. E. & Brotherhood, J. (1973). VI. Energy expenditure and food intake of individual men. *Br. J. Nutr.,* **9,** 286–300.

Edholm, O. G., Adam, J. M. & Best, T. W. (1974). Day-to-day weight changes in young men. *Ann. Hum. Biol.,* **3,** 3–12.

Edmonds, C. J., Smith, T., Griffiths, R. D., MacKenzie, J. & Edwards, R. H. T. (1985). Total body potassium and water and exchangeable sodium, in muscular dystrophy. *Clin. Sci.,* **68,** 379–85.

Edwards, D. A. W., Hammond, W. H., Healy, M. J. R., Tanner, J. M. & Whitehouse, R. H. (1955). Design and accuracy of calipers for measuring subcutaneous tissue thickness. *Br. J. Nutr.,* **2,** 133–43.

Egusa, G., Beltz, W. F., Grundy. S. M. & Howard, B. V. (1985). Influence of obesity on the metabolism of apolipoprotein B in humans. *J. Clin. Invest.,* **76,** 596–603.

Eiben, O. G. (1980). Recent data on variability in physique: some aspects of proportionality. In *Kinanthropometry II,* ed. M. Ostyn, G. Beunen & J. Simons, pp. 69–77. Baltimore: University Park Press.

Eiben, O. G., Sandor, G. Y. & Laszlo, J. (1977). The physique of patients suffering from Turner's syndrome. In *Growth and Development. Physique,* ed. O. G. Eiben, pp. 479–86. Budapest: Akademia Kiado.

Ekblöm, B. (1960). Effect of physical training on oxygen transport system in man. *Acta Physiol. Scand.,* Suppl. **328**, 1–45.

Elia, M., Carter, A. & Smith, R. (1979). The 3-methyl histidine content of human tissues. *Br. J. Nutr.,* **42**, 567–70.

Ellis, K. J. & Cohn, S. H. (1975). Correlation between skeletal calcium mass and muscle mass in man. *J. Appl. Physiol.,* **38**, 455–60.

Ellis, K. J., Shukla, K. K., Cohn, S. H. & Pierson, R. N. (1974). A predictor for total body potassium in man based on height, weight, sex and age: applications in metabolic disorders. *J. Lab. Clin. Med.,* **83**, 716–27.

Ellis, K. J., Vaswani, A., Zanzi, I. & Cohn, S. H. (1976). Total body sodium and chlorine in normal adults. *Metabolism,* **25**, 645–54.

Ellis, K. J., Yasumura, S., Vartsky, A. N. & Cohn, S. H. (1982). Total body nitrogen in health and disease: effects of age, weight, height and sex. *J. Lab. Clin. Med.,* **99**, 917–26.

Elson, D. (1965). Metabolism of nucleic acids (macromolecular DNA and RNA). *Ann. Rev. Biochem.,* **34**, 449–86.

Eriksson, A. W., Fellman, J., Forsius, H., Lehmann, W., Lewin, T. & Luukka, P. (1976). The origin of the Lapps in the light of recent genetic studies. in *Circumpolar Health,* ed. R. J. Shephard & S. Itoh, pp. 169–82. Toronto: University of Toronto Press.

Essén, B., Fohlin, L., Thorén, C. & Saltin, B. (1981). Skeletal muscle fiber types and sizes in anorexia nervosa. *Clin. Physiol.,* **1**, 395–403.

Etheridge, G. L. & Thomas, T. R. (1978). The effect of body position and water immersion on lung volumes of women. *Med. Sci. Sports Exerc.,* **10**, 61 (Abstract).

Evans, D. J., Hoffmann, R. G., Kalkhoff, R. K. & Kissebah, A. H. (1983). Relationship of androgenic activity to body fat topography, fat cell morphology, and metabolic aberrations in premenopausal women. *J. Clin. Endocrinol. Metab.,* **57**, 304–10.

Evans, D. J., Hoffman, R. G., Kalkhoff, R. K. & Kissebah, A. H. (1984). Relationship of body fat topography to insulin sensitivity and metabolic profiles in premenopausal women. *Metabolism,* **33**, 68–75.

Eveleth, P. B. (1979). Population differences in growth: environmental and genetic factors. In *Human Growth,* vol. 3, *Neurobiology and Nutrition,* ed. F. Falkner & J. M. Tanner, p. 381. London: Ballière, Tindall.

Eveleth, P. B. & Tanner, J. M. (1976). *Worldwide Variation in Human Growth.* Cambridge: Cambridge University Press.

Evers, S. E., Orchard, J. W. & Haddad, R. G. (1985). Bone density in post-menopausal North American Indian and Caucasian Females. *Hum. Biol.,* **57**, 719–726.

Exton-Smith, A. N., Millard, P. H., Payne, P. R. & Wheeler, E. F. (1969). Pattern of development and loss of bone with age. *Lancet,* (ii), 1154–7.

Ezrin, C., Godden, J. O., Volpe, R. & Wilson, R. (1973). *Systematic Endocrinology.* New York: Harper & Row.

Fairburn, C. G. & Garner, D. M. (1986). The diagnosis of bulimia nervosa. *Int. J. Eating Disorders,* **5**, 403–19.

Falk, J. R., Halmi, K. A. & Tyron, W. W. (1985). Activity measures in anorexia nervosa. *Arch. Gen. Psychiatr.,* **42**, 811–14.

Falkner, F. (1963). An air displacement method of measuring body volume in babies: a preliminary communication. *Ann. N.Y. Acad. Sci.*, **110**, 75–9.

Faller, I. L., Bond, E. E., Petty, D. & Pascale, L. R. (1955). The use of urinary deuterium oxide concentrations in a simple method for measuring total body water. *J. Lab. Clin. Med.*, **45**, 759–64.

Fanelli, M. R. & Kuczmarski, R. J. (1984). Ultrasound as an approach to assessing body composition. *Am. J. Clin. Nutr.*, **39**, 703–9.

Farr, V. (1966). Skinfold thickness as an indication of maturity of the newborn. *Arch. Dis. Childh.*, **41**, 301–8.

Faulhaber, J-D. (1974). Introduction to discussion. In *Lipid Metabolism, Obesity, and Diabetes Mellitus: Impact upon Atherosclerosis*, ed. H. Greten, R. Levine, E. F. Pfeiffer & A. E. Renold, pp. 83–8. Stuttgart: Georg Thieme.

Faust, I. M., Johnson, P. R. & Hirsch, J. (1977). Surgical removal of adipose tissue alters feeding behaviour and the development of obesity in rats. *Science*, **197**, 393–6.

Feitelberg, S., Epstein, S., Ismail, F. & D'Amanda, C. (1987). Deranged bone mineral metabolism in chronic alcoholism. *Metabolism*, **36**, 322–6.

Feldman, R., Sender, A. J., Siegelaub, A. B. & Oakland, M. S. (1969). Difference in diabetic and nondiabetic fat distribution patterns by skinfold measurements. *Diabetes*, **18**, 478–86.

Ferro-Luzzi, A. (1982). Meaning and constraints of energy-intake studies in free-living populations. In *Energy and Effort*, ed. G. A. Harrison, pp. 115–38. London: Taylor & Francis.

Ferro-Luzzi, A. (1988). Marginal energy malnutrition: some speculations on primary energy sparing mechanisms. In *Capacity for Work in the Tropics*, ed. K. J. Collins & D. F. Roberts, pp. 141–64. Cambridge: Cambridge University Press.

Fidanza, F., Keys, A. & Anderson, J. T. (1953). The density of body fat in man and other mammals. *J. Appl. Physiol.*, **6**, 252–6.

Fisler, J. S., Drenick, E. J., Blumfield, D. E. & Swendseid, M. E. (1982). Nitrogen economy during very low calorie reducing diets: quality and quantity of dietary protein. *Am. J. Clin. Nutr.*, **35**, 471–86.

Fitts, R. H. (1981). Aging and skeletal muscle. In *Exercise and Aging: The Scientific Basis*, ed. E. Smith & R. C. Serfass, pp. 31–44. Hillside, N.J.: Enslow Publishers.

Fletcher, D. & McNaughton, L. (1987). Three methods of assessing per cent body fat in elite cyclists. *J. Sports Med. Phys. Fitness*, **27**, 211–16.

Fletcher, R. F. (1962). The measurement of total body fat with skinfold calipers. *Clin. Sci.*, **22**, 333–46.

Flint, M. M., Drinkwater, B. L., Wells, C. L. & Horvath, S. M. (1977). Validity of estimating body fat of females: effect of age and fitness. *Hum. Biol.*, **49**, 559–72.

Flynn, M. A., Hanna, F. M. & Lutz, R. N. (1967). Estimation of body water compartments of preschool children. I. Normal children. *Am. J. Clin. Nutr.*, **20**, 1125–8.

Flynn, M. A., Woodruff, C. & Chase, G. (1972). Total body potassium in normal children. *Ped. Res.*, **6**, 239–45.

Fohlin, L. P. M. (1977). Exercise performance and body dimensions in anorexia nervosa before and after rehabilitation. *Acta Med. Scand.*, **204**, 61–5.

Fohlin, L. P. M. (1980). The effects of growth, body composition, and circulatory function of anorexia nervosa in adolescent patients. In *Children and Exercise* IX, ed. K. Berg & B. Eriksson, pp. 317–26. Baltimore: University Park Press.

Fohlin, L. P. M., Davies, C. T. M., Freyschuss, U., Bjarke, B. & Thorén, C. (1978). Body dimensions and exercise performance in anorexia nervosa patients. In *Pediatric Work Physiology,* ed. J. Borms & M. Hebbelinck, pp. 102–7. Basel: Karger.

Folk, G. E. (1974). *Textbook of Environmental Physiology*. Philadelphia: Lea & Febiger.

Fomon, S. J. (1978). *Nutritional Disorders of Children: Prevention, Screening and Follow-up*. Washington, DC: DHEW Publication HSA 78–5104.

Fomon, S. J., Jensen, R. L. & Owen, G. M. (1963). Determination of body volume of infants by a method of helium displacement. *Ann. N.Y. Acad. Sci.,* **110**, 80–90.

Fomon, S. J., Haschke, F., Ziegler, E. E. & Nelson, S. E. (1982). Body composition of reference children from birth to age 10 years. *Am. J. Clin. Nutr.,* **35**, 1169–75.

Forbes, G. B. (1964). Lean body mass and fat in obese children. *Pediatrics,* **34**, 308–14.

Forbes, G. B. (1974). Stature and lean body mass. *Am. J. Clin. Nutr.,* **27**, 595–602.

Forbes, G. B. (1978). Body composition in adolescence. In *Human Growth: An Advanced Treatise,* vol. II, ed. F. Falkner & J. Tanner, pp. 239–72. New York: Plenum Press.

Forbes, G. B. (1985). The effect of anabolic steroids on lean body mass: the dose response curve. *Metabolism,* **34**, 571–3.

Forbes, G. B. (1987). *Human Body Composition: Growth, Aging, Nutrition and Activity*. New York: Springer Verlag.

Forbes, G. B. & Amirhakimi, G. H. (1970). Skinfold thickness and body fat in children. *Hum. Biol.,* **42**, 401–18.

Forbes, G. B. & Bruining, G. J. (1976). Urinary creatinine excretion and lean body mass. *Am. J. Clin. Nutr.,* **29**, 1359–66.

Forbes, G. B. & Drenick, E. J. (1979). Loss of body nitrogen on fasting. *Am. J. Clin. Nutr.,* **32**, 1570–4.

Forbes, G. B. & Hursh, J. B. (1963). Age and sex trends in lean body mass calculated from K^{40} measurements, with a note on the theoretical basis for the procedure. *Ann. N.Y. Acad. Sci.,* **110**, 255–63.

Forbes, G. B. & McCoord, A. B. (1963). Changes in bone sodium during growth in the rat. *Growth,* **27**, 285–94.

Forbes, G. B. & McCoord, A. B. (1969). Long-term behaviour of radiosodium in bone: comparison with radiocalcium and effects of various procedures. *Calc. Tiss. Res. Internat.,* **4**, 113–28.

Forbes, G. B. & Perley, A. M. (1949). Determination of total body sodium in man with radiosodium. *J. Lab. Clin. Med.,* **34**, 1599–600.

Forbes, G. B. & Reina, J. C. (1970). Adult lean body mass declines with age: some longitudinal observations. *Metabolism,* **19**, 653–663.

Forbes, G. B., Kreipe, R. E., Lipinski, B. A. & Hodgman, C. H. (1984). Body composition changes during recovery from anorexia nervosa: comparison of two dietary regimes. *Am. J. Clin. Nutr.*, **40**, 1137–45.

Forbes, G. B., Brown, M. R., Welle, S. L. & Lipinski, B. A. (1986). Deliberate overfeeding in women and men: energy cost and composition of the weight gain. *Br. J. Nutr.*, **56**, 1–9.

Forbes, R. M., Cooper, A. R. & Mitchell, H. H. (1953). The composition of the adult human body as determined by chemical analysis. *J. Biol. Chem.*, **203**, 359–66.

Forster, R. E., Cohn, J. E., Briscoe, W. A., Blakemore, W. S. & Riley, R. L. (1955). A modification of the Krogh carbon monoxide breath-holding technique for estimating the diffusing capacity of the lung: a comparison with three other methods. *J. Clin. Invest.*, **34**, 1417–26.

Forsyth, H. L. & Sinning, W. E. (1973). The anthropometric estimation of body density and lean body weight of male athletes. *Med. Sci. Sports*, **5**, 174–80.

Forsyth, R., Plyley, M. J. & Shephard, R. J. (1988a). Reliability of circumferences versus skinfolds. Unpublished Report to Defence and Civil Institute of Environmental Medicine, Downsview, Out.

Forsyth, R., Plyley, M. & Shephard, R. J. (1988b). Residual volume as a tool in body fat prediction. *Ann. Nutr. Metab.*, **32**, 62–7.

Fortney, S. (1987). Thermoregulatory adaptations to inactivity. In *Adaptive Physiology to Stressful Environments*, ed. S. Samueloff & M. K. Yousef, pp. 75–83. Boca Raton: Chemical Rubber Company Press.

Foster, M. A., Hutchinson, J. M. S., Mallard, J. R. & Fuller, M. (1984). Nuclear magnetic resonance pulse sequence and discrimination of the high- and low-fat tissues. *Magnet. Reson. Imag.*, **2**, 187–92.

Frahm, J., Haase, A., Hänicke, W., Matthaei, D., Bomsdorf, H. & Helzel, T. (1985). Chemical shift selective MR imaging using a whole-body magnet. *Radiology*, **156**, 441–4.

Freedson, P., Sady, S., Katch, V., Reynolds, H. & Campaigne, B. (1979). Total body volume in females: validation of a theoretical model. *Hum. Biol.*, **51**, 499–505.

Freeman, S., Last, J. H., Petty, D. T. & Faller, I. L. (1955). *Total Body Water in Man: Adaptation of Measurement of Total Body Water to Field Studies.* Tech. Rep. EP-11. Natick, Mass.: Quartermaster Research and Development Center, Environmental Protection Division.

Freischlag, J. (1984). Weight loss, body composition, and health of high school wrestlers. *Phys. Sportsmed.*, **12** (1), 121–6.

Fried, T. & Shephard, R. J. (1970). Assessment of a lower extremity training programme. *Can. Med. Assoc. J.*, **103**, 260–6.

Friis-Hansen, B. (1961). Body water compartments in children: changes during growth and related changes in body composition. *Pediatrics*, **28**, 169–81.

Friis-Hansen, B. (1965). Hydrometry of growth and aging. In *Human Body Composition. Approaches and Applications*, ed. J. Brozek, pp. 191–209. Oxford: Pergamon Press.

Frisancho, A. R. (1978). Human growth and development among high-altitude populations. In *The Biology of High Altitude Peoples*, ed. P. T. Baker, pp. 117–71. Cambridge: Cambridge University Press.

Frisancho, A. R. (1981). New norms of upper limb fat and muscle areas for assessment of nutritional status. *Am. J. Clin. Nutr.*, **34**, 2540–5.

Frisancho, A. R. (1983). Perspectives on functional adaptation of the high altitude native. In *Hypoxia, Exercise and Altitude*, ed. J. R. Sutton, C. S. Houston & N. L. Jones. Proceedings of the Third Banff International Hypoxia Symposium, pp. 383–407. New York: Alan Liss.

Frisancho, A. R. (1984). New standards of weight and body composition by frame size and height for assessment of nutritional status of adults and the elderly. *Am. J. Clin. Nutr.*, **40**, 808–19.

Frisancho, A. R. & Flegel, P. N. (1982). Relative merits of old and new indices of body mass with reference to skinfold thickness. *Am. J. Clin. Nutr.*, **36**, 697–9.

Frisancho, A. R. & Flegel, P. N. (1983). Elbow breadth as a measure of frame size for US males and females. *Am. J. Clin. Nutr.*, **37**, 311–14.

Frisancho, A. R., Borkan, G. A. & Klayman, J. E. (1975). Patterns of growth of lowland and highland Peruvian Quechua of similar genetic composition. *Hum. Biol.*, **47**, 233–43.

Frisancho, A. R., Klayman, J. E. & Matos, J. (1977). Influence of maternal nutritional status on prenatal growth in a Peruvian urban population. *Am. J. Phys. Anthrop.*, **46**, 265–74.

Frisch, R. E. (1976). Fatness of girls from menarche to age 18 years, with a nomogram. *Hum. Biol.*, **48**, 353–9.

Frisch, R. E. (1983). Fatness and reproduction: delayed menarche and amenorrhea of ballet dancers and college athletes. In *Anorexia Nervosa: Recent Developments in Research*, ed. P. L. Darby, P. E. Garfinkel, D. M. Garner & D. V. Coscina, pp. 343–62. New York: A. Liss.

Frisch, R. E. & McArthur, J. W. (1974). Menstrual cycles: fatness as a determinant of minimum weight for height necessary for their maintenance or onset. *Science*, **185**, 949–51.

Frisch, R. E. & Revelle, R. (1970). Height and weight at menarche and a hypothesis of critical body weight and adolescent events. *Science*, **169**, 397–9.

Frohse, F. & Frankel, M. (1908). Die Muskeln des Menschlichen Armes. In *Handbuch der Anatomie des Menschen*, Tome II (2), ed. Karl von Bardeleben. Jena: G. Fischer.

Fryer, J. H. (1960). Specific gravity and total body water in men aged 60 or over. *Fed. Proc.*, **19**, 327.

Galton, D. J. (1966). An enzymatic defect in a group of obese patients. *Br. Med. J.*, (ii), 1498–500.

Gamble, J. L. (1962). The chloride space, arithmetic and deficit therapy. An evaluation of this parameter for estimating deficiencies of extracellular electrolytes in disease. *Pediatrics*, **30**, 990–4.

Gamble, J. L., Robertson, J. S., Hannigan, C. A., Foster, C. G. & Farr, L. E. (1953). Chloride, bromide, sodium and sucrose spaces in man. *J. Clin. Invest.*, **32**, 483–9.

Garfinkel, P. & Garner, D. M. (1982). *Anorexia Nervosa: A Multidimensional Perspective*. New York: Brunner & Mazel.

Garfinkel, P., Moldofsky, H. & Garner, D. (1980). The heterogeneity of anorexia nervosa. Bulimia as a distinct sub-group. *Arch. Gen. Psychiatr.*, **37**, 1036–40.

Garn, S. (1965). The applicability of North American growth standards in developing countries. *Can. Med. Assoc. J.*, **93**, 914–9.

Garn, S. M. (1956). Comparison of pinch caliper and X-ray measurements of skin plus subcutaneous fat. *Science*, **124**, 178–9.

Garn, S. M. (1957). Fat weight and fat placement in the female. *Science*, **12**, 1092–3.

Garn, S. M. (1961). Radiographic analysis of body composition. In *Techniques for Measuring Body Composition*, ed. J. Brozek & A. Henschel. Washington, DC: Nat. Acad. Sci. (Cited by Garn, 1963.)

Garn, S. M. (1963). Human biology and research in body composition. *Ann. N.Y. Acad. Sci.*, **110**, 429–46.

Garn, S. M. & Clark, D. C. (1976). Trends in fatness and the origin of obesity. *Pediatrics*, **57**, 443–56.

Garn, S. M. & LaVelle, M. (1983). Reproductive histories of low-weight girls and women. *Am. J. Clin. Nutr.*, **37**, 862–6.

Garn, S. M. & Nolan, P. (1963). A tank to measure body volume by water displacement (BOVOTA). *Ann. N.Y. Acad. Sci.*, **110**, 91–5.

Garn, S. M. & Pesick, S. D. (1982). Comparison of the Benn index and other body mass indices in nutritional assessment. *Am. J. Clin. Nutr.*, **36**, 573–5.

Garn, S. M. & Rohmann, C. G. (1966). Interaction of nutrition and genetics in timing of growth. *Pediatr. Clin. N. Amer.*, **13**, 353–79.

Garn, S. M. & Ryan, A. S. (1982). The effect of fatness on hemoglobin levels. *Am. J. Clin. Nutr.*, **36**, 189–91.

Garn, S. M., Poznanski, A. K. & Larson, K. (1976). Metacarpal lengths, cortical diameters and areas from the 10-state nutrition survey. In *Proceedings of the First Workshop on Bone Morphometry*, ed. C. F. G. Jaworski, pp. 367–91. Ottawa: University of Ottawa Press.

Garn, S. M., Cole, P. E. & Bailey, S. M. (1979). Living together as a factor in family line resemblances. *Hum. Biol.*, **51**, 565–87.

Garn, S. M., Ryan, A. S. & Robson, J. R. K. (1982). Fatness dependence and utility of the subscapular/triceps ratio. *Ecol. Food Nutr.*, **12**, 173–7.

Garn, S. M., Sullivan, T. V. & Hawthorne, V. M. (1987). Differential rates of fat change relative to weight change at different body sites. *Int. J. Obesity*, **11**, 519–26.

Garn, S. M., Sullivan, T. V. & Hawthorne, V. M. (1988). Persistence of relative fatness at different body sites. *Hum. Biol.*, **60**, 43–53.

Garner, D. M. (1985). Iatrogenesis in anorexia nervosa and bulimia nervosa. *Int. J. Eating Disorders*, **4**, 701–26.

Garrison, R. J., Feinleib, M., Castelli, W. P. & McNamara, P. M. (1983). Cigarette smoking as a confounder of the relationship between relative weight and long-term mortality: the Framingham study. *J. Amer. Med. Assoc.*, **249**, 2199–203.

Garrow, J. S. (1978). *Energy Balance and Obesity in Man*, 2nd edn. Amsterdam: North Holland Publishers.

Garrow, J. S. (1982). New approaches to body composition. *Am. J. Clin. Nutr.*, **35**, 1152–8.

Garrow, J. S. (1983). Indices of adiposity. *Nutr. Abs. Rev. Clin. Nutr.*, Series A, **53**, 697–708.

Garrow, J. S. & Warwick, P. M. (1978). Diet and obesity. In *The Diet of Man: Needs and Wants*, ed. J. Yudkin, pp. 127–44. Barking: Applied Science Publishers.

Garrow, J. S., Stalley, S., Diethelm, R., Pittet, P. H., Hesp, R. & Halliday, R. (1979). A new method for measuring body density of obese adults. *Br. J. Nutr.*, **42**, 173–83.

Genant, H. K., Cann, C. E. & Faul, D. D. (1982). Quantitative computed tomography for assessing vertebral bone mineral. In *Non-invasive Bone Measurements: Methodological Problems*, ed. J. Dequecker & C. C. Johnston, pp. 215–49. Oxford: IRL Press.

Gerven, D. P. van (1972). The contribution of size and shape variation to patterns of sexual dimorphism of the human femur. *Am. J. Phys. Anthrop.*, **37**, 49–60.

Ghesquière, J. L. A. (1971). Physical development and working capacity of Congolese. In *Human Biology and Environmental Change*, ed. R. Vorster, Malawi: International Biological Programme.

Ghesquière, J. L. A. & Eekels, R. (1984). Fitness of children in Kinshasa. In *Children and Sport*, ed. J. Ilmarinen & I. Välimäki, pp. 18–30. Berlin: Springer Verlag.

Ghesquière, J. L. A. & D'Hulst, C. (1988). Growth, stature and fitness of children in tropical areas. In *Capacity for Work in the Tropics*, ed. K. J. Collins & D. F. Roberts, pp. 165–80. Cambridge: Cambridge University Press.

Girandola, R. M., Wiswell, R. A., Mohler, J. G., Romero, G. T. & Barnes, W. S. (1977a). Effects of water immersion on lung volumes: implications for body composition analysis. *J. Appl. Physiol.*, **43**, 276–9.

Girandola, R. M., Wiswell, R. A. & Romero, G. T. (1977b). Body composition changes resulting from fluid ingestion and dehydration. *Res. Quart.*, **48**, 299–303.

Glick, Z. & Schvartz, E. (1974). Physical working capacity of young men of different ethnic groups in Israel. *J. Appl. Physiol.*, **37**, 22–6.

Glunky, P. (1983). Aging America renews its interest in osteoporosis. *Arch. Int. Med.*, **143**, 2055 (Note).

Gnaedinger, R. H., Reineke, E. P., Pearson, A. M., Van Huss, W. D., Wessel, J. A. & Montoye, H. J. (1963). Determination of body density by air displacement, helium dilution and underwater weighing. *Ann. N.Y. Acad. Sci.*, **110**, 96–108.

Godin, G. & Shephard, R. J. (1973). Body weight and the energy cost of activity. *Arch. Environ. Health*, **27**, 289–93.

Goldberg, A. L. & Goodman, H. M. (1969). Relationship between growth hormone and muscular work in determining muscle size. *J. Physiol.*, **200**, 655–66.

Goldberg, S. C., Halmi, K. A., Casper, R., Eckert, E. & Davis, J. M. (1977). Pretreatment predictors of weight change in anorexia nervosa. In *Anorexia Nervosa*, ed. R. A. Vigersky, pp. 31–41. New York: Raven Press.

Goldman, J. K., Bernardis, L. L. & Frohman, L. A. (1974). Food intake in hypothalamic obesity. *Am. J. Physiol.*, **227**, 88–91.

Goldman, R. F., Bullen, B. & Seltzer, C. (1963). Changes in specific gravity and body fat in overweight female adolescents as a result of weight reduction. *Ann. N.Y. Acad. Sci.,* **110**, 913–17.

Goldspink, G. (1972). Post-embryonic growth and differentiation of striated muscle. In *The Structure and Function of Muscle,* I *Structure,* Part I, ed. G. H. Bourne, pp. 179–236. New York: Academic Press.

Goldspink, G. & Howells, K. F. (1974). Work-induced hypertrophy in exercised normal muscles of different ages and the reversibility of hypertrophy after cessation of exercise. *J. Physiol.,* **239**, 179–93.

Gonyea, W. J. (1980). Muscle fiber splitting in trained and untrained animals. *Ex. Spts. Sci. Rev.,* **8**, 19–39.

Goodford, P. J. & Leach, E. H. (1966). The extracellular space of the smooth muscle of the guinea pig taenia coli. *J. Physiol.,* **186**, 1–10.

Göranzon, H. & Forsum, E. (1985). Effect of reduced energy intake versus increased physical activity on the outcome of nitrogen balance experiments in man. *Am. J. Clin. Nutr.,* **41**, 919–28.

Gotfredsen, A., Jensen, J., Borg, J. & Christiansen, C. (1986). Measurement of lean body mass and total body fat using dual photon absorptiometry. *Metabolism,* **35**, 88–93.

Graitcer, P. L. & Gentry, M. (1981). Measuring children: one reference. *Lancet,* (ii), 297–9.

Grand, T. I. (1977). Body weight: its relation to tissue composition, segment distribution and motor function. *Am. J. Phys. Anthrop.,* **45**, 101–8, 211–40.

Grande, F. (1961). In *Techniques for Measuring Body Composition,* ed. J. M. Brozek & A. Henschel, pp. 168–88. Washington, DC: US National Academy of Sciences – National Research Council.

Grant, J. P., Custer, P. B. & Thurlaw, J. (1981). Current techniques of nutritional assessment. *Surg. Clin. N. Am.,* **61**, 437–63.

Grant, M. W. (1964). Rate of growth in relation to birth rank and family size. *Br. J. Prev. Soc. Med.,* **18**, 35–42.

Grauer, W. O., Moss, A. A., Cann, C. E. & Goldberg, H. I. (1984). Quantification of body fat distribution in the abdomen using computed tomography. *Am. J. Clin. Nutr.,* **39**, 631–7.

Gray, G. E. & Gray, L. K. (1980). Anthropometric measurements and their interpretation: principles, practices and problems. *J. Am. Diet. Association.,* **77**, 534–9.

Green, H. (1986). Morphological, ultrastructural, and functional characteristics of the aging human skeletal muscle. In *Sports Medicine for the Mature Athlete,* ed. J. R. Sutton & R. M. Brock, pp. 17–26. Indianapolis: Benchmark Press.

Greene, J. A. (1939). Clinical study of the etiology of obesity. *Ann. Int. Med.,* **12**, 1797–803.

Greenleaf, J. E., Bernauer, E. M., Juhos, L. T., Young, H. L., Morse, J. T. & Staley, R. W. (1977). Effects of exercise on fluid balance and body composition of man during 14-day bed rest. *J. Appl. Physiol.,* **43**, 126–32.

Greenwald, A. J., Allen, T. H. & Bancroft, R. W. (1969). Abdominal gas volume at altitude and at ground level. *J. Appl. Physiol.,* **26**, 177–81.

Greenway, R. M., Houser, H. B., Lindau, O. & Weir, D. R. (1970). Long term changes in gross body composition of paraplegic and quadriplegic patients. *Paraplegia,* **7**, 301–8.

Greer, F. R., Searcy, J. E., Levin, S. R., Steichen, J. J., Asch, P. S. & Tsang, R. C. (1981). Bone mineral content and serum 25-hydroxyvitamin D concentration in breast-fed infants with and without supplemental vitamin D. *J. Pediatr.*, **98**, 696–701.

Greer, F. R., Lane, J., Weiner, S. & Mazess, R. B. (1983). An accurate and reproducible absorptiometric technique for determining bone mineral content in newborn infants. *Pediatr. Res.*, **17**, 259–62.

Greksa, L. P. (1988). Effect of altitude on the stature, chest depth and forced vital capacity of low to high-altitude migrant children of European ancestry. *Hum. Biol.*, **60**, 23–32.

Greksa, L. P. & Baker, P. T. (1982). Aerobic capacity of modernizing Samoan man. *Hum. Biol.*, **54**, 777–99.

Greksa, L. P., Spielvogel, H., Paredes-Fernandez, L., Paz-Zamora, M. & Caceres, E. (1984). The physical growth of urban children at high altitude. *Am. J. Phys. Anthrop.*, **65**, 315–22.

Greksa, L. P., Spielvogel, H. & Carceres, E. (1985). Effect of altitude on the physical growth of upper class children of European ancestry. *Ann. Hum. Biol.*, **12**, 225–32.

Griffith, E. R., Stonebridge, J. B., Piernick, D. & Lehman, J. F. (1973). Development of a method of X-ray densitometry for bone mineral measurement. *Am. J. Phys. Med.*, **52**, 128–49.

Griggs, R. C., Forbes, G. B., Moxley, R. T. & Herr, B. E. (1983). The assessment of muscle mass in progressive neuromuscular disease. *Neurology*, **33**, 158–65.

Grigoriev, A. I. & Kozlovskaya, I. B. (1988). Physiological responses of skeletomuscular system to muscle exercises under long-term hypokinetic conditions. *Physiologist*, **31**, S93–7.

Gruner, O. & Salmen, A. (1961). Vergleichende Korperwasserbestimmungen mit Hilfe von N-acetyl-4-aminoantipyrin und alkohol. *Klin. Wschr.*, **39**, 92–7.

Gryfe, C. I., Exton-Smith, A. N., Payne, P. R. & Wheeler, E. F. (1971). Pattern of development of bone in childhood and adolescence. *Lancet*, (i), 523–6.

Gundlach, B. L. & Visscher, G. J. W. (1986). The plethysmographic measurement of total body volume. *Hum. Biol.*, **58**, 783–99.

Gundlach, B. L., Nijikrake, H. G. M. & Hautvast, J. G. A. J. (1980). A rapid and simplified plethysmographic method for measuring body volume. *Hum. Biol.*, **52**, 23–33.

Gunther, B. (1975). Dimensional analysis and the theory of biological similarity. *Physiol. Rev.*, **55**, 659–99.

Guo, S., Roche, A. F., Chumlea, W. C., Miles, D. S. & Pohlman, R. L. (1987). Body composition predictions from impedance. *Hum. Biol.*, **59**, 221–33.

Gurney, S. M. & Jelliffe, D. B. (1973). Arm anthropometry in nutritional assessment: nomogram for rapid calculation of muscle-circumference and cross-sectional muscle and fat areas. *Am. J. Clin. Nutr.*, **26**, 912–15.

Gurr, M. I. & Kirtland, J. (1978). Adipose tissue cellularity: a review. 1. Techniques for studying cellularity. *Int. J. Obesity*, **2**, 401–27.

Gurr, M. I., Jung, R. T., Robinson, M. P. & James, W. P. T. (1982). Adipose tissue cellularity in man: the relationship between fat cell size and number:

the mass and distribution of body fat and the history of weight gain and loss. *Int. J. Obesity,* **6**, 419–36.

Gutmann, E. (1976). Neurotrophic relations. *Ann. Rev. Physiol.,* **38**, 177–216.

Gutmann, E. (1977). Muscle. In *Handbook of the Biology of Aging,* ed. C. Finch & L. Hayflick, New York: Van Nostrand, Reinhold.

Guyton, A. C., Harris, J. G. & Taylor, A. E. (1971). Interstitial fluid pressure. *Physiol. Rev.,* **51**, 527–63.

Gwinup, G. (1975). Effect of exercise alone on the weight of obese women. *Arch. Int. Med.,* **135**, 676–80.

Gwinup, G., Chelvam, R. & Steinberg, T. (1971). Thickness of subcutaneous fat and activity of underlying muscles. *Ann. Int. Med.,* **74**, 408–11.

Habicht, J-P. (1980). Some characteristics of indicators of nutritional status for use in screening and surveillance. *Am. J. Clin. Nutr.,* **33**, 531–5.

Habicht, J-P., Martorell, R., Yarbrough, C., Malina, R. M. & Klein, R. F. (1974). Height and weight standards for preschool children: how relevant are ethnic differences to growth potential? *Lancet,* (i), 611–15.

Hackett, P. H., Reeves, J. T., Reeves, C. D., Grover, R. F. & Rennie, D. (1980). Control of breathing in Sherpas at low and high altitude. *J. Appl. Physiol. Respirat. Exerc. Environ. Physiol.,* **49**, 374–9.

Hackney, A. C. & Deutsch, D. T. (1985). Accuracy of residual volume prediction – effects on body composition estimation in pulmonary dysfunction. *Can. J. Appl. Spt. Sci.,* **10**, 88–93.

Hagan, R. D., Upton, S. J., Wong, L. & Whittam, J. (1986). The effects of aerobic conditioning and/or caloric restriction in overweight men and women. *Med. Sci. Sports Exerc.,* **18**, 87–94.

Hager, A. (1981). Estimation of body fat in infants, children and adolescents. In *Adipose Tissue in Childhood,* ed. F. P. Bonnet, pp. 49–56. Boca Raton: CRC Press.

Hager, A., Sjöstrom, L., Arvidsson, B., Björntorp, P. & Smith, U. (1977). Body fat and adipose tissue cellularity in infants. A longitudinal study. *Metabolism,* **26**, 607–14.

Haggmark, T., Jansson, E. & Svane, B. (1978). Cross-sectional area of the thigh muscle in man measured by computed tomography. *Scand. J. Clin. Lab. Invest.,* **38**, 355–60.

Hall, A. (1987). Anorexia, bulimia and obesity. In *Are we Really What We Eat? Food Choice as a Basis for Better Health,* ed. J. Birkbeck, pp. 96–104. Auckland: Dairy Advisory Bureau.

Hall, D. A. (1973). Metabolic and structural aspects of aging. In *Textbook of Geriatric Medicine and Gerontology,* ed. J. C. Brocklehurst, pp. 17–45. Edinburgh: Churchill-Livingstone.

Hall, L. K. (1977). *Anthropometric estimations of body density of women athletes in selected activities.* Columbus, Ohio: Ohio State University, Ph.D. Thesis.

Halle (1797). Cited by Sheldon (1963).

Halmi, K. A., Goldberg, S. C., Casper, R., Eckert, E. & Davis, J. M. (1979). Pretreatment predictors in outcome in anorexia nervosa. *Br. J. Psychiatr.,* **143**, 71–8.

Hammel, H. T. (1964). Terrestrial animals in cold: recent studies of primitive man. In *Adaptation to the Environment,* ed. D. B. Dill, pp. 413–34. Washington, DC: American Physiological Society.

Hammond, W. H. (1955). Measurement and interpretation of subcutaneous fat, with norms for children and young males. *Br. J. Prev. Soc. Med., 9,* 201–11.

Hampton, M. C., Huenemann, R. L., Shapiro, L. R., Mitchell, B. W. & Behnke, A. R. (1966). A longitudinal study of gross body composition and body conformation and their association with food and activity in a teen-age population. *Am. J. Clin. Nutr., 19,* 422–35.

Hankin, M. E., Theile, H. M. & Steinbeck, A. W. (1972). Cortisol and aldosterone excretion and plasma cortisol concentrations in normal and obese female subjects. *Clin. Sci., 43,* 289–98.

Hansson, T. H., Roos, B. O. & Nachemson, A. (1975). Development of osteopenia in the fourth lumbar vertebra during prolonged bed rest after operation for scoliosis. *Acta Orthop. Scand., 46,* 621–30.

Hanzlickova, V. & Gutmann, E. (1975). Ultrastructural changes in senile muscle. *Adv. Exp. Med. Biol., 53,* 421–9.

Harris, C. I. (1981). Reappraisal of the quantitative importance of non-skeletal muscle source of N-methyl histidine in urine. *Biochem. J., 194,* 1011–14.

Harrison, G. A., Weiner, J. S., Tanner, J. M. & Barnicott, N. A. (1964). *Human Growth. An Introduction to Human Evolution, Variation and Growth.* Oxford: Clarendon Press.

Harrison, G. A., Kuchemann, C. F., Moore, M. A. S., Boyce, A. J., Baju, T., Mourant, A. E., Godber, M. J., Glasgow, B. G., Kopec, A. C., Tills, D. & Clegg, E. J. (1969). The effects of altitudinal variation in Ethiopian populations. *Phil. Trans. R. Soc. Lond., 256B,* 147–82.

Harrison, G. G. (1987). The measurement of total body electrical conductivity. *Hum. Biol., 59,* 311–17.

Harrison, J. E., Williams, W. C., Watts, J. & McNeill, K. G. (1975). A bone calcium index based on partial body calcium measurements by in vivo activation analysis. *J. Nucl. Med., 16,* 116–22.

Harrison, J. E., McNeill, K. G., Hitchman, A. J. & Britt, B. A. (1979). Bone mineral measurements of the central skeleton by in vivo neutron activation analysis for routine investigation of osteopenia. *Invest. Radiol., 14,* 27–34.

Harsha, D. W., Voors, A. W. & Berenson, G. S. (1980). Racial differences in subcutaneous fat patterns in children aged 7–15 years. *Am. J. Phys. Anthrop., 53,* 333–7.

Harvald, B. J. (1976). Current genetic trends in the Greenlandic population. In *Circumpolar Health,* ed. R. J. Shephard & S. Itoh, pp. 164–9. Toronto: University of Toronto Press.

Harvey, J. S. (1986). Overuse syndrome in young athletes. In *Sport for Children and Youths,* ed. M. R. Weiss & D. Gould, pp. 151–63. Champaign, Ill.: Human Kinetics Publishers.

Harvey, T. C., James, H. M. & Chettle, D. R. (1979). Birmingham Medical Research Expeditionary Society 1977 Expedition: effect of a Himalayan trek on whole-body composition, nitrogen and potassium. *Postgrad. Med. J., 55,* 475–7.

Haschke, F. (1983). Part I: Total body water in normal adolescent males. Part II: Body composition of the male reference adolescent. *Acta Paediatr, Scand., 307* (Suppl.), 1–23.

Hassi, J. (1977). *The Brown Adipose Tissue in Man.* Acta Universitatis Oulensis, Series D 21. Sweden: University of Oulu.

Hatfield, H. S. & Pugh, L. G. C. E. (1951). Thermal conductivity of human fat and muscle. *Nature* (Lond.), **168**, 918–19.

Hauspie, R., Susanne, C. & Alexander, F. (1977). Growth and maturation in children with chronic asthma. In *Growth and Development. Physique,* ed. O. G. Eiben, pp. 203–10. Budapest: Akademia Kiado.

Haverburg, L. N., Omstedt, P. T., Munro, H. N. & Young, V. R. (1975). N-methyl histidine content of mixed proteins in various rat tissues. *Biochim. Biophys. Acta,* **405**, 67–71.

Hawk, L. J. & Brook, C. G. D. (1979). Family resemblances of height, weight and body fatness. *Arch. Dis. Childh.,* **54**, 877–9.

Hawkins, T. & Goode, A. W. (1976). The determination of total body potassium using a whole-body monitor. *Phys. Med. Biol.,* **21**, 293–7.

Hawkins, W. W. (1964). Iron, copper and cobalt. In *Nutrition: A Comprehensive Treatise,* vol. I, ed. G. H. Beaton & E. W. McHenry, pp. 309–72. New York: Academic Press.

Hayes, P. A. (1986). Physiological aspects of survival clothing. *Joint International Conference on Escape, Survival and Rescue at Sea, Heathrow, London, 29–31 October, 1986.*

Hayes, P. A. (1988). The physiological basis for the development of immersion protective clothing. In *Environmental Ergonomics,* ed. I. B. Mekjavic, E. W. Banister & J. B. Morrison, pp. 221–39. London: Taylor & Francis.

Hayes, P. A., Smith, F. W. & Sowood, P. J. (1985). Distribution and quantification of human body fat using nuclear magnetic resonance (NMR) imaging: comparison with a skinfold caliper method. *J. Physiol.,* **369**, 160P.

Hayes, P. A., Sowood, P. J., Belyavin, A., Cohen, J. B. & Smith, F. W. (1988). Sub-cutaneous fat thickness measured by magnetic resonance imaging, ultrasound and calipers. *Med. Sci. Sports Exerc.,* **20**, 303–9.

Haymes, E. M. (1988). Temperature and exercise. In *Sport Science Perspectives for Women,* ed. J. Puhl, C. H. Brown & R. O. Voy, pp. 37–47. Champaign, Ill.: Human Kinetics Publishers.

Haymes, E. M., Lundegren, H. M., Loomis, J. L. & Buskirk, E. R. (1976). Validity of the ultrasonic technique as a method of measuring subcutaneous adipose tissue. *Ann. Hum. Biol.,* **3**, 245–51.

Heald, F. P., Hunt, E. E., Schwartz, R., Cook, C., Elliott, O. & Vajda, B. (1963). Measures of body fat and hydration in adolescent boys. *Pediatrics,* **31**, 226–39.

Heaney, R. P. (1986). Calcium intake, bone health, and aging. In *Nutrition, Aging and Health,* ed. E. A. Young, pp. 165–86. New York: A. R. Liss.

Heaney, R. P., Recker, R. R. & Saville, P. D. (1977). Calcium balance and calcium requirements in middle-aged women. *Am. J. Clin. Nutr.,* **30**, 1603–11.

Heaney, R. P., Recker, R. R. & Saville, P. D. (1978). Menopausal changes in bone remodelling, *J. Lab. Clin. Med.,* **92**, 964–70.

Heaney, R. P., Gallagher, J. C., Johnston, C. C., Neer, R., Parfitt, A. M. & Whedon, G. D. (1982). Calcium nutrition and bone health in the elderly. *Am. J. Clin. Nutr.,* **36**, 986–1013.

Heaton, J. M. (1972). The distribution of brown adipose tissue in the human. *J. Anat.*, **112**, 35–9.

Hebbelinck, M., Duquet, W. & Ross, W. D. (1973). A practical outline for the Heath-Carter somatotyping method applied to children. In *Pediatric Work Physiology*, ed. O. Bar-Or, pp. 71–9. Natanya, Israel: Wingate Institute.

Heldmaier, G. (1974). Temperature adaptation and brown adipose tissue in hairless and albino mice. *J. Comp. Physiol.*, **92**, 281–92.

Helenius, M. Y. T., Albanese, D., Micozzi, M. S., Taylor, P. R. & Heinonen, O. P. (1987). Studies of bioelectric resistance in overweight, middle-aged subjects. *Hum. Biol.*, **59**, 271–9.

Héroux, O. (1963). Patterns of morphological, physiological and endocrinological adjustments under different environmental conditions of cold. *Fed. Proc.*, **22**, 789–94.

Hetherington, A. W. & Ranson, S. W. (1942). The spontaneous activity and food intake of rats with hypothalamic lesions. *Am. J. Physiol.*, **136**, 609–17.

Hewitt, D. (1958). Sib resemblance in bone, muscle and fat measurements of the human calf. *Ann. Hum. Genet.*, **22**, 213–21.

Heymsfield, S. B. & McManus, C. B. (1985). Tissue components of weight loss in cancer patients. *Cancer*, **55**, (Suppl.), 238–49.

Heymsfield, S. B., Olafson, R. P., Kutner, M. H. & Nixon, D. W. (1979). A radiographic method of quantifying protein-calorie undernutrition. *Am. J. Clin. Nutr.*, **32**, 693–702.

Heymsfield, S. B., Stevens, V., Noel, R., McManus, C., Smith, J. & Nixon, D. (1982*a*). Biochemical composition of muscle in normal and semistarved human subjects: relevance to anthropometric measurements. *Am. J. Clin. Nutr.*, **36**, 131–42.

Heymsfield, S. B., McManus, C., Smith, J., Stevens, V. & Nixon, D. W. (1982*b*). Anthropometric measurement of muscle mass: revised equations for calculating bone-free arm muscle area. *Am. J. Clin. Nutr.*, **36**, 680–90.

Heymsfield, S. B., Arteaga, C., McManus, C., Smith, J. & Moffitt, S. (1983). Measurement of muscle mass in humans: validity of the 24 hour urinary creatinine method. *Am. J. Clin. Nutr.*, **37**, 478–94.

Heywood, P. F. (1974). *Malnutrition and productivity in Jamaican sugar cane cutters*. Ithaca: Cornell University, Ph.D. Dissertation.

Hiernaux, J. (1966). Peoples of Africa from 22° N to the equator. In *The Biology of Human Adaptability*, ed. P. T. Baker & J. S. Weiner, pp. 91–110. Oxford: Clarendon Press.

Hiernaux, J. (1977). Long-term biological effects of human migration from the African savanna to the equatorial forest: a case study of human adaptation to a hot and wet climate. In *Population Structure and Human Variation*, ed. G. A. Harrison, pp. 187–217. Cambridge: Cambridge University Press.

Hill, G. L., McCarthy, I. D., Collins, J. P. & Smith, A. H. (1978). A new method for the rapid measurement of body composition in critically ill surgical patients. *Br. J. Surg.*, **65**, 732–5.

Hill, G. L., Bradley, J. A. & Smith, R. C. (1979). Changes in body weight and body protein with intravenous nutrition. *J. Parent. Ent. Nutr.*, **3**, 215–18.

Himes, J. H. & Bouchard, C. (1985). Do the new Metropolitan Life Insurance weight–height tables correctly assess body frame and body fat relationships? *Am. J. Publ. Health*, **75**, 1076–9.

Himes, J. G., Roche, A. F. & Siervogel, R. M. (1979). Compressibility of skinfolds and the measurement of subcutaneous tissue thickness. *Am. J. Clin. Nutr.*, **32**, 1734–40.

Hines, H. M. & Knowlton, G. C. (1937). Electrolyte and water changes in muscle during atrophy. *Am. J. Physiol.*, **120**, 719–23.

Hirata, K-I. (1979). *Selection of Olympic Champions*. Santa Barbara: Institute of Environmental Stress, University of California.

Hirsch, J. & Gallian, E. (1968). Methods for the determination of adipose cell size in man and animals. *J. Lipid Res.*, **9**, 110–19.

Hirsch, J. & Goldrick, R. B. (1964). Lipogenesis and free fatty acid uptake and release in small aspirated samples of subcutaneous fat. *J. Clin. Invest.*, **43**, 1776–92.

Hodgdon, J. A. & Fitzgerald, P. I. (1987). Validity of impedance predictions at various levels of fatness. *Hum. Biol.*, **59**, 281–98.

Hoffer, E. C., Meador, C. K. & Simpson, D. C. (1969). Correlation of whole body impedance with total body water volume. *J. Appl. Physiol.*, **27**, 531–4.

Hollander, W., Chobanian, A. V. & Burrows, B. A. (1961). Body fluid and electrolyte composition in critically ill surgical patients. *Br. J. Surg.*, **65**, 732–5.

Hong, S. K., Rennie, D. W. & Park, Y. S. (1986). Cold acclimatization and deacclimatization of Korean diving women. *Exerc. Spt. Sci. Rev.*, **14**, 231–68.

Hoppeler, H., Howald, H., Conley, K., Lindstedt, S. L., Claasen, H., Vock, P. & Weibel, E. R. (1985). Endurance training in humans: aerobic capacity and structure of skeletal muscle. *J. Appl. Physiol.*, **59**, 320–7.

Horber, F. F., Scheidegger, J. R., Grünig, B. E. & Frey, F. J. (1985). Evidence that prednisone-induced myopathy is reversed by physical training. *J. Clin. Endocrinol. Metab.*, **61**, 83–8.

Horsman, A. & Simpson, M. (1975). The measurement of sequential changes in cortical bone geometry. *Br. J. Radiol.*, **48**, 471–6.

Houdas, Y. & Ring, E. F. J. (1982). *Human Body Temperature. Its Measurement and Regulation*. New York: Plenum Press.

Houston, M. E., Marrin, D. A., Green, H. J. & Thomson, J. A. (1981). The effect of rapid weight loss on physiological functions in wrestlers. *Phys. Sportsmed.*, **9** (11), 73–8.

Howell, J. A. (1917). An experimental study of the effect of stress and strain on bone development. *Anat. Rec.*, **13**, 233–52.

Howell, R. W. (1971). Obesity and smoking habits. *Br. Med. J.*, **4**, 625.

Howells, W. W. (1966). Variability in family lines vs population variability. *Ann. N.Y. Acad. Sci.*, **134**, 624–31.

Hsieh, S., Kline, G., Porcari, J. & Katch, F. I. (1985). Measurement of residual volume sitting and lying in air and water (and during underwater weighing) and its effects on computed body density. *Med. Sci. Sports Exerc.*, **17**, 204.

Hubbard, R. W., Smoake, J. A., Matther, W. T., Linduska, J. D. & Bowers, W. S. (1974). The effects of growth and endurance training on the protein and DNA content of rat soleus, plantaris, and gastrocnemius muscles. *Growth*, **38**, 171–85.

Hudash, G., Albright, J. P., McAuley, E., Martin, R. K. & Fulton, M. (1985).

Cross-sectional thigh components: computerized tomographic assessment. *Med. Sci. Sports Exerc.*, **17**, 417–21.

Huddleston, A. L., Rockwell, D., Kulund, D. N. & Harrison, R. B. (1980). Bone mass in lifetime tennis athletes. *J. Am. Med. Assoc.*, **244**, 1107–9.

Huenemann, R. (1972). Food habits of obese and non-obese adolescents. *Postgrad. Med.*, **51**, 99–105.

Hull, D. & Smales, O. R. C. (1978). Heat production in the newborn. In *Temperature Regulation and Energy Metabolism in the Newborn*, ed. J. C. Sinclair, pp. 129–56. New York: Grune & Stratton.

Hulley, S. B., Vogel, J. M. & Donaldson, C. L. (1971). The effect of supplemental oral phosphate on the bone mineral changes during prolonged bed rest. *J. Clin. Invest.*, **50**, 2506–18.

Hurst, W. W., Schemm, F. R. & Vogel, W. C. (1952). Urine blood ratios of deuterium oxide in man. *J. Lab. Clin. Med.*, **39**, 41–3.

Hurtado, A. (1964). Acclimatization to high altitudes. In *The Physiological Effects of High Altitude*, ed. W. H. Weihe, pp. 1–17. New York: Macmillan.

Huttunen, P., Hirvonen, J. & Kinnula, V. (1981). The occurrence of brown adipose tissue in outdoor workers. *Eur. J. Appl. Physiol.*, **46**, 339–45.

Hytten, F. E. (1980). Nutritional aspects of human pregnancy. In *Maternal Nutrition during Pregnancy and Lactation*, ed. H. Aebi & R. G. Whitehead, p. 27. Bern: Hans Huber.

Hytten, F. E. & Leitch, I. (1971). *The Physiology of Human Pregnancy*, 2nd edn. Oxford: Blackwell Scientific Publications.

Hytten, F. E., Taylor, K. & Taggart, N. (1966a). Measurement of total body fat in man by absorption of ^{85}K. *Clin. Sci.*, **31**, 111–19.

Hytten, F. E., Thomson, A. M. & Taggart, N. (1966b). Total body water in normal pregnancy. *J. Obstetr. Gyn. Brit. C'wlth*, **73**, 553–61.

Ikai, M. & Fukunaga, T. (1968). Calculation of muscle strength per unit cross-sectional area of human muscle by means of ultrasonic measurement. *Int. Z. Angew. Physiol.*, **26**, 26–32.

Iltis, P. W., Lesmes, G. R. & Schmidt, K. (1985). Immersion effects on hydrostatic weighing at various lung volumes. *Ann. Sports Med.*, **2**, 75–8.

Immink, M. D. C. & Viteri, F. E. (1981). Energy intake and productivity of Guatemalan sugarcane cutters: an empirical test of the efficiency wage hypothesis. Part I and II. *J. Develop. Econ.*, **9**, 251–87.

Ingram, D. L. & Weaver, M. E. (1969). A quantitative study of blood vessels of the pig's skin and the influence of environmental temperature. *Anat. Rec.*, **163**, 517–24.

Innes, J. A., Campbell, I. W., Campbell, C. J., Needle, A. L. & Munroe, J. F. (1974). Long-term follow up of therapeutic starvation. *Br. Med. J.*, (ii), 357–9.

Inokuchi, S., Ishikawa, H., Iwamoto, S. & Kimura, T. (1975). Age related changes in the histological composition of the rectus abdominis muscle of the adult human. *Hum. Biol.*, **47**, 231–49.

International Commission on Radiological Protection (1975). *Report of the Task Group on Reference Man*, No. 23, New York: Pergamon Press.

Itoh, S. (1974). *Physiology of Cold-Adapted Man*. Sapporo: Hokkaido University School of Medicine.

Itoh, S. (1975). Cold adaptability in relation to lipid metabolism. In *Physiological Adaptability and Nutritional Status of the Japanese. A. Thermal Adaptability of the Japanese and Physiology of the Ama*, ed. H. Yoshimura & S. Kobayashi, pp. 59–73. Tokyo: Japanese Committee for the International Biological Programme, Science Council of Japan.

Jackson, A. A., Picou, D. & Reeds, P. J. (1977). The energy cost of repleting tissue deficits during recovery from protein-energy malnutrition. *Am. J. Clin. Nutr., 30*, 1514–17.

Jackson, A. S. (1984). Research design and analysis of data procedures for predicting body density. *Med. Sci. Sports Exerc., 16*, 616–20.

Jackson, A. S. & Pollock, M. L. (1976). Factor analysis and multivariate scaling of anthropometric variables for the assessment of body composition. *Med. Sci. Sports, 8*, 196–203.

Jackson, A. S. & Pollock, M. L. (1978). Generalized equations for predicting body density of men. *Br. J. Nutr., 40*, 497–504.

Jackson, A. S. & Pollock, M. L. (1985). Practical assessment of body composition. *Phys. Sportsmed., 13*, 76–89.

Jackson, A. S., Pollock, M. L. & Gettman, L. R. (1978). Intertester reliability of selected skinfold and circumference and per cent fat estimates. *Res. Quart., 49*, 546–51.

Jackson, A. S., Pollock, M. L. & Ward, A. (1980). Generalized equations for predicting body density of women. *Med. Sci. Sports Exerc., 12*, 175–82.

Jacobson, P. C., Beaver, W., Grabb, S. A., Taft, T. N. & Talmage, R. V. (1984). Bone density in women: college athletes and older athletic women. *J. Orthop. Res., 2*, 328–32.

James, H. M., Dabek, J. T., Chettle, D. I., Dykes, P. W., Fremlin, J. H., Hardwicke, J., Thomas, B. J. & Vartsky, D. (1984). Whole body cellular and collagen nitrogen in healthy and wasted men. *Clin. Sci., 67*, 63–82.

James, W. P. & Trayhurn, P. (1976). An integrated view of the metabolic and genetic basis for obesity. *Lancet, 2* (7989), 770–3.

James, W. P. T. (1976). *Research on obesity*, p. 16. D.H.S.S./M.R.C. Group Report. London: Her Majesty's Stationery Office.

James, W. P. T. & Shetty, P. S. (1982). Metabolic adaptations and energy requirements in developing countries. *Hum. Nutr. Clin. Nutr., 36c*, 331–6.

Jamison, P. L. (1970). Growth of Wainwright Eskimos. Stature and weight. *Arctic Anthrop., 7*, 86–9.

Jamison, P. L. (1976). Growth of Eskimo children in northwestern Alaska. In *Circumpolar Health*, ed. R. J. Shephard & S. Itoh, pp. 223–9. Toronto: University of Toronto Press.

Janos, M., Janos, M,, Tamas, S. & Ivan, S. (1985). Assessment of biological development by anthropometric variables. In *Children and Exercise XI*, ed. R. A. Binkhorst, H. Kemper & W. H. M. Saris, pp. 341–5. Champaign, Ill.: Human Kinetics Publishers.

Jansky, L. (1973). Non-shivering thermogenesis and its thermoregulatory significance. *Biol. Rev., 48*, 85–132.

Jansky, L. (1976). Effects of cold and exercise on energy metabolism of small mammals. In *Progress in Animal Biometeorology*, ed. H. D. Johnson, Vol. I, Part I, pp. 239–58. Amsterdam: Swets & Zeitlinger.

Jarrett, A. S. (1965). Effect of immersion on intrapulmonary pressure. *J. Appl. Physiol.*, **20**, 1261–6.

Jarrett, R. J. (1986). Is there an ideal body weight? *Br. Med. J.*, **293**, 493–5.

Jelliffe, D. B. (1966). *The Assessment of the Nutritional Status of the Community.* Monograph 53. Geneva: WHO.

Jelliffe, D. B. & Jelliffe, E. F. P. (1982). *Assessment of Nutritional Status.* Oxford: Oxford University Press.

Jensen, P. S., Orphanoudakis, S. C., Rauschkolb, E. N., Baron, R., Lang, R. & Rasmussen, H. (1980). Assessment of bone mass in the radius by computed tomography. *Am. J. Roentg.*, **134**, 285–92.

Jéquier, E. & Schutz, Y. (1983). Long-term measurements of energy expenditure in humans using a respiration chamber. *Am. J. Clin. Nutr.*, **38**, 989–98.

Jéquier, E., Pittet, P. & Gygax, P. H. (1978). Thermic effect of glucose and thermal body insulation in lean and obese subjects: a calorimetric approach. *Proc. Nutr. Soc.*, **37**, 45–53.

Jéquier, J-C., LaBarre, R., Rajic, M., Beaucage, C., Shephard, R. J. & Lavallée, H. (1977). Le postulat de Normalité dans les études longitudinales. In *Frontiers of Activity and Child Health*, ed. H. Lavallée & R. J. Shephard, pp. 55–65. Québec City: Editions du Pélican.

Johnston, F. E. (1982). Relationships between body composition and anthropometry. *Hum. Biol.*, **54**, 221–45.

Johnston, F. E. (1985). Systematic errors in the use of the Mellits-Cheek equation to predict body fat in lean females. *New Engl. J. Med.*, **312**, 588–9.

Johnston, F. E., Hamill, P. V. V. & Lemeshow, S. (1974). Skinfold thicknesses in a national probability sample of US males and females 6 through 17 years. *Am. J. Phys. Anthropol.*, **40**, 321–4.

Johnston, F. E., Bogin, B., MacVean, R. B. & Newman, B. C. (1984). A comparison of international standards versus local reference data for the triceps and subscapular skinfolds of Guatemalan children and youth. *Hum. Biol.*, **56**, 157–71.

Jones, D. J., Fox, M. M., Babigan, H. H. & Hutton, H. E. (1980). Epidemiology of anorexia nervosa in Monroe County, New York. *Psychosom. Med.*, **42**, 551–8.

Jones, H. H., Priest, J. D., Hayes, W. C., Tichenor, C. C. & Nagel, D. H. (1977). Humeral hypertrophy in response to exercise. *J. Bone Joint Surg.*, **59**A, 204–8.

Jones, P. R. M. & Pearson, J. (1969). Anthropometric determination of leg fat and muscle plus bone volumes in young male and female adults. *J. Physiol.*, **204**, 63P.

Joos, S. K., Mueller, W. H., Hanis, C. L. & Schull, W. J. (1984). Diabetes alert study: weight history and upper body obesity in diabetic and non-diabetic Mexican American adults. *Ann. Hum. Biol.*, **11**, 167–71.

Kachadorian, W. A., Johnson, R. E. & Buffington, R. E. (1972). Nitrogen excretion as a measure of protein metabolism in man under different conditions of renal function. *Int. Z. Angew. Physiol.*, **30**, 309–13.

Kalkhoff, R. K., Hartz, A. H., Rupley, D., Kissebah, A. H. & Kelber, S. (1983). Relationship of body fat distribution to blood pressure, carbohydrate intolerance, and plasma lipids in healthy, obese women. *J. Lab. Clin. Med.*, **102**, 621–7.

Kaplan, M. L. & Leveille, G. A. (1974). Calorigenic effects of high protein meal in obese and non-obese subjects. *Fed. Proc.*, **33**, 701 (Abstract).

Kaplansky, A. S., Savina, E. A., Portugalov, V. V., Ilyina-Kakueva, E. I., Alexeyev, E. I., Durnova, G. N., Pankova, A. S., Plakshuta-Plakutina, G. I., Shvets, V. N. & Yakovleva, V. I. (1980). Results of morphological investigations aboard Biosatellites Cosmos. *Physiologist*, **23** (6), S51–4.

Kaplowitz, H. J., Mueller, W. H., Selwyn, B. J., Malina, R. M., Bailey, D. A. & Mirwald, R. L. (1987). Sensitivities, specificities and positive predictive values of simple indices of body fat distribution. *Hum. Biol.*, **59**, 809–25.

Katch, F. I. (1969). Practice curves and errors of measurement in estimating underwater weight by hydrostatic weighing. *Med. Sci. Sports*, **1**, 212–20.

Katch, F. I. (1983). Individual differences of ultrasound assessment of subcutaneous fat: effects of body position. *Hum. Biol.*, **55**, 789–95.

Katch, F. I. & Katch, V. L. (1984). The body composition profile. Techniques of measurement and applications. *Clin. Sports Med.*, **3**, 31–63.

Katch, F. I. & McArdle, W. (1973). Prediction of body density from simple anthropometric measurements in college age women and men. *Hum. Biol.*, **45**, 445–54.

Katch, F. I. & McArdle, W. D. (1983). *Nutrition, Weight Control and Exercise*, 2nd edn, p. 332. Philadelphia: Lea & Febiger.

Katch, F. I. & Michael, E. D. (1968). Prediction of body density from skinfold and girth measurements of college females. *J. Appl. Physiol.*, **25**, 92–94.

Katch, F. I., Michael, E. D. & Horvath, S. M. (1967). Estimation of body volume by underwater weighing: description of a simple method. *J. Appl. Physiol.*, **23**, 811–13.

Katch, F. I., Behnke, A. R. & Katch, V. L. (1979). Estimation of body fat from skinfolds and surface area. *Hum. Biol.*, **51**, 411–24.

Katch, V. L. & Freedson, P. S. (1982). Body size and shape: derivation of the 'HAT' frame size model. *Am. J. Clin. Nutr.*, **36**, 699–75.

Katch, V. L. & Katch, F. I. (1973). A simple anthropometric method for calculating segmental leg limb volume. *Res. Quart.*, **45**, 211–14.

Katch, V. L. & Michael, E. D. (1973). The relationship between various segmental leg measurements, leg strength and relative endurance performance of college females. *Hum. Biol.*, **45**, 371–83.

Katch, V. L., Michael, E. D. & Amuchie, F. A. (1973). The use of body weight and girth measurements in predicting segmental leg volume of females. *Hum. Biol.*, **45**, 293–303.

Katch, V. L., Campaigne, B., Freedson, P., Katch, F. I. & Behnke, A. R. (1980). Contribution of breast volume and weight to body fat distribution in females. *Am. J. Phys. Anthrop.*, **53**, 93–100.

Katch, V. L., Freedson, P. S., Katch, F. I. & Smith, L. (1982). Body frame size: validity of self-appraisal. *Am. J. Clin. Nutr.*, **36**, 676–9.

Kato, S. & Ishiko, T. (1966). Obstructed growth of children's bones due to excessive labour in remote corners. In *Proceedings of International Congress of Sport Sciences*, ed. K. Kato, p. 479. Tokyo: Japanese Union of Sport Sciences.

Katsuki, S., Shibayama, H., Ikai, M. & Kondo, S. (1965). An ultrasonic apparatus for measurement of body composition. *Jap. J. Phys. Educ.*, **40**, 34–41.

Katz, J. J. (1960). Chemical and biological studies with deuterium. *Amer. Sci.*, **48**, 544–80.

Kaufman, L. & Wilson, C. J. (1973). Determination of extracellular fluid volume by fluorescent excitation analysis of bromine. *J. Nucl. Med.*, **14**, 812–15.

Kavanagh, T., Yacoub, M. H., Mertens, D. J., Kennedy, R. B. & Campbell, R. B. (1988). Cardiorespiratory response to exercise training after orthotopic cardiac transplantation. *Circulation*, **77**, 162–71.

Keatinge, W. R., Coleshaw, S. R. K., Millard, C. E. & Axelsson, S. (1986). Exceptional case of survival in cold water. *Br. Med. J.*, **292**, 171–2.

Kemper, H. C. G. & Vershuur, R. (1974). Relationship between biological age, habitual physical activity and morphological, physiological characteristics of 12 and 13-year-old boys. *Acta Paediatr. Scand.*, **28**, Suppl., 191–203.

Kemsley, W. F. F., Billewicz, W. Z. & Thomson, A. M. (1962). A new weight-for-height standard based on British anthropometric data. *Br. J. Prev. Soc. Med.*, **16**, 189–95.

Kendall, R. E., Hall, D. J., Hailey, A. & Babigan, H. M. (1973). The epidemiology of anorexia nervosa. *Psychiatr. Med.*, **3**, 200–3.

Kennedy, N. S. J., Eastell, R., Smith, M. A. & Tothill, P. (1983). Normal levels of total body sodium and chlorine by neutron activation analysis. *Phys. Med. Biol.*, **28**, 215–21.

Kerpel-Fronius, E. & Kovach, S. (1948). The volume of extracellular body fluids in malnutrition. *Pediatrics*, **2**, 21–3.

Keys, A. (1956). Recommendations concerning body measurements for the characterization of nutritional status. *Hum. Biol.*, **28**, 111–23.

Keys, A. & Brozek, J. (1953). Body fat in adult man. *Physiol. Rev.*, **33**, 245–345.

Keys, A., Brozek, J., Henschel, A., Mickelsen, O. & Taylor, H. L. (1950). *Biology of Human Starvation*. Minneapolis: University of Minnesota Press.

Keys, A., Anderson, J. T. & Brozek, J. (1955). Weight gain from simple over-eating. *Metabolism*, **4**, 427–32.

Keys, A., Fidanza, F., Karvonen, M. J., Kimura, N. & Taylor, H. L. (1972). Indices of relative weight and obesity. *J. Chron. Dis.*, **25**, 329–43.

Keys, A., Taylor, H. L. & Grande, F. (1973). Basal metabolism and age of adult man. *Metabolism*, **22**, 579–87.

Khairi, M. R. A., Cronin, J. H., Robb, J. A., Smith, D. M., Yu, P. L. & Johnston, C. C. (1976). Femoral trabecular pattern index and bone mineral content measurement by photon absorptiometry in senile osteoporosis. *J. Bone Joint Surg.*, **58** (Section A), 221–6.

Khosla, T. & Billewicz, W. Z. (1964). Measurement of change in body weight. *Br. J. Nutr.*, **18**, 227–39.

Khosla, T. & Lowe, C. R. (1967). Indices of obesity derived from body weight and height. *Br. J. Prev. Soc. Med.*, **21**, 122–8.

Khosla, T. & Lowe, C. R. (1971). Obesity and smoking habits. *Br. Med. J.*, **4**, 10–13.

Kielmann, A. A. & McCord, C. (1978). Weight for age as an index of risk of death in children. *Lancet*, (i), 1247–50.

Kiiskinen, A. & Heikkinen, E. (1978). Physical training and connective tissues in young mice: biochemistry of long bones. *J. Appl. Physiol.*, **44**, 50–4.

King, J. W., Brelsford, J. H. & Tullos, H. S. (1969). Analysis of the pitching arm of the professional baseball pitcher. *Clin. Orthop.*, **67**, 116–23.

Kirton, A. H. & Pearson, A. M. (1963). Relationships between potassium content and body composition. *Ann. N.Y. Acad. Sci.*, **110**, 221–8.

Kissebah, A. H., Vydelingum, N., Murray, R., Evans, D. J., Hartz, A. J., Kalkhoff, R. K. and Adams, P. W. (1982). Relation of body fat distribution to metabolic complications of obesity. *J. Clin. Endocrinol. Metab.*, **54**, 254–60.

Kleiber, M. (1932). Body size and metabolism. *Hilgardia*, **6**, 315–53.

Kleiber, M. (1947). Body size and metabolic rate. *Physiol. Rev.*, **27**, 511–41.

Kleiber, M. (1975). *The Fire of Life: An Introduction to Animal Energetics.* Huntingdon, N.Y.: Robert E. Kreiger.

Klemm, T., Banzer, D. H. & Schneider, U. (1976). Bone mineral content of the growing skeleton. *Am. J. Roentg.*, **126**, 1283–4.

Kley, H. K., Deselaers, T., Peerenboom, H. & Krüskemper, H. L. (1980). Enhanced conversion of androstenedione to estrogens in obese males. *J. Clin. Endocrinol. Metab.*, **51**, 1128–32.

Klibanski, A., Neer, R. M., Beitins, I. Z., Ridgway, E. C., Zervas, N. T. & McArthur, J. W. (1980). Decreased bone density in hyperprolactinemic women. *New Engl. J. Med.*, **303**, 1511–14.

Klish, W. J., Forbes, G. B., Gordon, A. & Cochran, W. J. (1984). New method for the estimation of lean body mass in infants (EMME instrument): validation in nonhuman models. *J. Pediatr. Gastroenterol. Nutr.*, **3**, 199–204.

Klish, W. J., Cochran, W. J., Fiorotto, M. L., Wong, W. W. & Klein, P. D. (1987). The bioelectrical measurement of body composition during infancy. *Hum. Biol.*, **59**, 319–27.

Knight, G. S., Beddoe, A. H., Streat, S. J. & Hill, G. L. (1986). Body composition of two human cadavers by neutron activation and chemical analysis. *Am. J. Physiol.*, **251**, E179–85.

Knittle, J. L. & Hirsch, J. (1968). Effect of early nutrition on the development of rat epididymal fat pads. Cellularity and metabolism. *J. Clin. Invest.*, **47**, 2091–8.

Knussman, R. (1967). Interkorrelationen im Hautleistensystem des Menschen und ihre faktoranalytische Ausertung. *Humangenetik*, **4**, 221–43.

Knittle, J. L., Timmers, K., Ginsberg-Fellner, F., Brown, R. E. & Katz, D. P. (1979). The growth of adipose tissue in children and adolescents: cross-sectional and longitudinal studies of adipose cell number and size. *J. Clin. Invest.*, **63**, 239–46.

Knowler, W. C. & Garrow, J. (1982). Obesity indices derived from weight and height. *Int. J. Obesity*, **6**, 241–3.

Koch, G. & Röcker, L. (1980). Total amount of hemoglobin, plasma and blood volumes, and intravascular protein masses in trained boys. In *Children and Exercise* IX, ed. K. Berg & B. Eriksson, pp. 109–15. Baltimore: University Park Press.

Kochakian, C. D. (1976). *Anabolic-androgenic Steroids.* New York: Springer-Verlag.

Kohara, Y. (1975). Anthropometric observations on the Japanese Ama, with special reference to the thickness of subcutaneous fat. In *Physiological Adaptability and Nutritional Status of the Japanese. A. Thermal Adaptability of the Japanese and Physiology of the Ama,* ed. H. Yoshimura & S.

Kobayashi, pp. 211–18. Tokyo: Japanese Committee for the International Biological Program, Science Council of Japan.

Kohlrausch, W. (1929). Zusammenhange von Korperform und Leistung. Ergebnisse der anthropometrische Messungen an den Athleten der Amsterdammer Olimpiade. *Arbeitsphysiologie*, **2**, 187–204.

Konishi, F. (1967). Urinary creatinine as a possible index of muscular activity. *Res. Quart.*, **38**, 398–404.

Korecky, B. & Rakusan, K. (1978). Normal and hypertrophic growth of the rat heart: Changes in cell dimensions and number. *Am. J. Physiol.*, **234**, H123–8.

Kotulan, J., Reznickova, M. & Placheta, Z. (1980). Exercise and growth. In *Youth and Physical Activity*, ed. Z. Placheta, pp. 61–117. Brno: J. E. Purkynje University.

Kow, I. M., Wilkinson, P. M. & Chornock, F. W. (1967). Axonal delivery of neuroplasmic components to muscle cells. *Science*, **155**, 342–55.

Kowalski, C. J. (1972). A commentary on the use of multivariate methods in anthropometric research. *Am. J. Phys. Anthrop.*, **36**, 119–32.

Krabbe, S., Christensen, T., Worm, J., Christiansen, C. & Transböl, I. (1978). Relationship between haemoglobin and serum testosterone in normal children and adolescents and in boys with delayed puberty. *Acta Paediatr. Scand.*, **67**, 655–8.

Krabbe, S., Hummer, L. & Christiansen, C. (1984). Longitudinal study of calcium metabolism in male puberty. II. Relationship between mineralization and serum testosterone. *Acta Paediatr. Scand.*, **73**, 750–5.

Kreisberg, R. A., Bowdoin, B. & Meador, C. K. (1970). Measurement of muscle mass in humans by isotopic dilution of creatine ^{14}C. *J. Appl. Physiol.*, **28**, 264–7.

Kretschmer, E. & Enke, W. (1936). *Die Personlichkeit der Athletiker*. Leipzig: Thieme.

Krogh, A. (1916). *The Respiratory Exchange of Animals and Man*. London: Longmans.

Krogman, W. M. (1970). Growth of head, face, trunk and limbs in Philadelphia White and Negro children of elementary and high school age. *Mon. Soc. Res. Child Develop.*, **35**, 3 (Serial 136).

Krølner, B. & Pors-Nielsen, S. (1980). Measurement of bone mineral content of the lumbar spine I. Theory and application of a new two-dimensional dualphoton attenuation method. *Scand. J. Clin. Lab. Invest.*, **40**, 653–63.

Krølner, B. & Toft, B. (1983). Vertebral bone loss: an unheeded side effect of therapeutic bed rest. *Clin. Sci.*, **64**, 537–40.

Krølner, B., Toft, B. & Nielsen, S. P. (1983). Physical exercise as prophylaxis against involutional vertebral bone loss: a controlled trial. *Clin. Sci.*, **64**, 541–6.

Krotkiewski, M., Aniansson, A., Grimby, G., Björntorp, P. & Sjöstrom, L. (1979). The effect of unilateral isokinetic strength training on local adipose and muscle tissue morphology, thickness and enzymes. *Eur. J. Appl. Physiol.*, **42**, 271–81.

Krotkiewski, M., Björntorp, P., Sjöstrom, L. & Smith, U. (1983). Impact of obesity on metabolism in men and women. Importance of regional adipose tissue distribution. *J. Clin. Invest.*, **72**, 1150–62.

Krzywicki, H. J. & Consolazio, C. F. (1968). Body composition methodology in military nutrition survey. In *Body Composition in Animals and Man*, pp. 492–511. Washington, DC: Nat. Acad. Sci. US.

Krzywicki, H. J., Ward, G. M., Rahman, D. P., Nelson, R. A. & Consolazio, C. F. (1974). A comparison of methods for estimating human body composition. *Am. J. Clin. Nutr.*, **27**, 1380–5.

Kulwich, R., Feinstein, L., Golumbic, C., Hiner, R. L., Seymour, W. R. & Kauffman, W. R. (1961). Relationship of gamma ray measurements to the lean content of hams. *J. Animal Sci.*, **20**, 497–502.

Kushner, R. F. & Schoeller, D. A. (1986). Estimation of total body water by bioelectrical impedance analysis. *Am. J. Clin. Nutr.*, **44**, 417–24.

Kvist, H., Sjöstrom, L. & Tylen, U. (1986). Adipose tissue volume determinations in women by computed tomography: technical considerations. *Int. J. Obesity*, **10**, 53—67.

Kyere, K., Oldroyd, B., Oxby, C. B., Burkinshaw, L., Ellis, R. E. & Hill, G. L. (1982). The feasibility of measuring total body carbon by counting neutron inelastic scatter gamma rays. *Phys. Med. Biol.*, **6**, 805–17.

Ladell, W. S. S. (1964). Terrestrial animals in humid heat: man. In *Handbook of Physiology*, section 4, ed. D. B. Dill, pp. 625–60. Washington, DC: American Physiological Society.

Lafontan, M. & Berlan, M. (1981). Alpha-adrenergic receptors and the regulation of lipolysis in adipose tissue. *Trends Pharmacol. Sci. May*, 126–9.

Lafontan, M., Dang-Tran, L. & Berlan, M. (1979). Alpha-adrenergic antilipolytic effect of adrenaline in human fat cells of the thigh: comparison with adrenaline responsiveness of different fat deposits. Eur. J. Clin. Invest., 9, 261–6.

Lamarck, J. B. P. A. (1809). *Histoire naturelle des animaux sans vertèbres*. Paris: Verdière.

Lambert, T. & Teissier, G. (1927). Théorie de la similitude biologique. *Ann. Physiol.*, **3**, 212–46.

Lamki, L., Ezrin, C., Koven, I. & Steiner, G. (1973). L-thyroxine in the treatment of obesity without increase in loss of lean body mass. *Metabolism*, **22**, 617–22.

Landau, R. L. (1973). The metabolic influence of progesterone. In *Handbook of Physiology. Endocrinology* II, Part I, ed. R. O. Greep, pp. 573–89. Washington, DC: American Physiological Society.

Landry, F., Carrière, S., Poirier, L., LeBlanc, C., Gaudreau, J., Moisau, A., Carrier, R. & Potvin, R. (1980). Observations sur la condition physique des Québecois. *Union Méd. Can.*, **109**, 1–6.

Lane, H. W., Roessler, G., Nelson, E. W. & Cerda, J. J. (1977). Whole body counter measurements of total body potassium before and after exercise. *Fed. Proc.*, **36**, 433.

Lane, H. W., Roessler, G. S., Nelson, E. W. & Cerda, J. J. (1978). Effect of physical activity on human potassium metabolism in a hot and humid environment. *Am. J. Clin. Nutr.*, **31**, 838–43.

Lanyon, L. E. & Rubin, C. T. (1983). Regulation of bone mass in response to physical activity. In *Osteoporosis, a Multidisciplinary Problem*, ed. A. St J. Dixon, R. G. G. Russell & T. C. B. Stamp, pp. 51–61. London: Royal Society of Medicine.

Lanza, E. (1983). Determination of moisture, protein, fat and calories in raw pork and beef by near infra-red spectroscopy. *J. Food Sci.,* **48**, 471–4.

Lapidus, L., Bengtsson, C., Larsson, B., Pennert, K., Rybo, E. & Sjöstrom, L. (1984). Distribution of adipose tissue and risk of cardiovascular disease and death: a 12-year follow up of participants in the population study of women in Gothenburg, Sweden. *Br. Med. J.,* **289**, 1257–61.

Larivière, G., Lavallée, H. & Shephard, R. J. (1974). See Shephard, Lavallée & Larivière (1978).

Larsson, B., Svardsudd, K., Welin, L., Wilhelmsen, L., Björntorp, L. & Tibblin, G. (1984). Abdominal adipose tissue distribution, obesity, and risk of cardiovascular disease and death: 13-year follow up of participants in the study of men born in 1913. *Br. Med. J.,* **288**, 1401–4.

Larsson, L., Grimby, G. & Karlsson, J. (1979). Muscle strength and speed of movement in relation to age and muscle morphology. *J. Appl. Physiol. Resp. Environ. Exerc. Physiol.,* **46**, 451–6.

Lasker, G. & Womack, H. (1975). An anatomical view of demographic data: biomass, fat mass and lean body mass of the United States and Mexican human populations. In *Biosocial Interactions in Populations,* ed. E. S. Watts, pp. 43–53. The Hague: Mouton.

Latimer, H. B. & Lowrance, E. W. (1965). Bilateral asymmetry in weight and length of human bones. *Anat. Rec.,* **152**, 217–224.

Latin, R. W. & Ruhling, R. O. (1986). Total lung capacity, residual volume and predicted residual volume in a densitometric study of older men. *Br. J. Sports Med.,* **20**, 66–8.

Lavietes, P. H., Bourillon, J. & Klinghoffer, K. H. (1936). The volume of the extracellular fluids of the body. *J. Clin. Invest.,* **15**, 261–8.

Laven, J. T. (1983). Skinfold compressibility – a measure of body composition and a source of error. *Am. J. Clin. Nutr.,* **37**, 727 (Abstract).

Lawlor, M. R., Crisman, R. P. & Hodgson, J. A. (1985). Bioelectrical impedance analysis as a method to assess body composition. *Med. Sci. Sports Exerc.,* **17**, 271 (Abstract).

Le Bideau, G. & Rivolier, J. (1958). La mesure du pli cutané. *Ann. Nutr. Alim.,* **12**, 121–6.

Lee, J., Kolonel, L. N. & Hinds, M. W. (1981). Relative merits of the weight-corrected-for-height indices. *Am. J. Clin. Nutr.,* **34**, 2521–9.

Lee, J., Kolonel, L. N. & Hinds, M. W. (1982). The use of an inappropriate weight–height derived index of obesity can produce misleading results. *Int. J. Obesity,* **6**, 233–9.

Lee, M. M. C. (1957). Physical and structural age changes in the human skin. *Anat. Rec.,* **129**, 473–88.

Lee, M. M. C. (1959). Thickening of the subcutaneous tissues in paralyzed limbs in chronic hemiplegia. *Hum. Biol.,* **31**, 187–93.

Lee, M. M. C. & Ng, C. (1965). Post-mortem studies of skinfold caliper measurements and actual thickness of skin and subcutaneous tissue. *Hum. Biol.,* **37**, 91–103.

Leger, L. A., Lambert, J. & Martin, P. (1982). Validity of plastic skinfold caliper measurements. *Hum. Biol.,* **54**, 667–75.

Leibman, J., Gotch, F. A. & Edelman, I. S. (1960). Tritium assay by liquid scintillation spectrometry. Comparison of tritium and deuterium oxides as tracers for body water. *Circ. Res.*, **8**, 907–12.

Leithead, C. S. & Lind, A. R. (1964). *Heat Stress and Heat Disorders*. London: Cassell.

LeMaho, Y., Van Kha, H. V., Koubi, H., Dewasmes, G., Girard, J., Ferré, P. & Cagnard, M. (1981). Body composition, energy expenditure and plasma metabolites in long-term fasting geese. *Am. J. Physiol.*, **241**, E342–54.

Lentz, T. L. (1971). Nerve trophic function: in vitro assay of effects of nerve tissue on muscle cholinesterase activity. *Science*, **171**, 187–9.

Leonard, J. I., Leach, C. S. & Rambaut, P. C. (1983). Quantitation of tissue loss during prolonged space flight. *Am. J. Clin. Nutr.*, **38**, 667–79.

Lepkovsky, S. (1973). Newer concepts in the regulation of food intake. *Am. J. Clin. Nutr.*, **26**, 271–84.

Lerner, A., Heitlinger, L. A. & Rossi, T. M. (1985). Midarm muscle circumference as indicator of muscle mass. *J. Pediatr.*, **106**, 168–9.

Lesser, G. T. & Markofsky, J. (1979). Body water compartments with human aging using fat-free mass as the reference standard. *Am. J. Physiol.*, **236**, R215–20.

Lesser, G. T. & Zak, G. (1963). Measurement of total body fat in man by the simultaneous absorption of two inert gases. *Ann. N.Y. Acad. Sci.*, **110**, 40–52.

Lesser, G. T., Blumberg, A. G. & Steele, J. M. (1952). Measurement of total body fat in living rats by the absorption of cyclopropane. *Am. J. Physiol.*, **169**, 545–53.

Lesser, G. T., Perl, W. & Steele, J. M. (1960). Determination of total body fat by absorption of an inert gas; measurements and results in normal human subjects. *J. Clin. Invest.*, **39**, 1791–806.

Lesser, G. T., Deutsch, S. & Markofsky, J. (1971). Use of independent measurement of body fat to evaluate overweight and underweight. *Metabolism*, **20**, 792–804.

Leusink, J. A. (1974). A comparison of the body composition estimated by densitometry and total body potassium measurement in trained and untrained subjects. *Pflüg. Archiv.*, **348**, 357–62.

Levin, M. E., Boisseau, V. C. & Avioli, L. V. (1976). Effects of diabetes mellitus on bone mass in juvenile and adult-onset diabetes. *New Engl. J. Med.*, **294**, 241–5.

Lewis, D. S., Rollwitz, W. L., Bertrand, H. A. & Masoro, E. J. (1986). Use of NMR for measurement of total body water and estimation of body fat. *J. Appl. Physiol.*, **60**, 836–40.

Li, C. C. (1961). *Human Genetics. Principles and Methods*. New York: McGraw-Hill.

Lim, T. P. K. (1963). Critical evaluation of the pneumatic method for determining body volume; its history and technique. *Ann. N.Y. Acad. Sci.*, **110**, 72–4.

Lindsay, R., Hart, D. M., Aitken, J. M., MacDonald, E. B., Anderson, J. B. & Clarke, A. C. (1976). Long-term prevention of postmenopausal osteoporosis by estrogen. *Lancet*, (i), 1038–40.

Linnell, S. L., Stager, J. M., Blue, P. W., Oyster, N. & Robertshaw, D. (1984).

Bone mineral content and menstrual regularity in female runners. *Med. Sci. Sports Exerc.*, **16**, 343–8.

Ljunggren, H. (1957). Studies on body composition with special reference to the composition of obesity tissue and non-obesity tissue. *Acta Endocrinol.*, Suppl. 33, 1–58.

Ljunggren, H. (1965). Sex differences in body composition. In *Human Body Composition. Approaches and Applications*, ed. J. Brozek. Oxford: Pergamon Press.

Ljunggren, H., Ikkos, D. & Luft, R. (1961). Basal metabolism in women with obesity and anorexia nervosa. *Br. J. Nutr.*, **15**, 21–34.

Lloyd, R. D. & Mays, C. W. (1987). A model for human body composition by total body counting. *Hum. Biol.*, **59**, 7–30.

Lloyd, R. D., Mays, C. W. & Taysum, D. H. (1979). Gamma-ray spectrometry of humans at the University of Utah. *Radiat. Environ. Biophys.*, **16**, 157–75.

Loeb, J. N. (1976). Corticosteroids and growth. *New Engl. J. Med.*, **295**, 547–52.

Loeppky, J. A., Myhre, L. G., Venters, M. D. & Luft, U. C. (1977). Total body water and lean body mass estimated by ethanol dilution. *J. Appl. Physiol.*, **42**, 803–8.

Lohman, T. (1984). Research progress in validation of laboratory methods of assessing body composition. *Med. Sci. Sports Exerc.*, **16**, 596–603.

Lohman, T. (1986). Applicability of body composition techniques in children and youths. *Adv. Pediatr. Sport Sci.*, **2**, 29–57.

Lohman, T. & Pollock, M. L. (1981). Which caliper? How much training? *JOHPER*, **52**, 27–9.

Lohman, T., Wilmore, J. H., Friestad, G. & Slaughter, M. H. (1983). *AAHPERD Research Abstracts*, p. 122.

Lohman, T. G. (1981). Skinfolds and body density and their relation to body fatness: a review. *Hum. Biol.*, **53**, 181–225.

Lohman, T. G., Boileau, R. A. & Massey, B. H. (1975). Prediction of lean body mass in young boys from skinfold thickness and body weight. *Hum. Biol.*, **47**, 245–62.

Lohman, T. G., Slaughter, M. H., Selinger, A. & Boileau, R. A. (1978). Relationship of body composition to somatotype in college age men. *Hum. Biol.*, **50**, 147–9.

Lohman, T. G., Boileau, R. A. & Slaughter, M. H. (1984a). Body composition in children and youth. In *Advances in Pediatric Sport Sciences*, ed. R. A. Boileau, pp. 29–57. Champaign, Ill.: Human Kinetics Publishers.

Lohman, T. G., Slaughter, M. H., Boileau, R. A., Bunt, J. & Lussier, L. (1984b). Bone mineral measurements and their relation to body density in children, youth and adults. *Hum. Biol.*, **56**, 667–79.

Lohman, T. G., Pollock, M. L., Slaughter, M. H., Brandon, L. J. & Boileau, R. A. (1984c). Methodological factors and the prediction of body fat in female athletes. *Med. Sci. Sports Exerc.*, **16**, 92–6.

Londeree, B. R. & Forkner, L. (1978). Changes in ^{40}K counts with exercise. *Res. Quart.*, **49**, 95–100.

Lopes, J., Russell, D. M., Whitwell, J. & Jeejebhoy, K. N. (1982). Skeletal muscle function in malnutrition. *Am. J. Clin. Nutr.*, **36**, 602–10.

Lukaski, H. C. & Johnson, P. E. (1985). A simple inexpensive method of determining total body water using a tracer dose of D_2O and infrared absorption of biological fluids. *Am. J. Clin. Nutr.*, **41**, 363–70.

Lukaski, H. C. & Mendez, J. (1980). Relationship between fat-free weight and urinary 3-methyl histidine excretion in man. *Metabolism*, **29**, 758–61.

Lukaski, H. C., Mendez, J., Buskirk, E. R. & Cohn, S. H. (1981). Relationship between endogenous 3-methylhistidine excretion and body composition. *Am. J. Physiol.*, **240**, E302–7.

Lukaski, H. C., Johnson, P. E., Bolunchuk, W. W. & Lyken, G. I. (1985). Assessment of fat free mass using bioelectrical impedance measurement of the human body. *Am. J. Clin. Nutr.*, **41**, 810–17.

Lukaski, H. C., Bolonchuk, W. W., Hall, C. B. & Siders, W. A. (1986). Validation of tetrapolar bioelectrical impedance method to assess human body composition. *J. Appl. Physiol.*, **60**, 1327–32.

Lykken, G. I., Jacob, R. A., Munoz, P. M. & Sandstead, H. H. (1980). A mathematical model of creatine metabolism in normal males – comparison between theory and experiment. *Am. J. Clin. Nutr.*, **33**, 2674–85.

Lykken, G. I., Lukaski, H. C., Bolonchuk, W. W. & Sanstead, H. H. (1983). Potential errors in body composition as estimated by whole body scintillation counting. *J. Lab. Clin. Med.*, **101**, 651–8.

Lynch, T. N., Jensen, R. L., Stevens, P. M., Johnson, R. L. & Lamb, L. E. (1967). Metabolic effects of prolonged bed rest: their modification by simulated altitude. *Aerospace Med.*, **38**, 10–20.

Maas, G. D. (1974). *The Physique of Athletes*. Leiden: Leiden University Press.

MacDougall, J. D., Ward, G. R., Sale, D. G. & Sutton, J. R. (1977). Biochemical adaptation of human skeletal muscle to heavy resistance training and immobilization. *J. Appl. Physiol.*, **43**, 700–3.

MacDougall, J. D., Wenger, H. A. & Green, H. J. (1982). *Physiological Testing of the Elite Athlete*. Ottawa: Canadian Association of Sport Sciences.

Mack, P., Brown, W. & Trapp, H. (1949). The quantitative evaluation of bone density. *Am. J. Roentg.*, **61**, 808–25.

Mack, P., La Chance, P., Vose, G. & Vogt, F. (1967). Bone demineralization of foot and hand of Gemini-Titan IV, V and VII astronauts during orbital flight. *Radium Ther. Nucl. Med.*, **100**, 503–11.

MacMahon, S. W., Wilcken, D. E. L. & MacDonald, G. J. (1986). The effect of weight reduction on left ventricular mass. *New Engl. J. Med.*, **314**, 334–9.

MacMillan, M. G., Reid, C. M., Shirling, D. & Passmore, R. (1965). Body composition, resting oxygen consumption and urinary creatinine in Edinburgh students. *Lancet*, (i), 728–9.

Madsen, M. (1977). Vertebral and peripheral bone mineral content by photon absorptiometry. *Invest. Radiol.*, **12**, 185–8.

Mahalko, J. R. & Johnson, L. K. (1980). Accuracy of prediction of long-term energy needs. *J. Am. Diet. Assoc.*, **77**, 557–61.

Malhotra, M. S. (1966). People of India, including primitive tribes – a survey on physiological adaptation, physical fitness and nutrition. In *The Biology of Human Adaptability*, ed. P. T. Baker & J. S. Weiner, pp. 329–56. Oxford: Clarendon Press.

Malina, R. M. (1971). Skinfolds in American Negro and White children. *J. Am. Diet. Assoc.*, **59**, 34–39.

Malina, R. M. (1973). Biological substrata. In *Comparative Studies of Blacks and Whites in the United States,* ed. K. S. Miller & R. W. Dreger, pp. 53–123. New York: Seminar Press.

Malina, R. M. (1978). Growth of muscle tissue and muscle mass. In *Human Growth,* Vol. 2, Postnatal Growth, ed. F. Falkner & J. M. Tanner, pp. 273–94. New York: Plenum Press.

Malina, R. M. (1979). Secular changes in growth, maturation and physical performance. *Ex. Spts. Sci. Rev.,* **6**, 203–55.

Malina, R. M. (1980). Physical activity, growth and functional capacity. In *Human Physical Growth and Maturation,* ed. F. E. Johnston, A. F. Roche & C. Susanne, pp. 307–27. New York: Plenum Press.

Malina, R. M. (1984). Human growth, maturation and regular physical activity. In *Advances in Pediatric Sport Sciences,* ed. R. A. Boileau, pp. 59–83. Champaign, Ill.: Human Kinetics Publishers.

Malina, R. M. & Little, B. B. (1985). Body composition, strength and motor performance in undernourished boys. In *Children and Exercise* XI, ed. R. A. Binkhorst, H. C. G. Kemper & W. H. M. Saris, pp. 293–300. Champaign, Ill.: Human Kinetics Publishers.

Malina, R. M., Brown, K. H. & Zavaleta, A. N. (1987). Relative lower extremity length in Mexican American and in American Black and White youth. *Am. J. Phys. Anthrop.,* **72**, 89–94.

Mallard, J., Hutchinson, J. M. S., Edelstein, W. A., Lung, C. R., Foster, M. A. & Johnson, G. (1980). In vivo n.m.r. imaging in medicine; the Aberdeen approach both physical and biological. *Phil. Trans. R. Soc. Lond.,* **289**, 519–33.

Mansour, A. M. & Nass, M. M. K. (1970). In vivo cortisol action on RNA synthesis in rat liver nuclei and mitochondria. *Nature,* **228**, 665–7.

Manzke, E., Chestnut, C. H., Wergedal, J. E., Baylink, D. J. & Nelp, W. B. (1975). Relationship between local and total bone mass in osteoporosis. *Metabolism,* **24**, 605–15.

Marcus, R. (1986). Osteoporosis alert: clues to thinning bone. *Geriatrics,* (Feb/Mar), 26–40.

Maresh, M. M. (1963). Tissue changes in the individual during growth from X-rays of the extremities. *Ann. N.Y. Acad. Sci.,* **110**, 465–74.

Maresh, M. M. (1966). Changes in tissue widths during growth. *Am. J. Dis. Childr.,* **111**, 142–55.

Margen, S., Chu, J. Y., Kaufman, N. A. & Calloway, D. H. (1974). Studies in calcium metabolism. I. The calciuretic effect of dietary protein. *Am. J. Clin. Nutr.,* **27**, 584–9.

Marks, C. & Katch, V. (1986). Biological and technological variability of residual lung volume and the effect on body fat calculations. *Med. Sci. Sports. Exerc.,* **18**, 485–8.

Martin, A. D. (1984). *An Anatomical Basis for Assessing Human Body Composition: evidence from 25 dissections.* Burnaby, BC: Simon Fraser University, Ph.D. Thesis.

Martin, A. D., Ross, W. D., Drinkwater, D. T. & Clarys, J. P. (1985). Prediction of body fat by skinfold caliper: assumptions and cadaver evidence. *Int. J. Obesity,* **9**, Suppl. 1, 31–9.

Martin, J. D. (1981). Health care in northern Canada – a historical perspective. In *Circumpolar Health* 81, ed. B. Harvald & J. P. Hart Hansen, pp. 80–7. Copenhagen: Nordic Council for Arctic Medical Research.

Martin, R. E. & Wool, I. G. (1968). Formation of active hybrids from subunits of muscle ribosomes from normal and diabetic rats. *Proc. Nat. Acad. Sci. (USA)*, **60**, 569–74.

Martin, R. K., Albright, J. P., Clarke, W. R. & Niffenegger, J. A. (1981). Load-carrying effects on the adult beagle tibia. *Med. Sci. Sports Exerc.*, **13**, 343–9.

Martin, S., Neale, G. & Elia, M. (1985). Factors affecting maximal momentary grip strength. *Hum. Nutr. Clin. Nutr.*, **39C**, 137–47.

Martin, T. G., Kessler, W. V., Stant, E. G., Christian, J. E. & Andrews, F. N. (1963). Body composition of calves and pigs measured by large volume liquid scintillation counting and conventional chemical analyses. *Ann. N.Y. Acad. Sci.*, **110**, 213–20.

Martorell, R. (1985). Child growth retardation: a discussion of its causes and its relationship to health. In *International Symposium on Nutritional Adaptation in Man*, ed. J. C. Waterlow, pp. 13–29. London: John Libbey.

Martorell, R. & Arroyave, G. (1988). Malnutrition, work output and energy needs. In *Capacity for Work in the Tropics*, ed. K. Collins & D. Roberts, pp. 57–75. Cambridge: Cambridge University Press.

Martorell, R. & Habicht, J-P. (1986). Growth in early childhood in developing countries. In *Human Growth: A Comprehensive Treatise*, ed. F. Falkner & J. M. Tanner, pp. 241–62. New York: Plenum Press.

Martorell, R., Malina, R. M., Castillo, R. O., Mendoza, F. S. & Pawson, I. G. (1988). Body proportions in three ethnic groups: children and youths 2–17 years in NHANES II and HHANES. *Hum. Biol.*, **60**, 205–22.

Matiegka, J. (1921). The testing of physical efficiency. *Am. J. Phys. Anthrop.*, **4**, 223–30.

Maughan, R. J., Watson, J. S. & Weir, J. (1983). Relationship between muscle strength and muscle cross-sectional area in male sprinters and endurance runners. *Eur. J. Appl. Physiol.*, **50**, 309–18.

Maughan, R. J., Watson, J. S. & Weir, J. (1984). The relative proportion of fat, muscle and bone in the normal human forearm as determined by computed tomography. *Clin. Sci.*, **66**, 683–9.

Mayer, J. (1960). Exercise and weight control. In *Science and Medicine of Exercise and Sports*, ed. W. E. Johnson, pp. 301–10. New York: Harper and Row.

Mayer, J. (1972). *Human Nutrition. Its Physiological, Medical and Social Aspects*. Springfield, Ill.: C. C. Thomas.

Mayhew, J. L. & Gross, P. M. (1974). Body composition changes in young women with high resistance weight training. *Res. Quart.*, **45**, 433–40.

Mayhew, J. L., Piper, F. C. & Holmes, M. A. (1981). Prediction of body density, fat weight and lean body mass in male athletes. *J. Spts Med. Phys. Fitness*, **21**, 383–9.

Mayhew, J. L., Piper, F. C., Kass, J. A. & Montaldi, D. H. (1983). Prediction of body composition in female athletes. *J. Spts Med. Phys. Fitness*, **23**, 333–40.

Mayhew, J. L., Clark, B. A., McKeown, B. C. & Montaldi, D. H. (1985). Accuracy of anthropometric equations for estimating body composition of female athletes. *J. Spts Med. Phys. Fitness*, **25**, 120–6.

Mazess, R. B. (1971). Estimation of bone and skeletal weight by direct photon absorptiometry. *Invest. Radiol.*, **6**, 52–60.

Mazess, R. B. & Cameron, R. J. (1972). Direct readout of bone mineral content using radionuclide absorptiometry. *Int. J. Appl. Radiat. Isot.*, **23**, 471–9.

Mazess, R. B. & Mather, W. (1973). Bone mineral content in normal US whites. In *International Conference on Bone Mineral Measurement*, ed. R. B. Mazess, Washington, DC: Dept. of HEW Publication NIH 75-863.

Mazess, R. B. & Mather, W. (1974). Bone mineral content of North Alaska Eskimos. *Am. J. Clin. Nutr.*, **27**, 916–25.

Mazess, R. B. & Mather, W. (1975). Bone mineral content in Canadian Eskimos. *Hum. Biol.*, **47**, 45–63.

Mazess, R. B., Cameron, J. R., O'Connor, R. & Knutzen, D. (1964). Accuracy of bone mineral measurement. *Science*, **145**, 388–9.

Mazess, R. B., Peppler, W. W., Chestnut, C. H., Nelp, W. B., Cohn, S. H. & Zanzi, I. (1981). Total body bone mineral and lean body mass by dual photon absorptiometry. II. Comparison with total body calcium by neutron activation analysis. *Calcif. Tiss. Internat.*, **33**, 361–3.

Mazess, R. B., Peppler, W. W. & Gibbons, M. (1984). Total body composition by dual-photon (^{153}Gd) absorptiometry. *Am. J. Clin. Nutr.*, **40**, 834–9.

McArdle, J. J. & Sansone, F. M. (1977). Re-innervation of fast and slow twitch muscle following nerve crush at birth. *J. Physiol.*, **271**, 567–86.

McCance, R. A. & Widdowson, E. M. (1951). Composition of the body. *Br. Med. Bull.*, **7**, 296–303.

McCully, K. K., Kent, J. A. & Chance, B. (1988). Application of ^{31}P magnetic resonance spectroscopy to the study of athletic performance. *Sports Med.*, **5**, 312–21.

McGowan, A., Jordan, M. & MacGregor, J. (1975). Skinfold thicknesses in neonates. *Biol. Neon.*, **25**, 66–84.

McKeown, T. & Record, R. G. (1952). Observations on foetal growth in multiple pregnancies in man. *J. Endocrinol.*, **8**, 386–401.

McKeran, R. O., Halliday, D. & Purkiss, P. (1978). Comparison of human myofibrillar protein catabolic rate derived from 3-methylhistidine excretion with synthetic rate from muscle biopsies during L-alpha^{15}N lysine infusion. *Clin. Sci. Mol. Med.*, **54**, 471–5.

McMahon, T. (1975). Using body size to understand the structural design of animals. Quadrupedal locomotion. *J. Appl. Physiol.*, **39**, 619–27.

McNair, J. R., Brown, B. S., Weber, J. & McGowan, R. (1984). The clinical evaluation of body fat. *J. Arkansas Med. Soc.*, **80**, 425–30.

McNeill, K. G. & Green, R. M. (1959). Measurements with a whole body counter. *Can. J. Physics*, **37**, 683–9.

McNeill, K. G. & Harrison, J. E. (1981). Partial body neutron activation-truncal. In *Non-invasive Measurements of Bone Mass and Their Clinical Application*, ed. S. H. Cohn, pp. 165–90. Boca Raton: CRC Press.

McNeill, K. G., Mernagh, J. R., Jeejebhoy, K. N., Wolman, S. L. & Harrison, J. E. (1979a). In vivo measurements of body protein based on the determination of prompt gamma analysis. *Am. J. Clin. Nutr.*, **32**, 1955–61.

McNeill, K. G., Mernagh, J. R., Jeejebhoy, K. N. & Harrison, J. E. (1979*b*). In *Nuclear Activation Techniques in the Life Sciences, 1978*. Vienna: International Atomic Energy Agency.

McNeill, K. G., Harrison, J. E., Mernagh, J. R., Stewart, J. S. & Jeejebhoy, K. N. (1982). Changes in body protein, body potassium and lean body mass during total parental nutrition. *J. Parent. Enterol. Nutr.,* **6**, 106–8.

Meador, C. K., Kreisberg, R. A., Friday, J. P., Bowdoin, B., Coan, P., Armstrong, J. & Hazelrig, J. B. (1968). Muscle mass determination by isotopic dilution of creatine[14]C. *Metabolism,* **17**, 1104–8.

Meleski, B. W., Shoup, R. F. & Malina, R. M. (1982). Size, physique and body composition of competitive female swimmers 11 through 20 years of age. *Hum. Biol.,* **54**, 609–25.

Mellits, E. D. & Cheek, D. B. (1970). The assessment of body water and fatness from infancy to adulthood. *Monogr. Soc. Res. Child Dev.,* **35**, 12–26.

Mendez, J. & Lukaski, H. C. (1981). Variability of body density in ambulatory subjects measured on different days. *Am. J. Clin. Nutr.,* **34**, 78–81.

Mendez, J., Keys, A., Anderson, T. & Grande, F. (1960). Density of fat and bone mineral of mammalian body. *Metabolism,* **9**, 472–77.

Mendez, J., Prokop, E., Picon-Reategui, E., Akers, R. & Buskirk, E. R. (1970). Total body water by D_2O dilution using saliva samples and gas chromatography. *J. Appl. Physiol.,* **28**, 354–7.

Mendez, J., Lukaski, H. C. & Buskirk, E. R. (1984). Fat-free mass as a function of max. O_2 consumption and 24-hour creatinine and 3-methyl histidine excretion. *Am. J. Clin. Nutr.,* **39**, 710–5.

Meneely, G. R., Ball, C. O. T., Ferguson, J. L., Payne, D. D., Lorimer, A. R., Weiland, R. L., Rolf, M. L. & Heyssel, R. M. (1962). Use of computers in measuring body electrolytes by gamma spectrometry. *Circ. Res.,* **11**, 539–48.

Merklin, R. J. (1974). Growth and distribution of human fetal brown fat. *Anat. Rec.,* **178**, 637–46.

Mernagh, J. R., Harrison, J. E. & McNeill, D. G. (1977). In vivo determination of nitrogen using Pu-Be sources. *Phys. Med. Biol.,* **22**, 831–5.

Mernagh, J. R., Harrison, J. E., Krondl, A., McNeill, K. G. & Shephard, R. J. (1986). Composition of lean tissue in healthy volunteers for nutritional studies in health and disease. *Nutr. Res.,* **6**, 499–507.

Merz, A. L., Trotter, M. & Peterson, R. R. (1956). Estimation of skeletal weight in the living. *Am. J. Phys. Anthrop.,* **14**, 589–609.

Mészaros, J., Szmodis, I., Mohacsi, J. & Szabo, T. (1984). Prediction of final stature at the age of 11–13 years. In *Children and Sport,* ed. J. Ilmarinen & I. Valimaki, pp. 30–6. Berlin: Springer Verlag.

Mészaros, J., Mohacsi, J., Szabo, T. & Szmodis, I. (1985). Assessment of biological development by anthropometric variables. In *Children and Exercise XI,* ed. R. A. Binkhorst, H. C. G. Kemper & W. H. M. Saris, pp. 341–5. Champaign, Ill.: Human Kinetics Publishers.

Miall, W. E., Ashcroft, M. T., Lovell, H. G. & Moore, F. (1967). A longitudinal study of the decline in adult height with age in two Welsh communities. *Hum. Biol.,* **39**, 445–54.

Metropolitan Life Insurance Company (1983). 1983 Metropolitan height and weight tables. *Stat. Bull. Metropolitan Life Insurance Co.*, **64** (Jan–June), 64.

Mettau, J. W. (1978). *Measurement of Total Body Fat in Low Birth Weight Infants.* Rotterdam: Bronder Offset BV.

Mettau, J. W., Degenhart, N. J., Visser, H. K. A. & Holland, W. P. S. (1977). Measurement of total body fat in newborns and infants by absorption and desorption of non-radioactive xenon. *Pediatr. Res.*, **11**, 1097–101.

Meyer, R. A., Kushmerick, M. J. & Brown, T. R. (1982). Application of ^{31}P NMR spectroscopy to the study of striated muscle metabolism. *Am. J. Physiol.*, **242**, C1–C11.

Miller, A. T. & Blyth, C. S. (1952). Estimation of lean body mass and body fat from basal oxygen consumption and creatinine excretion. *J. Appl. Physiol.*, **5**, 73–8.

Miller, C. E. & Remenchik, A. P. (1963). Problems involved in accurately measuring the K content of the human body. *Ann. N.Y. Acad. Sci.*, **110**, 175–88.

Miller, D. S. & Mumford, P. (1967). Gluttony: I. An experimental study of low or high protein diets. II. Thermogenesis in overeating man. *Am. J. Clin. Nutr.*, **20**, 1212–22; 1223–29.

Miller, M. E. & Cappon, C. J. (1984). Anion-exchange chromatographic determination of bromide in serum. *Clin. Chem.*, **30**, 781–3.

Miller, M. E., Ward, L., Thomas, B. J., Cooksley, W. G. E. & Shepard, R. W. (1982). Altered body composition and muscle protein degradation in nutritionally growth-retarded children with cystic fibrosis. *Am. J. Clin. Nutr.*, **36**, 492–9.

Millward, D. J., Bates, P. C., Grimble, G. K., Brown, J. G., Nathan, M. & Rennie, M. J. (1980). Quantitative importance of non-skeletal muscle sources of N-methylhistidine in urine. *Biochem. J.*, **190**, 225–8.

Milne, J. S. (1985). *Clinical Effects of Ageing: A Longitudinal Study.* London: Croom Helm.

Minokoshi, Y., Saito, M. & Shimazu, T. (1988). Sympathetic activation of lipid synthesis in brown adipose tissue in the rat. *J. Physiol.*, **398**, 361–70.

Mirwald, R. L. & Bailey, D. A. (1986). *Maximal Aerobic Power.* London, Ont.: Sport Dynamics.

Mitchell, H. H., Hamilton, T. S., Streggerda, F. R. & Bean, H. W. (1945). The chemical composition of the adult human body and its bearing on the biochemistry of growth. *J. Biol. Chem.*, **158**, 625–37.

Mittelman, K., Crawford, S., Bhakthan, G., Gutman, G. & Holliday, S. (1986). Anthropometric changes in older cyclists: effects of a trans-Canada bicycle tour. In *Perspectives in Kinanthropometry*, ed. J. A. P. Day, pp. 107–13. Champaign, Ill.: Human Kinetics Publishers.

Moleschott, J. (1859). *Physiologie der Nahrungsmittel – Ein Handbuch der Diatetik.* Giessen: Ferbersche Universitatsbuchhandlung.

Möller, P., Bergström, J., Eriksson, S., Furst, P. & Hellström, K. (1979). Effect of aging on free amino acids and electrolytes in leg skeletal muscle. *Clin. Sci.*, **56**, 427–32.

Mollison, P. O., Veall, N. & Cutbush, M. (1950). Red cell volume and plasma volume in newborn infants. *Arch. Dis. Childh.*, **25**, 242–53.

Montegriffo, V. M. E. (1971). A survey of the incidence of obesity in the United Kingdom. *Postgrad. Med. J.,* **47** Suppl., 418–22.

Montoye, H. J. (1984). Exercise and osteoporosis. In *Exercise and Health,* ed. H. Eckert & H. J. Montoye, pp. 59–73. Champaign, Ill.: Human Kinetics Publishers.

Montoye, H. J. & Gayle, R. (1978). Familial relationships in maximal oxygen uptake. *Hum. Biol.,* **50**, 241–50.

Moore, F. D., Olesen, K. H., McMurray, J. D., Parker, H. V., Ball, M. R. & Boyden, C. M. (1963). *The Body Cell Mass and the Supporting Environment.* Philadelphia: W. B. Saunders.

Moore-Ede, M. C., Brennan, M. F. & Ball, M. R. (1975). Circadian variation of intercompartmental potassium fluxes in man. *J. Appl. Physiol.,* **38**, 163–70.

Moran, L., Custer, P. & Murphy, G. (1980). Nutritional assessment of lean body mass. *J. Parent. Ent. Nutr.,* **4**, 595 (Abstract).

Morey-Holton, E. R. & Arnaud, S. B. (1985). Spaceflight and calcium metabolism. *Physiologist,* **28** (6), S9–12.

Morgan, D. B. & Burkinshaw, L. (1983). Estimation of non-fat body tissues from measurements of skinfold thickness, total body potassium and total body nitrogen. *Clin. Sci.,* **65**, 407–14.

Mostardi, R., Porterfield, J. A. & Greenberg, B. (1982). The physiology of ballet dancing. *Fed. Proc.,* **41**, 1676.

Motil, K. J., Grande, R. J., Mathews, D. E., Bier, D. M., Maletskos, C. J. & Young, V. R. (1982). Whole body leucine metabolism in adolescents with Crohn's disease and growth failure during nutritional supplementation. *Gastroenterology,* **82**, 1359–68.

Moulton, C. R. (1923). Age and chemical development in mammals. *J. Biol. Sci.,* **57**, 79–97.

Mount, L. E. (1979). *Adaptation to Thermal Environment. Man and his Productive Animals.* Baltimore: University Park Press.

Moynihan, C. M., Stark, O. & Peckham, C. S. (1986). Obesity in 16-year-olds assessed by relative weight and doctors' rating. *Int. J. Obesity,* **10**, 3–10.

Mueller, W. H. (1978). Transient environmental changes and age-limited genes as causes of variation in sib–sib and parent–offspring correlations. *Ann. Hum. Biol.,* **5**, 395–8.

Mueller, W. H. (1983). The genetics of human fatness. *Yb. Phys. Anthrop.,* **26**, 215–30.

Mueller, W. H. & Reid, R. M. (1979). A multivariate analysis of fatness and relative fat patterning. *Am. J. Phys. Anthrop.,* **50**, 199–208.

Mueller, W. H. & Stallones, L. (1981). Anatomical distribution of subcutaneous fat: skinfold site choice and construction of indices. *Hum. Biol.,* **53**, 321–35.

Mueller, W. H. & Wohlieb, J. C. (1981). Anatomical distribution of subcutaneous fat and its description by multivariate methods. How valid are principal components? *Am. J. Phys. Anthrop.,* **54**, 25–35.

Mueller, W. H., Ven, F., Soto, P., Schull, V. N., Rothhammer, F. & Schull, W. J. (1979). A multinational Andean genetic and health program. VIII. Lung function changes with migration between altitudes. *Am. J. Phys. Anthrop.,* **51**, 183–96.

Mueller, W. H., Joos, S. K., Hanis, C. L., Zavaleta, A. N., Eichner, J. & Schull, W. J. (1984). The Diabetes Alert study: Growth, fatness and fat patterning, adolescence through adulthood in Mexican Americans. *Am. J. Phys. Anthrop.*, **64**, 389–99.

Mueller, W. H., Deutsch, M. I., Malina, R. M., Bailey, D. A. & Mirwald, R. L. (1986). Subcutaneous fat topography: age changes and relationship to cardiovascular fitness in Canadians. *Hum. Biol.*, **58**, 955–73.

Mukherjee, D. & Roche, A. F. (1984). The estimation of per cent body fat, body density and total body fat by maximum R^2 regression equations. *Hum. Biol.*, **56**, 79–109.

Müller, A. (1911). Stoffwechsel- und Respirationsversuche zur Frage der Eiweissmast. *Z. Physiol. Pathol. Stoffwechsels*, **6**, 617–29.

Munro, A., Joffe, A., Ward, J. S., Wyndham, C. H. & Fleming, P. W. (1966). An analysis of the errors in certain anthropometric measurements. *Int. Z. Angew. Physiol.*, **23**, 93–106.

Murray, S. & Shephard, R. J. (1984). Relationships between neck and waist girths, skinfold readings and hydrostatic estimates of body fat in young women. *Can. J. Appl. Spt. Sci.*, **9**, 6P.

Murray, S. & Shephard, R. J. (1988). Possible anthropometric alternatives to skinfold measurements. *Hum. Biol.*, **60**, 273–82.

Murray, S. J., Shephard, R. J., Greaves, S., Allen, C. & Radomski, M. (1986). Effects of cold stress and exercise on fat loss in females. *Eur. J. Appl. Physiol.*, **55**, 610–18.

Myhre, L. G. & Kessler, N. V. (1966). Body density and potassium 40 measurements of body compositon as related to age. *J. Appl. Physiol.*, **21**, 1251–5.

Nadeshdin, W. A. (1932). Zur Untersuchung der Minderwertigkeit der Organe an Leichen. *Deutsche Z. ges. gerichtl. Med.*, **18**, 426–31.

Naeye, R. L. & Roode, P. (1970). The sizes and numbers of cells in the visceral organs in human obesity. *Am. J. Clin. Pathol.*, **54**, 251–3.

Naeye, R. L., Benirschke, K., Hagstrom, J. W. C. & Marcus, C. C. (1966). Intrauterine growth of twins as estimated from liveborn birthweight data. *Pediatrics*, **37**, 409–16.

Nagamine, S. (1975). Bone formation and nutritional status in Japanese children and adults. In *Human Adaptability*, Vol. 3. *Physiological Adaptability and Nutritional Status of the Japanese. B. Growth, Work Capacity and Nutrition of Japanese*, ed. K. Asahian & R. Shigiya, pp. 94–9. Tokyo: Japanese Committee for the International Biological Programme, Science Council of Japan.

Narici, M. V., Roi, G. S. & Landoni, L. (1988). Force of knee extensor and flexor muscles and cross-sectional area determined by nuclear magnetic resonance imaging. *Eur. J. Appl. Physiol.*, **57**, 39–44.

Nathan, D. G., Piomelli, S., Cummins, J. F., Gardner, F. H. & Limauro, A. L. (1963). The effect of androgen on some aspects of body composition and erythropoiesis in octagenarian males. *Ann. N.Y. Acad. Sci.*, **110**, 965–77.

National Institutes of Health (1986). Consensus development panel on the health implications of obesity. *Ann. Int. Med.*, **103**, 1073–7.

Needham, J. (1950). *Biochemistry and Morphogenesis*. Cambridge: Cambridge University Press.

Neel, J. M. (1962). Diabetes mellitus: a thrifty genotype rendered detrimental by 'progress'. *Am. J. Hum. Genet.*, **14**, 354–62.

Nelp, W. B., Palmer, H. E., Murano, R., Pailthorp, K., Hinn, G. M., Rich, C., Williams, J. L., Rudd, T. G. & Denney, J. D. (1970). Measurement of total body calcium (bone mass) in vivo with the use of total body neutron activation and analysis. *J. Lab. Clin. Med.*, **76**, 151–62.

Nelp, W. B., Denney, J. D., Murano, R., Hinn, G. M., Williams, J. L., Rudd, T. G. & Palmer, H. E. (1972). Absolute measurement of total body calcium (bone mass) in vivo. *J. Lab. Clin. Med.*, **79**, 430–8.

Nelson, E. A. & Craig, A. B. (1978). Physiologic responses to a transcontinental bicycle trip. *Phys. Sportsmed.*, **6**, 83–93.

Nelson, J. K. & Nelson, K. R. (1986). Skinfold profiles of black and white boys and girls ages 11–13. *Hum. Biol.*, **58**, 379–90.

Nelson, M. E., Fisher, E. C., Catsos, P. D., Meredith, C. N., Turksoy, R. N. & Evans, W. J. (1986). Diet and bone status in amenorrheic runners. *Am. J. Clin. Nutr.*, **43**, 910–16.

Nelson, R. A. (1982). Nutrition and physical performance. *Phys. Sportsmed.*, **10** (4), 55–63.

Neumann, G., Jelliffe, D. B., Zerfas, A. J. & Jelliffe, E. F. (1982). Nutritional assessment of the child with cancer. *Cancer Res.*, **42**, 699S–712S.

Newman, M. T. (1960). Adaptations in the physique of American Aborigines to nutritional factors. *Hum. Biol.*, **32**, 288–313.

Newham, M. T. (1961). Biological adaptation of man to his environment: heat, cold, altitude and nutrition. *Ann. N.Y. Acad. Sci.*, **91**, 617–33.

Nichols, B. L., Alleyne, G. A. O., Barnes, D. J. & Hazlewood, C. F. (1969). Relationship between muscle potassium and total body potassium in infants with malnutrition. *J. Pediatr.*, **74**, 49–57.

Nilsson, B. E. & Westlin, N. E. (1971). Bone density in athletes. *Clin. Orthop. Rel. Res.*, **77**, 179–82.

Nishizawa, N., Noguchi, T. & Hareyama, S. (1977). Fractional flux rates of N-methylhistidine in skin and gastrointestine. The contribution of these tissues to urinary excretion of N-methyl histidine in the rat. *Br. J. Nutr.*, **38**, 149–51.

Noble, B. J. (1986). *Physiology of Exercise and Sport*. St. Louis: Times–Mirror/Mosby College.

Nobmann, E. D. (1981). Health and nutrition attitudes, knowledge and practices of Southwest Alaskan Eskimo adolescents. In *Circumpolar Health* 81, ed. B. Harvald & J. P. Hart Hansen, pp. 40–6. Copenhagen: Nordic Council for Arctic Medical Research.

Noppa, H., Andersson, M., Bengtsson, C., Bruce, A. & Isaksson, B. (1979). Body composition in middle-aged women with special reference to the correlation between body fat mass and anthropometric data. *Am. J. Clin. Nutr.*, **32**, 1388–95.

Nordin, B., Barnett, E., MacGregor, J. & Nisbet, J. (1962). Lumbar spine densitometry. *Br. Med. J.*, **1**, 1793–6.

Norgan, N. G. (1982). Human energy stores. In *Energy and Effort*, ed. G. A. Harrison, pp. 139–58. London: Taylor & Francis.

Norgan, N. G. & Durnin, J. V. G. A. (1980). The effects of 6 weeks of over-feeding on the body weight, body composition and energy metabolism of young men. *Am. J. Clin. Nutr.*, **33**, 978–88.

Norgan, N. G. & Ferro-Luzzi, A. (1982). Weight–height indices as estimators of fatness in men. *Hum. Nutr. Clin. Nutr.*, **36C**, 363–72.

Norgan, N. G. & Ferro-Luzzi, A. (1985). The estimation of body density in men. Are equations general? *Ann. Hum. Biol.*, **12**, 1–15.

Novak, L. P. (1963). Age and sex differences in body density and creatinine excretion of high school children. *Ann. N.Y. Acad. Sci.*, **110**, 545–76.

Novak, L. P. (1970). Comparative study of body composition of American and Filipino women. *Hum. Biol.*, **42**, 206–16.

Novak, L. P. (1972). Aging, total body potassium, fat-free mass and cell mass in males and females between ages 18 and 85 years. *J. Gerontol.*, **27**, 438–43.

Novak, L. P., Tauxe, W. N. & Orvis, A. L. (1973). Estimation of total body potassium in normal adolescents by whole body counting: age and sex differences. *Med. Sci. Sports*, **5**, 147–55.

Novotny, V. (1981). Veranderungen des Knochenalters im Verlauf einer mehrjahrigen sportlicher Belastung. *Med. Sport*, **21**, 44–7.

Noyons, A. K. M. & Jongbloed, J. (1935). Uber die Bestimmung des wahren Volumens und des spezifischen Gewichtes von Mensch und Tier mit Hilfe von Luftdruckveranderung. *Pflüg. Arch. ges. Physiol.*, **235**, 588–96.

Nukada, A. (1975). Industrialization as a factor for secular increase in physiques of school children in Japan. In *Human Adaptability*, Vol. 3. *Physiological Adaptability and Nutritional Status of the Japanese. B. Growth, Work Capacity and Nutrition of Japanese*, ed. K. Asahina & R. Shigiya, pp. 107–11. Tokyo: Japanese Committee for the International Biological Programme, Science Council of Japan.

Nunneley, S. A., Wissler, E. H. & Allan, J R. (1985). Immersion cooling. Effect of clothing and skinfold thickness. *Aviat. Space Environ. Med.*, **56**, 1177–82.

Nyboer, J. (1970). Electrorheometric properties of tissues and fluids. *Ann. N.Y. Acad. Sci.*, **170**, 410–20.

Nylander, I. (1971). The feeling of being fat and dieting in a school population. *Acta Socio-med. Scand.*, **1**, 17–26.

Oakley, J. R., Parsons, R. J. & Whitelaw, A. G. L. (1977). Standards for skinfold thickness in British newborn infants. *Arch. Dis. Childh.*, **52**, 287–90.

Oberhausen, E. & Onstead, C. O. (1965). Relationship of potassium content of man with age and sex. In *Radioactivity in Man*, II, ed. G. R. Meneely & S. M. Linde, pp. 179–85. Springfield, Ill.: C. C. Thomas.

Odeblad, E. (1966). Micro NMR in high permanent magnetic fields. *Acta Obstetr. Gynecol. Scand.*, **45**, 1–88.

Oganov, V. S., Skuratova, S. A., Potapov, A. N. & Shirvinskaya, M. A. (1980). Physiological mechanisms of adaptation of rat skeletal muscles to weightlessness and similar functional requirements. *Physiologist*, **23** (6), S16–21.

O'Hara, W. J., Allen, C. & Shephard, R. J. (1977a). Loss of body fat during an arctic winter expedition. *Can. J. Physiol. Pharm.*, **55**, 1235–41.

O'Hara, W. J., Allen, C. & Shephard, R. J. (1977b). Loss of weight and fat during exercise in a cold chamber. *Eur. J. Appl. Physiol.*, **37**, 205–18.

O'Hara, W. J., Allen, C. & Shephard, R. J. (1977c). Treatment of obesity by exercise in the cold. *Can. Med. Assoc. J.*, **117**, 773–86.

O'Hara, W., Allen, C., Shephard, R. J. & Allen, G. (1979). Fat loss in the cold: a controlled study. *J. Appl. Physiol.*, **46**, 872–7.

Oppliger, R. A., Looney, M. A. & Tipton, C. M. (1987). Reliability of hydrostatic weighing and skinfold measurements of body composition using a generalizability study. *Hum. Biol.*, **59**, 77–96.

Orchard, J. W., Evers, S. E. & Haddad, R. G. (1984). Identifying postmenopausal patients at risk of significant bone loss. *Can. Fam. Phys.*, **30**, 2503–8.

Orpin, M. J. & Scott, P. H. (1984). Estimation of total body fat using skinfold caliper measurements. *N.Z. Med. J.*, **63**, 501.

Osborn, R. H. & DeGeorge, F. V. (1959). *Genetic Basis of Morphological Variation*. Cambridge, Mass.: Harvard University Press.

Oscai, L. B. & Holloszy, J. O. (1969). Effects of weight changes produced by exercise, food restriction or overeating on body composition. *J. Clin. Invest.*, **48**, 2124–8.

Oscai, L. B., Babirak, S. P., McGarr, J. A. & Spirakis, C. N. (1974). Effect of exercise on adipose tissue cellularity. *Fed. Proc.*, **33**, 1956–8.

Osserman, E. F., Pitts, G. C., Welham, W. C. & Behnke, A. R. (1950). In vivo measurement of body fat and body water in a group of normal men. *J. Appl. Physiol.*, **2**, 633–9.

Osterback, L. L. & Viitsalo, J. (1986). Growth selection of young boys participating in different sports. In *Children and Exercise* XII, ed. J. Rutenfranz, R. Mocellin & F. Klimt, pp. 373–80. Champaign, Ill.: Human Kinetics Publishers.

Ostrove, S. M. & Vaccaro, P. (1982). Effect of immersion on RV in young women: implications for measurement of body density. *Int. J. Spts Med.*, **3**, 220–3.

Pallier, E., Max, J. P., Burlet, C. & Debry, G. (1985). Human and animal fat cell size determinations using an image analysing computer. *Am. J. Clin Nutr.*, **41**, 818–20.

Palomino, H., Mueller, W. H. & Schull, W. J. (1979). Altitude heredity, and body proportions in northern Chile. *Am. J. Phys. Anthrop.*, **50**, 39–50.

Palsson, J. O. P. (1981). Anthropometric studies of Icelandic women. In *Circumpolar Health* 81, ed. B. Harvald & J. P. Hart Hansen, pp. 194–200. Copenhagen: Nordic Council for Medical Research.

Panksepp, J. (1974). Hypothalamic regulation of energy balance and feeding behaviour. *Fed. Proc.*, **33**, 1150–65.

Parfitt, A. M. & Whedon, G. D. (1982). Calcium nutrition and bone health in the elderly. *Am. J. Clin. Nutr.*, **36**, 986–1013.

Parizkova, J. (1974). Particularities of lean body mass and fat development in growing boys as related to their motor activity. *Acta Paediatr. Scand.*, **28**, Suppl., 233–44.

Parizkova, J. (1977). *Body Fat and Physical Fitness*. The Hague: Martinus Nijhoff.

Parizkova, J. & Goldstein, H. (1970). A comparison of skinfold thickness using the Best and Harpenden calipers. *Hum. Biol.*, **42**, 436–41.

Parizkova, J. & Roth, Z. (1972). The assessment of depot fat in children from skinfold thickness measurements by Holtain (Tanner/Whitehouse) caliper. *Hum. Biol.*, **44**, 613–620.

Parizkova, J., Vaneckova, M., Sprynarova, S. & Vamberova, M. (1971). Body composition and fitness in obese children before and after special treatment. *Acta Paediatr. Scand.*, **217**, 80–5.

Parizkova, J., Merhautova, J. & Prokopec, M. (1972). Comparaison entre la croissance des jeunes Tunisiens et celle des jeunes Tchéques âgés de 11 à 12 ans. *Biométrie Humaine* (Paris), **7**, 1–10.

Parnell, R. W. (1965). Human size, shape and composition. In *Human Body Composition. Approaches and Applications*, ed. J. Brozek, pp. 61–72. Oxford: Pergamon Press.

Parra, A., Argote, R. M., Garcia, G., Cervantes, C., Alatorre, S. & Pérez-Pastén, E. (1979). Body composition in hypopituitary dwarfs before and during human growth hormone therapy. *Metabolism*, **28**, 851–7.

Pascale, L. B., Frankel, T., Freeman, S., Faller, I. L. & Bond, E. E. (1954). *A Means of Measuring Total Body Water In Humans without Venipuncture.* US Army Med. Nutr. Lab. Rep. Denver, 135.

Pascale, L. R., Grossman, M. I., Sloane, H. S. & Frankel. T. (1956). Correlations between thickness of skinfolds and body density in 88 soldiers. *Hum. Biol.*, **28**, 165–76.

Passmore, R., Strong, J. A., Swindells, Y. E. & Eldin, N. (1963). The effect of overfeeding on two fat young women. *Br. J. Nutr.*, **17**, 373–83.

Patterson-Buckendahl, P., Arnaud, S. B., Mechanic, G. L., Martin, R. B., Grindeland, R. E. & Cann, C. E. (1987). Fragility and composition of growing rat bone after one week of space flight. *Am. J. Physiol.*, **262**, R240–6.

Pawan, G. E. S. & Clode, M. (1960). The gross chemical composition of subcutaneous adipose tissue in the mean and obese human subject. *J. Biochem.*, **74**, 9P.

Pawson, I. G. (1977). Growth characteristics of a population of Tibetan origin in Nepal. *Am. J. Phys. Anthropol.*, **47**, 473–82.

Pearson, A. M. (1963). Implications of research on body composition for animal biology: An introductory statement. *Ann. N.Y. Acad. Sci.*, **110**, 291–301.

Peppler, W. W. & Mazess, R. B. (1981). Total body bone mineral and lean body mass by dual-photon absorptiometry. I. Theory and measurement procedure. *Calcif. Tiss. Internat.*, **33**, 353–9.

Perry, L. & Leornard, B. (1963). Obesity and mental health. *J. Am. Med. Assoc.*, **183**, 807–8.

Pertschuk, M. J., Corsby, L. O. & Mullen, J. L. (1983). Non-linearity of weight gain and nutrition intake in anorexia nervosa. In *Anorexia Nervosa: Recent Developments in Research*, ed. P. L. Darby, P. E. Garfinkel, D. M. Garner & D. V. Coscina, pp. 301–10. New York: A. Liss.

Pérusse, L., Lortie, G., Leblanc, C., Tremblay, A., Thériault, G. & Bouchard, C. (1987). Genetic and environmental sources of variation in physical fitness. *Ann. Hum. Biol.*, **14**, 425–34.

Pérusse, L., LeBlanc, C. & Bouchard, C. (1988*a*). Inter-generation transmission of physical fitness in the Canadian population. *Can. J. Spt. Sci.*, **13**, 8–14.

Pérusse, L., Tremblay, A., Leblanc, C. & Bouchard, C. (1988*b*). Genetic and environmental influences on level of habitual physical activity and exercise participation. *Am. J. Epidemiol.*, **129**, 1012–22.

Pérusse, L., Tremblay, A., Leblanc, C., Cloninger, C. R., Reich, T., Rice, J. &

Bouchard, C. (1988c). Familial resemblance in energy intake: contribution of genetic and environmental factors. *J. Clin. Nutr.*, **47**, 629–35.

Peterson, R. E., O'Toole, J. J., Kirkendall, W. W. & Kempthorn, O. (1959). The variability of extracellular fluid space (sucrose) in man during a 24 hour period. *J. Clin. Invest.*, **38**, 1644–58.

Pethig, R. (1979). *Dielectric and Electronic Properties of Biological Materials*. Chichester: John Wiley.

Pett, L. B. & Ogilvie, G. F. (1956). The Canadian weight–height survey. *Hum. Biol.*, **28**, 177–88.

Pfau, A., Kallistratos, G. & Schröder, J. (1961). Zur Bestimmung des Fleischgehaltes im Schweineschincken mit Hilfe von 40-K-Gammaaktivitätsmessungen. *Atompraxis*, **8**, 279–84.

Picou, D., Halliday, D. & Garrow, J. S. (1966). Total body protein, collagen and non-collagen protein in infantile protein malnutrition. *Clin. Sci.*, **30**, 345–51.

Pierson, R. N., Lin, D. H. Y. & Phillips, R. A. (1974). Total body potassium in health: effects of age, sex, height and fat. *Am. J. Physiol.*, **226**, 206–12.

Pierson, R. N., Wang, J., Colt, E. W. & Neumann, P. (1982). Body composition measurements in normal man: the potassium, sodium, sulfate and tritium spaces in 58 adults. *J. Chron. Dis.*, **35**, 418–28.

Pierson, W. R. (1963). A photogrammetric technique for the estimation of surface area and volume. *Ann. N.Y. Acad. Sci.*, **110**, 109–12.

Pincherle, G. (1971). Obesity and smoking habits. *Br. Med. J.*, **4**, 298.

Pinson, E. A. (1952). Water exchanges and barriers as studied by the use of hydrogen isotopes. *Physiol. Rev.*, **32**, 123–34.

Pirke, K. M., Pahl, J., Schweiger, U. & Warnhoff, M. (1985). Metabolic and endocrine indices of starvation in bulimia: a comparison with anorexia nervosa. *Psychiatr. Res.*, **15**, 33–9.

Pittet, Ph., Chappins, Ph., Acheson, K., de Techtermann, F. & Jequier, E. (1976). Thermic effect of glucose in obese subjects studied by direct and indirect calorimetry. *Br. J. Nutr.*, **35**, 281–92.

Plato, C. C., Wood, J. L. & Norris, A. H. (1980). Bilateral asymmetry in bone measurements of the hand and lateral hand dominance. *Am. J. Phys. Anthrop.*, **52**, 27–31.

Plocher, T. A. & Powley, T. L. (1976). Effect of hypophysectomy on weight gain and body composition of the genetically obese yellow (A^y/a) mouse. *Metabolism*, **25**, 593–602.

Pocock, N. A., Eisman, J. A., Yeates, M. G., Sambrook, P. N. & Eberl, S. (1986). Physical fitness is a major determinant of femoral neck and lumbar spine bone mineral density. *J. Clin. Invest.*, **78**, 618–21.

Polgar, G. & Promadht, V. (1971). *Pulmonary Function Testing in Children: Techniques and Standards*, p. 123. Philadelphia: W. B. Saunders.

Pollock, M. & Jackson, A. S. (1984). Research progress in validation of clinical methods of assessing body composition. *Med. Sci. Sports Exerc.*, **16**, 606–13.

Pollock, M. L., Cureton, T. K. & Greninger, L. (1969). Effects of frequency of training on working capacity, cardiovascular function, and body composition of adult men. *Med. Sci. Sports Exerc.*, **1**, 70–4.

Pollock, M. L., Broida, J., Kendrick, Z., Miller, H. S., Janeway, R. & Linnerud, A. C. (1972). Effects of training two days per week at different intensities on middle-aged men. *Med. Sci. Sports Exerc.*, **4**, 192–7.

Pollock, M. L., Getman, L. R., Jackson, A., Ayres, J., Ward, A. & Linnerup, A. (1977). Body composition of elite class distance runners. *Ann. N.Y. Acad. Sci.*, **301**, 361–70.

Pollock, M. L., Wilmore, J. H. & Fox, S. M. (1984). *Exercise in Health and Disease*. Philadelphia: W. B. Saunders.

Postl, B. (1981). Six year follow up of Keewatin Inuit children born 1973–1974. In *Circumpolar Health 81*, ed. B. Harvald & J. P. Hart Hansen, pp 175–9. Copenhagen: Nordic Council for Arctic Medical Research.

Poulos, P. P., Poulos, J. G., Pifer, M., Van Woert, W. & Parks, M. E. (1956). A constant change (single-injection) method for the estimation of the volume of distribution of substances in body fluid compartments. *J. Clin. Invest.*, **35**, 921–33.

Poznanski, A. K., Kuhns, L. R. & Guire, K. E. (1980). New standards of cortical mass in the humerus of neonates; a means of evaluating bone loss in the premature infant. *Radiology*, **134**, 639–44.

Prahl-Andersen, B., Kowalski, C. J., Heyendael, P. (1979). *A Mixed-Longitudinal Interdisciplinary Study of Growth and Development*. London: Academic Press.

Prefaut, C., Lupi, E. & Anthonisen, N. R. (1976). Human lung mechanics during water immersion. *J. Appl. Physiol.*, **40**, 320–3.

Prentice, A. M. (1984). Adaptations to long-term low energy intake. In *Energy Intake and Activity*, ed. E. Pollitt & P. Amante, pp. 3–31. New York: Liss.

Prentice, A. M., Black, A. E., Coward, W. A., Davies, H. L., Goldberg, G. R., Murgatroyd, P. R. Ashford, J., Sawyer, M. & Whitehead, R. G. (1986). High levels of energy expenditure in obese women. *Br. Med. J.*, **292**, 983–7.

Presta, E., Segal, K. R., Gutin, B., Harrison, G. G. & Van Itallie, T. B. (1983a). Comparison in man of total body electrical conductivity and lean body mass derived from body density: validation of a new body composition method, *Metabolism*, **32**, 524–7.

Presta, E., Wang, J., Harrison, G. G. Björntorp, P., Harker, H. H. & Van Itallie, T. B. (1983b). Measurement of total body electical conductivity; a new method for estimation of body composition. *Am. J. Clin. Nutr.*, **37**, 735–9.

Preston, T., Reeds, P. J., East, B. W. & Holmes, P. H. (1985). A comparison of body protein determinations in rats by in vivo neutron activation and carcass analysis. *Clin. Sci.*, **68**, 349–55.

Price, P. A., Parthemore, J. G. & Deftos, L. J. (1980). New biochemical marker for bone metabolism. Measurement by radioimmunoassay of bone Gla protein in the plasma of normal subjects and patients with bone disease. *J. Clin. Invest.*, **66**, 878–83.

Price, R. R., Wagner, J., Larson, K. H., Patton, J. A., Touya, J. J. *et al.* (1977). Techniques for measuring regional and total bone mineral mass to bone function ratios. *Proceedings of Symposium on Radionuclide Imaging, Los Angeles, 1976*, pp 145–8.

Prosser, C. L. (1964). Perspectives of adaptation. In *Adaptation to Environment*, ed. D. B. Dill, pp. 11–25. Washington, DC: American Physiological Society.

Pryor, H. B. (1940). *Width–Weight Tables for Boys and Girls from 1 to 17 years – for Men and Women from 18 to 41+ years*. Stanford: Stanford University Press.

Pullar, J. D. & Webster, A. J. F. (1974). Heat loss and energy retention during growth in congenitally obese and lean rats. *Br. J. Nutr.*, **31**, 377–92.

Pullar, J. D. & Webster, A. J. F. (1977). The energy cost of fat and protein deposition in the rat. *Br. J. Nutr.*, **37**, 355–63.

Purves, D. (1976). Longterm regulation in the vertebrate peripheral nervous system. In *International Review of Physiology 10. Neurophysiology II*, ed. R. Porter, pp. 125–77. Baltimore: University Park Press.

Quetelet, A. (1835). *Sur l'homme et le développement de ses facultés*. Paris: Bachélier Imprimeur-Libraire.

Raja, C., Singh, R., Bharadwaj, H. & Singh, I. P. (1978). Relationship of extremity volumes with total body volume and body density in young Indian girls. *Hum. Biol.*, **50**, 103–13.

Rambaut, P., Dietlein, L., Vogel, J. & Smith, M. (1972). Comparative study of two direct methods of bone mineral measurement. *Aerospace Med.*, **43**, 646–50.

Rambaut, P. C., Leach, C. S. & Johnson, P. C. (1975). Calcium and phosphorus changes of the Apollo 17 crew members. *Nutr. Metabol.*, **18**, 62–9.

Ramirez, M. E. (1987). Biological variability in a migrating isolate: Tokelau effects of migration on fat patterning in adults. *Hum. Biol.*, **59**, 901–9.

Ramirez, M. E. & Mueller, W. H. (1980). The development of obesity and fat patterning in Tokelau children. *Hum. Biol.*, **52**, 675–87.

Rathbun, E. N. & Pace, N. (1945). Studies on body composition. I. Determination of body fat by means of the body specific gravity. *J. Biol. Chem.*, **158**, 667–76.

Ravussin, E., Schutz, Y., Acheson, K. J., Dusmet, M., Bourquin, L. & Jéquier, E. (1985). Short-term, mixed diet overfeeding in man: no evidence for 'luxuskonsumption'. *Am. J. Physiol.*, **249**, E470–7.

Reeve, J., Veall, N. & Wooton, R. (1978). Problems in the analysis of dynamic tracer studies. *Clin. Sci. Mol. Med.*, **55**, 225–30.

Refsum, H. E. & Strømme, S. B. (1974). Urea and creatinine production and excretion in urine during and after prolonged heavy exercise. *Scand. J. Clin. Lab. Invest.*, **33**, 247–54.

Riechley, K. B., Mueller, W. H., Harris, C. L., Tulloch, B. R., Barton, S. & Schull, W. J. (1987). Centralized obesity and cardiovascular disease risk in Mexican Americans. *Am. J. Epidemiol.*, **125**, 373–86.

Reid, I. R., Mackie, M. & Ibbertson, H. K. (1986). Bone mineral content in Polynesian and white New Zealand women. *Br. Med. J.*, **292**, 1547–8.

Reid, J. T., Bensdadoun, A., Paladines, O. L. & Van Niekerk, B. D. H. (1963). Body water estimations in relation to body composition and indirect calorimetry in animals. *Ann. N.Y. Acad. Sci.*, **110**, 327–42.

Revicki, D. A. & Israel, R. G. (1986). Relationship between body mass indices and measures of body adiposity. *Am. J. Publ. Health*, **76**, 992–4.

Reynolds, E. L. & Grote, P. (1948). Sex differences in the distribution of tissue components in the human leg from birth to maturity. *Anat. Rec.*, **102**, 45–53.

Rice, J. (1981). Cited by Pérusse *et al.* (1988).

Rice, J., Cloninger, R. C. & Reich, T. (1978). Multifactorial inheritance with cultural transmission and assortative mating. I. Description and properties of the unitary models. *Am. J. Hum. Genet.*, **30**, 618–43.

Riggs, B. L., Wahner, H. W., Melton, L. J., Richelson, L. S., Judd, H. L. & Offord, K. P. (1986). Rates of bone loss in the appendicular and axial skeletons of women. Evidence of substantial vertebral bone loss before menopause. *J. Clin. Invest.*, **77**, 1487–91.

Rigotti, N. A., Nussbaum, S. R., Herzog, D. B. & Neer, R. M. (1984). Osteoporosis in women with anorexia nervosa. *N. Engl. J. Med.*, **311**, 1601–6.

Ringe, J. D., Rehpenning, W. & Kuhlencordt, F. (1977). Physiologie Anderung des Mineralgehalts von Radius und Ulna in Abhangigkeit von Lebensalter und Geschlecht. *Fortschr. Geb. Rontgenstr. Nuklearmed. Erganzungband*, **126**, 376–80.

Robbins, N., Karpati, G. & Engel, W. K. (1969). Histochemical and contractile properties in the cross-innervated guinea pig soleus muscle. *Arch. Neurol.*, **20**, 318–29.

Roberts, D. F. (1953). Body weight, races and climate. *Am. J. Phys. Anthrop.*, **11**, 533–58.

Roberts, D. F. (1988). Genetics of working capacity. In *Capacity for Work in the Tropics*, ed. K. J. Collins & D. F. Roberts, pp. 181–92. Cambridge: Cambridge University Press.

Roberts, M. A., Andrews, G. A. & Caird, F. I. (1975). Skinfold thickness on the dorsum of the hand in the elderly. *Age and Ageing*, **4**, 8–15.

Robertson, C. H., Engle, C. M. & Bradley, M. E. (1978). Lung volumes in man immersed to the neck: dilution and plethysmographic techniques. *J. Appl. Physiol.*, **44**, 679–82.

Robinson, M. F. & Watson, P. E. (1965). Day-to-day variations in body-weight of young women. *Br. J. Nutr.*, **19**, 225–35.

Robson, J. R. K., Bazin, M. & Soderstrom, R. (1971). Ethnic differences in skinfold thickness. *Am. J. Clin. Nutr.*, **29**, 864–8.

Roche, A. F. (1984). Research progress in the field of body composition. *Med. Sci. Sports Exerc.*, **16**, 579–83.

Roche, A. F. (1987). Some aspects of the criterion methods for the measurement of body composition. *Hum. Biol.*, **59**, 209–20.

Roche, A. F., Siervogel, R. M., Chumlea, W. C. & Webb, F. (1981). Grading body fatness from limited anthropometric data. *Am. J. Clin. Nutr.*, **34**, 2831–8.

Rode, A. & Shephard, R. J. (1971). Cardiorespiratory fitness of an arctic community. *J. Appl. Physiol.*, **31**, 519–26.

Rode, A. & Shephard, R. J. (1973a). Growth, development and fitness of the Canadian Eskimo. *Med. Sci. Sports*, **5**, 161–9.

Rode, A. & Shephard, R. J. (1973b). Fitness of the Canadian Eskimo: the influence of season. *Med. Sci. Sports*, **5**, 170–3.

Rode, A. & Shephard, R. J. (1984). Ten years of 'civilisation': fitness of Canadian Inuit. *J. Appl. Physiol. Resp. Environ. Exerc. Physiol.*, **56**, 1472–7.

Roessler, G. S. & Dunavant, B. G. (1967). Comparative evaluation of a whole body counter potassium 40 method for measuring lean body mass. *Am. J. Clin. Nutr.*, **20**, 1171–8.

Rohrer. (1908). Eine Neue Formel zum Bestimmung der Korperfulle. M.m.W. cited by Hirata (1979).

Rolland-Cachera, M. F., Sempe, M., Guilloud-Bataille, M., Patios, E., Pequignot-Guggenbuhl, F. & Fautrad, V. (1982). Adiposity indices in children. *Am. J. Clin. Nutr.*, **36**, 178–84.

Roos, B. O. & Skoldborn, H. (1974). Dual photon absorptiometry in lumbar vertebrae. I. Theory and method. *Acta Radiol.*, **13**, 226–80.

Rosenbaum, S. & Skinner, R. K. (1985). A survey of heights and weights of adults in Great Britain, 1980. *Ann. Hum. Biol.*, **12**, 115–27.

Ross, W. D. & Marfell-Jones, M. J. (1982). Kinanthropometry. In *Physiological testing of the elite athlete*, ed. J. D. MacDougall, H. A. Wenger & H. J. Green, pp. 76–115. Ottawa: Canadian Association of Sports Sciences.

Ross, W. D. & Ward, R. (1984). Proportionality of Olympic athletes. In *Physical Structure of Olympic Athletes*. Part II. *Kinanthropometry of Olympic Athletes*, ed. J. E. L. Carter, pp. 110–143. Basel: Karger.

Ross, W. D., Drinkwater, D. T., Bailey, D. A., Marshall, G. R. & Leahy, R. M. (1980). Kinanthropometry: traditions and new perspectives. In *Kinanthropometry II*, ed. M. Ostyn, G. Beunen & J. Simons, pp. 3–27. Baltimore: University Park Press.

Ross, W. D., Ward, R., Leahy, R. M. & Day, J. A. P. (1982). Proportionality of Montreal athletes. In: *Physical Structure of Olympic Athletes*. Part I. *The Montreal Olympic Games Anthropological Project*, ed. J. E. L. Carter, pp. 81–106. Basel: Karger.

Ross, W. D., Eiben, O. G., Ward, R., Martin, A. D., Drinkwater, D. T. & Clarys, J. P. (1986). Alternatives for the conventional methods of human body composition and physique assessment. In *Perspectives in Kinanthropometry*, ed. J. A. P. Day, pp. 203–20. Champaign, Ill.: Human Kinetics Publishers.

Rother, P., Hunger, H., Vahle, H. & Rother, B. (1973). Uber die Rekonstruktion der Korperhohe aus den Massen langer Rohrenknochen sowie uber den Einfluss des Alterns und der Akzeleration auf die Korperhohe und die vertikalen Proportionen des Menschen. *Gegenbaurs morph. Jb.*, **119**, 767.

Rothwell, N. J. & Stock, M. J. (1979). Regulation of energy balance in two models of reversible obesity in the rat. *J. Comp. Physiol. Psychol.*, **93**, 1024–34.

Rothwell, N. J. & Stock, M. J. (1981). Thermogenesis: comparative and evolutionary considerations. In *The Body Weight Regulatory System: Normal and Disturbed Mechanisms*, ed. L. A. Cioffi, W. P. T. James & T. B. Van Itallie, New York: Raven Press.

Rowe, J. W. (1985). Alterations in renal function. In *Principles of Geriatric Medicine*, ed. R. Andres, E. L. Bierman & W. R. Hazzard, pp. 319–24. New York: McGraw-Hill.

Rowe, J. W., Andres, R., Tobin, J. D., Norris, A. H. & Shock, N. W. (1976). The effect of age on creatinine clearance in men: a cross-sectional and longitudinal study. *J. Gerontol.*, **31**, 155–63.

Royal College of Physicians (1983). Obesity: a report of the Royal College of Physicians. *J. Roy. Coll. Phys. Lond.*, **17**, 3–58.

Ruiz, L., Colley, J. R. T. & Hamilton, P. J. S. (1971). Measurement of triceps skinfold thickness. An investigation of sources of variation. *Br. J. Prev. Soc. Med.,* **25**, 165–7.

Ruff, C. B. & Jones, H. H. (1981). Bilateral asymmetry in cortical bone of the humerus and tibia – sex and age factors. *Hum. Biol.,* **53**, 69–86.

Russell, D., Prendergast, P. J., Darby, P., Garfinkel, P. E., Whitwell, J. & Jeejebhoy, K. N. (1983). A comparison between muscle function and body composition in anorexia nervosa: the effect of re-feeding. *Am. J. Clin. Nutr.,* **38**, 229–37.

Russell, G. F. M. & Mezey, A. G. (1962). An analysis of weight gain in patients with anorexia nervosa treated with high calorie diets. *Clin. Sci.,* **23**, 449–61.

Rutenfranz, J., Lange Andersen, K., Seliger, V., Ilmarinen, J., Klimmer, F., Kylian, H. & Ruppel, M. (1984). Maximal aerobic power affected by maturation and body growth during childhood and adolescence. In *Children and Sport,* ed. J. Ilmarinen & I. Valimaki, pp. 67–85. Berlin: Springer Verlag.

Rutledge, M. M., Clark, J., Woodruff, C., Krause, G. & Flynn, M. A. (1976). A longitudinal study of total body potassium in normal breastfed and bottle-fed infants. *Pediatr. Res.,* **10**, 114–17.

Ryan, R. J., Williams, J. D., Ansell, B. M. & Bernstein, L. M. (1957). The relationship of body composition to oxygen consumption and creatinine excretion in healthy and wasted men. *Metabolism,* **6**, 365–77.

Sady, S., Freedson, P., Katch, V. L. & Reynolds, H. M. (1978). Anthropometric model of total body volume for males of different sizes. *Hum. Biol.,* **50**, 529–40.

Sagild, U. (1956). Total exchangeable potassium in normal subjects, with special reference to changes with age. *Scand. J. Clin. Lab. Invest.,* **i**, 44–50.

Salans, L. B., Horton, E. S. & Sims, E. A. H. (1971). Experimental obesity in man: cellular character of the adipose tissue. *J. Clin. Invest.,* **50**, 1005–11.

Salans, L. B., Gushman, S. W. & Weismann, R. E. (1973). Adipose cell size and number in non-obese and obese patients. *J. Clin. Invest.,* **52**, 929–41.

Samueloff, S. (1987). Population studies on human adaptability. In *Adaptive Physiology to Stressful Environments,* ed. S. Samueloff & M. K. Yousef, pp. 27–34. Boca Raton: Chemical Rubber Company Press.

Samueloff, S. (1988). Environmental, genetic and leg mass influences on energy expenditure. In *Capacity for Work in the Tropics,* ed. K. J. Collins & D. F. Roberts, pp. 205–14. Cambridge: Cambridge University Press.

Samus & Rameaux (1839). Mémoire addressé à l'Académie Royale. *Bulletin de l'Académie Royale,* **3**, 1094–100 (Cited by Mount, 1979).

Saris, W. H. M., Binkhorst, R. A., Cramwinckel, A. B., Van Waesberghe, F. & Van der Veen-Hezemans, A. M. (1980). The relationship between working performance, daily physical activity, fatness, blood lipids, and nutrition in schoolchildren. In *Children and Exercise IX,* ed. K. Berg & B. Eriksson, pp. 166–74. Baltimore: University Park Press.

Satwanta, K., Bharadwaj, H,. & Singh, I. P. (1977). Relationship of body density to body measurement in young Punjabi women: applicability of body composition prediction equations developed for women of European descent. *Hum. Biol.,* **49**, 203–13.

Satwanta, K., Sing, I. P. & Bharadwaj, H. (1980). Body fat from skinfold thicknesses and weight–height indices: a comparison. *Z. Morph. Anthrop.*, **71**, 93–100.

Satyanarayana, K., Nadamuni Naidu, A., Chatterjee, B. & Narasinga Rao, B. S. (1977). Body size and work output. *Am. J. Clin. Nutr.*, **30**, 322–5.

Satyanarayana, K., Naidu, N. & Rao, N. (1980). Adolescent growth spurt among rural Indian boys in relation to their nutritional status in early childhood. *Ann. Hum. Biol.*, **7**, 359–65.

Savard, R., Bouchard, C., LeBlanc, C. & Tremblay, A. (1978). Familial resemblance in fatness indicators. *Ann. Hum. Biol.*, **10**, 111–18.

Sawka, M. N., Weber, H. & Knowlton, R. G. (1978). The effect of total body submersion on residual lung volume and body density measurements in man. *Ergonomics*, **21**, 89–94.

Schantz, P., Randall-Fox, E., Hutchison, W., Tyden, A. & Åstrand, P. O. (1983). Muscle fibre type distribution, muscle cross-sectional area and maximal voluntary strength in humans. *Acta Physiol. Scand.*, **117**, 219–26.

Schemmel, R. (1980). Assessment of obesity. In *Nutrition, Physiology and Obesity,* ed. R. Schemmel. Boca Raton: CRC Press.

Schlechte, J. A., Sherman, B. & Martin, R. (1983). Bone density in amenorrheic women with and without hyperprolactinemia. *J. Clin. Endocrin. Metab.*, **56**, 1120–3.

Schleimer, K. (1983). Dieting in teenage schoolgirls. A longitudinal prospective study. *Acta Paediatr. Scand.*, Suppl., 312.

Schloerb, P. R., Friis-Hansen, B. J., Edelman, I. S., Solomon, A. K. & Moore, F. D. (1950). The measurement of total body water in the human subject by deuterium oxide dilution. *J. Clin. Invest.*, **29**, 1296–310.

Schmid, P. & Schlick, W. (1976). Das Prinzip der Messung des luftfreien Korpervolumens mit hilfe einer Druckdifferenz-Tauchsonde. *Eur. J. Appl. Physiol.*, **35**, 59–67.

Schoeller, D. A., Van Sauten, D. W., Peterson, D. W., Jaspan, J. & Klein, P. D. (1980). Total body water measurements in humans with ^{18}O and ^{2}H labelled water. *Am. J. Clin. Nutr.*, **33**, 2686–93.

Schoeller, D. A., Kushner, R. F., Taylor, P., Dietz, W. H. & Bandini, L. (1985). Measurement of total body water: isotope dilution techniques. In *Body Composition Assessment in Youth and Adults,* ed. A. F. Roche, pp. 24–29. Columbus, Ohio: Ross Laboratories.

Scholander, P. F., Hammel, H. T., Hart, J. S., Le Messurier, D. H. & Steen, J. (1958a). Cold adaptation in Australian aborigines. *J. Appl. Physiol.*, **13**, 211–18.

Scholander, P. F., Hammel, H. T., Andersen, K. L. & Loyning, Y. (1958b). Metabolic adaptation to cold in man. *J. Appl. Physiol.*, **12**, 1–8.

Schreider, E. (1951). Anatomical factors of body heat regulation. *Nature*, **167**, 823–4.

Schreider, E. (1957). Ecological rules and body heat regulation in man. *Nature*, **179**, 915–16.

Schull, W. J. (1986). Genetics of health related fitness. In *Sport and Human Genetics,* ed. R. Malina & C. Bouchard, pp. 89–104. Champaign, Ill.: Human Kinetics Publishers.

Schultz, A. H. (1926). Foetal growth of man and other primates. *Quart. Rev. Biol.*, **1**, 465–521.

Schutte, J. E., Longhurst, J. C., Gaffney, F. A., Bastian, B. C. & Blomqvist, C. G. (1981). Total plasma creatinine: an accurate measure of total striated muscle. *J. Appl. Physiol.*, **51**, 762–66.

Schutte, J. E., Lilljeqvist, R. E. & Johnson, R. L. (1983). Growth of lowland children of European ancestry during sojourn at high altitude (3200 m). *Am. J. Phys. Anthrop.*, **61**, 221–6.

Schutte, J. E., Townsend, E. J., Hugg, J., Shoup, R. F., Malina, R. M. & Blomqvist, C. G. (1984). Density of lean body mass is greater in Blacks than in Whites. *J. Appl. Physiol.*, **56**, 1647–9.

Schutz, Y., Bessard, T. & Jéquier, E. (1984). Diet-induced thermogenesis measured over a whole day in obese and non-obese women. *Am. J. Clin. Nutr.*, **40**, 542–52.

Scott, E. C. & Johnston, F. E. (1982). Critical fat, menarche, and the maintenance of menstrual cycles: a critical review. *J. Adolesc. Health Care*, **2**, 249–60.

Seckler, D. (1980). 'Malnutrition', an intellectual Odyssey. *West. J. Agric. Econ.*, **5**, 219–27.

Segal, K. R., Gutin, B., Presta, E., Wang, J. & Van Itallie, T. B. (1985). Estimation of human body composition by electrical impedance methods: a comparative study. *J. Appl. Physiol.*, **58**, 1565–71.

Selinger, A. (1977). The body as a three component system. Ph.D. Thesis, University of Illinois, Urbana, Ill. *Dissert. Abstr.*, **38** (10), 5996A.

Selye, H. (1950). *Stress*. Montreal: Acta Medical Publishers.

Shaffer, P. A. & Coleman, W. (1909). Protein metabolism in typhoid fever. *Arch. Int. Med.*, **4**, 538–600.

Shaw, J. C. L. (1973). Parental nutrition in the management of sick low birth weight infants. *Pediatr. Clin. N. Am.*, **20**, 333–58.

Sheldon, W. H. (1963). *The Varieties of Human Physique. An Introduction to Constitutional Psychology*. New York: Hafner.

Sheng, H. P. & Huggins, R. A. (1979). A review of body composition studies, with emphasis on total body water and fat. *Am. J. Clin. Nutr.*, **32**, 630–47.

Shephard, R. H. & Memma, H. E. (1967). Skin thickness in endocrine disease, a roentgenographic study. *Ann. Int. Med.*, **66**, 531–9.

Shephard, R. J. (1956). The influence of age on the haemoglobin level in congenital heart disease. *Br. Heart J.*, **18**, 49–54.

Shephard, R. J. (1971). The working capacity of schoolchildren. In *Frontiers of Fitness*, ed. R. J. Shephard, pp. 319–44. Springfield, Ill.: C. C. Thomas.

Shephard, R. J. (1974). *Men at Work. Applications of Ergonomics to Performance and Design*. Springfield, Ill.: C. C. Thomas.

Shephard, R. J. (1977). *Endurance Fitness,* 2nd edn. Toronto: University of Toronto Press.

Shephard, R. J. (1978a). *Human Physiological Work Capacity*. Cambridge: Cambridge University Press.

Shephard, R. J. (1978b). Fitness, obesity and health. In *Proceedings of the first RSG4 Physical Fitness Symposium with Special Reference to Military Forces*, ed. C. Allen, pp. 238–61. Downsview, Ont.: Defence and Civil Institute of Environmental Medicine.

Shephard, R. J. (1982a). *Physiology and Biochemistry of Exercise*. New York: Praeger.

Shephard, R. J. (1982b). *Physical Activity and Growth*. Chicago, Ill.: Year Book Publishers.

Shephard, R. J. (1985a). Adaptation to exercise in the cold., *Sports Med.*, 2, 59–71.

Shephard, R. J. (1985b). The value of exercise in preventive medicine. In *The Value of Preventive Medicine*, ed. D. Evered & J. Whelan, pp. 164–82. London: Pitman.

Shephard, R. J. (1986). *Fitness of a Nation. Lessons from the Canada Fitness Survey*, Basel: Karger.

Shephard, R. J. (1987). *Physical Activity and Aging*, 2nd edn. London: Croom Helm Publishing.

Shephard, R. J. (1988). Work capacity: methodology in a tropical environment. In *Capacity for Work in the Tropics*, ed. K. J. Collins & D. F. Roberts, pp. 1–30. Cambridge: Cambridge University Press.

Shephard, R. J. & Rode, A. (1973). Fitness for arctic life: the cardiorespiratory status of the Canadian Eskimo. In *Polar Human Biology*, ed. O. G. Edholm & E. K. E. Gunderson, pp. 216–239. London: Heinemann.

Shephard, R. J. & Rode, A. (1985). Acculturation and the biology of aging. In *Circumpolar Health* 1984, ed. R. Fortuine, pp. 45–8. Seattle: University of Washington Press.

Shephard, R. J. & Seliger, V. (1969). On the estimation of total lung capacity from chest X-rays. Radiographic and helium dilution estimates on children aged 10–12 years. *Respiration*, 26, 327–36.

Shephard, R. J., Carey, G. C. R. & Phair, J. J. (1958). Evaluation of a portable box-bag for field testing of pulmonary diffusion. *J. Appl. Physiol.*, 12, 79–85.

Shephard, R. J., Allen, C., Bar-Or. O., Davies, C. T. M., Degré, S., Hedman, R., Ishii, K., Kaneko, M., LaCour, R., di Prampero, P. & Seliger, V. (1969a). The working capacity of Toronto schoolchildren. *Can. Med. Assoc. J.*, 100, 560–6.

Shephard, R. J., Jones, G., Ishii, K., Kaneko, M. & Olbrecht, A. J. (1969b). Factors affecting body density and thickness of subcutaneous fat. Data on 518 Canadian city dwellers. *Am. J. Clin. Nutr.*, 22, 1175–89.

Shephard, R. J., Kaneko, M. & Ishii, K. (1971). Simple indices of obesity. *J. Spts Med. Phys. Fitness*, 11, 154–61.

Shephard, R. J., Hatcher, J. & Rode. A. (1973). On the body composition of the eskimo. *Eur. J. Appl. Physiol.*, 10, 1–13.

Shephard, R. J., Lavallée, H., Larivière, G., Rajic, M., Brisson, G., Beaucage, C., Jéquier, J. C. & LaBarre, R. (1975). La capacité physique des enfants Canadiens: une comparaison entre les enfants Canadiens-Francais, Canadiens-Anglais et Esquimaux. II Anthropométrie et volumes pulmonaires. *Union Méd. Can.* 104, 259–69.

Shephard, R. J., Killinger, D. & Fried, T. (1977). Responses to sustained use of anabolic steroid. *Br. J. Spts Med.*, 11, 170–3.

Shephard, R. J., Lavallée, H. & Larivière, G. (1978a). Competitive selection among age-class ice-hockey players. *Brit. J. Spts Med.*, 12, 11–13.

Shephard, R. J., Lavallée, H., Rajic, M., Jéquier, J-C., Brisson, G. & Beaucage, C. (1978b). Radiographic age in the interpretation of physiological and anthropological data. In *Pediatric Work Physiology*, ed. J. Borms & M. Hebbelinck, pp. 124–33. Basel: Karger.

Shephard, R. J., Lavallée, H., Jéquier, J-C., LaBarre, R. & Rajic, M. (1979). A community approach to assessments of exercise tolerance in health and disease. *J. Spts Med. Phys. Fitness*, **19**, 296–304.

Shephard, R. J., Lavallée, H., LaBarre, R., Rajic, M., Jéquier, J-C. & Volle, M. (1984a). Body dimensions of Québecois children. *Ann. Hum. Biol.*, **11**, 243–52.

Shephard, R. J., Goodman, J., Rode, A. & Schaefer, O. (1984b). Snowmobile use and decrease of stature among the Inuit. *Arctic Med. Res.*, **38**, 32–6.

Shephard, R. J., LaBarre, R., Jéquier, J-C., Lavallée, H., Rajic, M. and Volle, M. (1985a). The 'unisex phantom', sexual dimorphism, and proportional growth assessment. *Am. J. Phys. Anthrop.*, **67**, 403–12.

Shephard, R. J., Kofsky, P. R., Harrison, J. E., McNeill, K. G. & Krondl, A. (1985b). Body composition of older female subjects. New approaches and their limitations. *Hum. Biol.*, **57**, 671–86.

Shephard, R. J., Bouhlel, E., Vandewalle, H. & Monod, H. (1988). Muscle mass as a factor limiting physical work. *J. Appl. Physiol.*, **64**, 1472–9.

Shetty, P. S., Jung, R. T. & James, W. P. T. (1979). Reduced dietary-induced thermogenesis in obese subjects before and after weight loss. *Proc. Nutr. Soc.*, **38**, 87A.

Shirai, I. (1975). Nutrition and physical fitness. In *Human Adaptability*, Vol. 3. *Physiological Adaptability and Nutritional Status of the Japanese. B. Growth, Work Capacity and Nutrition of Japanese*, ed. K. Asahina & R. Shigiya, pp. 100–7. Tokyo: Japanese Committee for the International Biological Programme, Science Council of Japan.

Shizgal, H. M. (1981). The effect of malnutrition on body composition. *Surg. Gyn. Obstetr.*, **152**, 22–6.

Shizgal, H. M. (1985). Body composition of patients with malnutrition and cancer. *Cancer*, **55**, 250–3.

Shizgal, H. M., Forse, R. A., Spanier, A. H. & MacLean, L. D. (1979). Protein malnutrition following intestinal bypass for morbid obesity. *Surgery*, **86**, 60–8.

Shukla, K. K., Ellis, K. J., Dombrowski, C. S. & Cohn, S. H. (1973). Physiological variation of total body potassium in man. *Am. J. Physiol.*, **224**, 271–4.

Sidney, K. H. & Shephard, R. J. (1973). Physiological characteristics and performance of the whitewater paddler. *Int. Z. Angew. Physiol.*, **32**, 55–70.

Sidney, K. H., Shephard, R. J. & Harrison, J. (1977). Endurance training and body composition of the elderly. *Am. J. Clin. Nutr.*, **30**, 326–33.

Sievert, R. M. (1956). Untersuchungen über die Gammastrahlung des menschlichen Korpers. *Strahlentherapie*, **99**, 185–95.

Simpson, N. E, & McAlpine, P. J. (1976). A comparison of genetic markers in the blood of circumpolar populations. In *Circumpolar Health*, ed. R. J. Shephard & S. Itoh, pp. 219–20. Toronto, Ontario: University of Toronto Press.

Sims, E. A. H., Bray, G. A., Danforth, E., Glennen, J. A., Horton, E. S., Salans, L. B. & O'Connell, M. (1974). Experimental obesity in man. In

Lipid Metabolism, Obesity and Diabetes Mellitus: Impact upon Atherosclerosis, ed. H. Greten, R. Levine, E. F. Pfeiffer & A. E. Renold, pp. 70–6. Stuttgart: Georg Thieme.

Sinaki, M., McPhee, M. C., Hodgson, S. F., Merritt, J. M. & Offord, K. P. (1986). Relationship between bone mineral density and strength of back extension in healthy post-menopausal women. *Proc. Mayo Clin.,* **61**, 116–22.

Singer, F. R. (1981). Metabolic bone disease. In *Endocrinology and Metabolism,* ed. P. Felig, J. D. Baxter, A. E. Broadus & L. A. Frohman, pp. 1081–118. New York: McGraw-Hill.

Singer, K. (1986). Injuries and disorders of the epiphyses in young athletes. In *Sport for Children and for Youths,* ed. M. R. Weiss & D. Gould, pp. 141–50. Champaign, Ill.: Human Kinetics Publishers.

Sinning, W. E., (1974). Body composition assessment of college wrestlers. *Med. Sci. Sports,* **6**, 139–45.

Sinning, W. E. (1978). Anthropometric estimation of body density, fat and lean body weight in women gymnasts. *Med. Sci. Sports Exerc.,* **10**, 243–9.

Sinning, W. E. & Hackney, A. C. (1986). Body composition estimation by girths and skeletal dimensions in male and female athletes. In *Perspectives in Kinanthropometry,* ed. J. A. P. Day, pp. 239–44. Champaign, Ill.: Human Kinetics Publishers.

Sinning, W. E. & Wilson, J. R. (1984). Validity of generalized equations for body composition in women athletes. *Res. Quart.,* **55**, 153–60.

Sinning, W. E., Dolny, D. G., Little, K. D., Cunningham, L. N., Racaniello, A., Siconolfi, S. F. & Sholes, J. L. (1985). Validity of generalized equations for body composition analysis in male athletes. *Med. Sci. Sports Exerc.,* **17**, 124–30.

Siri, W. E. (1956a). Apparatus for measuring human body volume. *Rev. Sci. Instrum.,* **27**, 729–38.

Siri, W. E. (1956b). The gross composition of the body. *Adv. Biol. Med. Phys.,* **4**, 239–80.

Siri, W. E. (1961). Body composition from fluid spaces and density. In *Techniques for Measuring Body Composition,* ed. J. Brozek & A. Henschel, pp. 108–17. Washington, DC: National Academy of Sciences/National Research Council.

Sippell, W. G., Dörr, H. G., Bidlingmaier, F. & Knorr, D. (1980). Plasma levels of aldosterone, corticosterone, 11-deoxycorticosterone, progesterone, 17-hydroxyprogesterone, cortisol and cortisone during infancy and childhood. *Pediatr. Res.,* **14**, 39–46.

Sjöstrom, L., Björntorp, P. & Vrana, J. (1971). Microscopic fat cell size measurements on frozen cut adipose tissue in comparison with automatic determinants of osmium-fixed fat cells. *J. Lipid Res.,* **12**, 521–30.

Sjöstrom, L., Kvist, H., Cederblad, A. & Tylen, U. (1986). Determination of total adipose tissue and body fat in women by computed tomography. ^{40}K and tritium. *Am. J. Physiol.,* **250**, E736–45.

Škerlj, B. (1958). Age changes in fat distribution in the female body. *Acta Anatomica,* **38**, 56–63.

Škerlj, B. (1959). Towards a systematic morphology of the human body. *Acta Anatomica,* **39**, 220–43.

Škerlj, B. (1961). Ein Beitrag zur Grossgewebenanalyse am lebenden Menschen. *Acta Anatomica*, **44**, 131–6.

Skrabal, F., Arnot, R. N., Joplin, G. F. & Fraser, T. R. (1972). The effect of glucocorticoid withdrawal on body water and electrolytes in hypopituitary patients with adrenocortical insufficiency as investigated with ^{77}Br, ^{43}K, ^{24}Na and ^{3}H$_2$O. *Clin. Sci.*, **43**, 79–90.

Skreslet, S. & Aarefjord, F. (1968). Acclimatisation to cold in man induced by frequent scuba diving in cold water. *J. Appl. Physiol.*, **24**, 177–81.

Skrobak-Kaczynski, J. & Lewin, T. (1976). Secular changes in Lapps of Northern Finland. In *Circumpolar Health*, ed. R. J. Shephard & S. Itoh. Toronto: University of Toronto Press.

Slaughter, M. H., Lohman, T. G., Boileau, R. A., Stillman, R. J., Van Loan, M., Horswill, C. A. & Wilmore, J. H. (1984). Influence of maturation on relationship of skinfolds to body density: a cross-sectional study. *Hum. Biol.*, **56**, 681–9.

Sloan, A. W. (1967). Estimation of body fat in young men. *J. Appl. Physiol.*, **23**, 311–15.

Sloan, A. W. & Bredell, G. A. G. (1973). A comparison of two simple methods of determining pulmonary residual volume. *Hum. Biol.*, **45**, 23–9.

Sloan, A. W. & Koeslag, J. H. (1973). A trial of the ponderax skinfold caliper. *S. Afr. Med. J.*, **47**, 125–7.

Sloan, A. W. & Shapiro, M. (1972). A comparison of skinfold measurements with three standard calipers. *Hum. Biol.*, **44**, 29–36.

Smith, D. M., Nance, W. E., Kang, K. W., Christian, J. C. & Johnston, C. C. (1973). Genetic factors in determining bone mass. *J. Clin. Invest.*, **52**, 2800–8.

Smith, D. M., Norton, J. A., Khairi, R. & Johnston, C. C. (1976a). The measurement of rates of mineral loss with aging. *J. Lab. Clin. Med.*, **87**, 882–92.

Smith, D. M., Khairi, M. R. A., Norton, J. & Johnston, C. C. (1976b). Age and activity effects on the rate of bone mineral loss. *J. Clin. Invest.*, **58**, 716–21.

Smith, E. L. (1982). Exercise for prevention of osteoporosis: a review. *Phys. Sportsmed.*, **10** (3), 72–9.

Smith, E. L., Sempos, C. T. & Purvis, R. W. (1981a). Bone mass and strength decline with age. In *Exercise and Aging: The Scientific Basis*, ed. E. L. Smith & R. C. Serfass, pp. 59–88, Hillside NJ: Enslow Publishers.

Smith, E. L., Reddan, W. & Smith, P. E. (1981b). Physical activity and calcium modalities for bone mineral increase in aged women. *Med. Sci. Sports Exerc.*, **13**, 60–4.

Smith, N. J. (1980). Excessive weight loss and food aversion in athletes simulating anorexia nervosa. *Pediatrics*, **66**, 139–42.

Smith, T., Hesp, R. & MacKenzie, J. (1979). Total body potassium calibrations for normal and obese subjects to two types of whole body counter. *Phys. Med. Biol.*, **24**, 171–5.

Smith, U. (1983). Regional differences and effect of cell size on lipolysis in human adipocytes. In *The Adipocyte and Obesity*, ed. A. Angel, pp. 245–50. New York: Raven Press.

Smith, U., Hammersten, J., Björntorp, P. & Kral, J. G. (1979). Regional differences and effect of weight reduction on human fat cell metabolism. *Eur. J. Clin. Invest.,* **9**, 327–32.

Snyder, A. C., Wenderoth, M. P., Johnston, C. C. & Hui, S. L. (1986). Bone mineral content of elite lightweight amenorrheic oarswomen. *Hum. Biol.,* **58**, 863–9.

Soberman, R., Brodie, B. B., Levy, B. B., Axelrod, J., Hollander, V. & Steele, J. M. (1949). The use of antipyrine in the measurement of total body water in man. *J. Biol. Chem.,* **179**, 31–42.

Society of Actuaries (1959). *Build and Blood Pressure Study.* Chicago: Society of Actuaries.

Society of Actuaries (1980). *Build Study, 1979.* Chicago: Society of Actuaries.

Soedjatmoko, K. (1981). The challenge of world hunger. In *Nutrition in Health and Disease and International Development,* ed. A. E. Harper & G. K. Davis, pp. 1–16. New York: A. R. Liss.

Sohar, E. & Sneh, E. (1973). Follow-up of obese patients: 14 years after a successful reducing diet. *Am. J. Clin. Nutr.,* **26**, 845–8.

Solomon, A. K., Edelman, I. S. & Soloway, S. (1950). The use of the mass spectrometer to measure deuterium in body fluids. *J. Clin. Invest.,* **29**, 1311–19.

Sommer, H. M. (1985). The influence of exercise and training on the locomotor system of children: a longitudinal study of adolescent tennis players. In *Children and Exercise* XI, ed. R. A. Binkhorst, H. C. G. Kemper & W. H. M. Saris, pp. 308–18. Champaign, Ill.: Human Kinetics Publishers.

Southgate, D. A. T. & Hey, E. N. (1976). Chemical and biochemical development of the human fetus. In *The Biology of Human Fetal Growth,* ed. D. R. Roberts & A. M. Thompson, pp. 195–209. London: Taylor & Francis.

Spady, D. W., Payne, P. R., Picou, D. & Waterlow, J. C. (1976). Energy balance during recovery from malnutrition. *Am. J. Clin. Nutr.,* **29**, 1073–88.

Spencer, T. & Heywood, P. (1983). Seasonability, subsistence agriculture and nutrition in a lowland community of Papua New Guinea. *Ecol. Food Nutr.,* **13**, 221–9.

Spengler, D. M., Morey, E. R., Carter, D. R., Turner, R. T. & Baylink, D. J. (1979). Effect of spaceflight on bone strength. *Physiologist,* **22**, 118.

Spray, C. M. & Widdowson, E. M. (1950). The effect of growth and development on the composition of mammals. *Br. J. Nutr.,* **4**, 332–53.

Spurr, G. B. (1988). Effects of child malnutrition. In *Capacity for Work in the Tropics,* ed. K. J. Collins & D. F. Roberts, pp. 107–40. Cambridge: Cambridge University Press.

Spurr, G. B., Barac-Nieto, M. & Maksud, M. G. (1977). Productivity and maximal oxygen consumption in sugar cane cutters. *Am. J. Clin. Nutr.,* **30**, 316–21.

Spurr, G. B., Barac-Nieto, M., Lotero, H. & Dahners, H. W. (1981). Comparisons of body fat estimated from total body water and skinfold thicknesses of undernourished men. *Am. J. Clin. Nutr.,* **34**, 1944–53.

Srivastava, S. S., Mani, K. V., Soni, C. M. & Bhati, J. (1957). Effect of muscular exercise on urinary excretion of creatine and creatinine. *Ind. J. Med. Res.,* **55**, 953–60.

Stallones, R. A. (1969). Population patterns of disease and body weight. In *Obesity*, ed. N. L. Wilson, pp. 109–18. Philadelphia: F. A. Davis.

Standard, K. L., Wills, V. G. & Waterlow, J. C. (1959). Indirect indicators of muscle mass in malnourished infants. *Am. J. Clin. Nutr.*, **7**, 271–9.

Stansell, M. J. & Hyder, A. R. (1976). Simplified body composition analysis using deuterium dilution and deuteron photodisintegration. *Aviat. Space Environ Med.*, **47**, 839–45.

Stearns, G. (1939). Mineral metabolism of normal infants. *Physiol. Rev.*, **19**, 415–30.

Stefanik, P. A., Heald, F. P. & Mayer, J. (1959). Caloric intake in relation to energy output of obese and non-obese adolescent boys. *Am. J. Clin. Nutr.*, **7**, 55–62.

Stein, T. P., Schulter, M. D. & Diamond, C. E. (1983). Nutrition, protein turnover and physical activity in young women. *Am. J. Clin. Nutr.*, **38**, 223–8.

Steinberg, I. & Zaske, D. E. (1985). Body composition and pharmacokinetics. In *Body Composition Assessment in Youth and Adults*, ed. A. F. Roche, pp. 96–101. Columbus, Ohio: Ross Laboratories.

Steinhaus, A. H. (1933). Chronic effects of exercise. *Physiol. Rev.*, **13**, 103–47.

Steinkamp, R., Cohen, N. L., Gaffey, W. R., McKay, T., Bron, G., Siri, W. E., Sargeant, T. W. & Isaacs, E. (1965). Measures of body fat and related factors in normal adults – I. Introduction and methodology. II. A simple clinical method to estimate body fat and lean body mass. *J. Chron. Dis.*, **18**, 1279–90; 1291–307.

Stern, J. S. & Greenwood, M. R. C. (1974). A review of the development of adipose cellularity in man and animals. *Fed. Proc.*, **33**, 1952–5.

Sterner, T. G. & Burke, E. J. (1986). Body fat assessment: a comparison of visual estimation and skinfold techniques. *Phys. Sportsmed.*, **14**, 101–7.

Stinson, S. (1982). The effect of high altitude on the growth of children of high socio-economic status in Bolivia. *Am. J. Phys. Anthrop.*, **59**, 61–71.

Stokes, M. & Young, A. (1986). Measurement of quadriceps cross-sectional area by ultrasonography: a description of the technique and its application in physiotherapy. *Physiother. Practice*, **2**, 31–6.

Stordy, B. J., Marks, V., Kalucy, R. S. & Crisp, A. H. (1977). Weight gain, thermic effect of glucose and resting metabolic rate during recovery from anorexia nervosa. *Am. J. Clin. Nutr.*, **30**, 138–46.

Stouffer, J. R. (1963). Relationship of ultrasonic measurements and X-rays to body composition. *Ann. N.Y. Acad. Sci.*, **110**, 31–9.

Strong, L. H., Gee, G. K. & Goldman, R. F. (1985). Metabolic and vasomotor insulative responses occurring on immersion in cold water. *J. Appl. Physiol.*, **58**, 964–77.

Stuart, H. C. & Reed, R. B. (1951). Certain technical aspects of longitudinal studies of child health and development. *Am. J. Publ. Health*, **41**, 85–90 (Pt. II, Supp.).

Stunkard, A. J. & Albaum, J. M. (1981). The accuracy of self-reported weights. *Am. J. Clin. Nutr.*, **34**, 1593–9.

Stunkard, A. J., Sorensen, T. I. A., Hanis, C., Teasdale, T. W., Chakraborty, R., Schull, W. J. & Shulsinger, F. (1986). An adoption study of human obesity. *New Engl. J. Med.*, **314**, 193–8.

Stupakov, G. P. (1988). Biomechanical characteristics of bone. Structure changes following real and simulated weightlessness. *Physiologist*, **31**, S4–7.

Sukhatme, P. V. & Margen, S. (1982). Autoregulatory homeostatic nature of energy balance. *Am. J. Clin. Nutr.*, **35**, 355–65.

Suzuki, S. (1970). Experimental studies on factors on growth. *Monogr. Soc. Res. Child. Dev.*, **35**, 6–11.

Suzuki, S. (1975). Experimental studies on the interrelationships among nutrition, physical exercise and health components. In *Human Adaptability*, Vol. 3. *Physiological Adaptability and Nutritional Status of the Japanese. B. Growth, Work Capacity and Nutrition of the Japanese*, ed. K. Asahina & R. Shigiya, pp. 83–94. Tokyo: Japanese Committee for the International Biological Programme, Science Council of Japan.

Suzuki, S., Ohta, F., Tsuji, K., Oshima, S., Tsuji, E., Ohta, H. & Suzuki, H. (1969). *Experimental Studies on the Interrelationships of Nutrition, Physical Exercise and Health Components*. Report II. *Influence on Bone Development*. Tokyo: Japanese Institute of Nutrition, Annual Report.

Svoboda, M. D. & Query, L. M. (1986). Hydrostatic weighing of women throughout the menstrual cycle. In *Perspectives in kinanthropometry*, ed. J. A. P. Day, pp. 245–50. Champaign, Ill.: Human Kinetics Publishers.

Swan, R. C., Madisso, H. & Pitts, R. F. (1954). Measurement of extracellular fluid volume in nephrectomized dogs. *J. Clin. Invest.*, **33**, 1447–56.

Szabo, S., Doka, J., Apor, P. *et al.* (1972). Die Beziehungen zwischen Knochenlebensalter, funktionellen anthropometrischen Daten und der aeroben Kapazitat. *Schw. Z. Sportmed.*, **20**, 109–15.

Szakall, A. (1943). Maximale Leistung und maximale Arbeit. *Arbeitsphysiologie*, **13**, 9.

Szathmary, E. J. E. & Holt, N. (1983). Hyperglycemia in Dogrib Indians of the North West Territories, Canada: association with age and a centripetal distribution of body fat. *Hum. Biol.*, **55**, 493–515.

Szilagyi, T., Hideg, J. Szöör, A., Berényi, E., Rapcsák, M. & Pozsgai, A. (1981). The effect of hypokinesis and hypoxia on the function of muscles. *Physiologist*, **24** (6), S11–12.

Taggart, N. (1962). Diet, activity and body weight. A study of variations in a woman. *Br. J. Nutr.*, **16**, 223–5.

Talso, P. J., Miller, C. E., Carballo, A. J. & Vasquez, I. (1960). Exchangeable potassium as a parameter of body composition. *Metabolism*, **9**, 456–71.

Tanner, J. M. (1962). *Growth at Adolescence*. Oxford: Blackwell Scientific Publications.

Tanner, J. M. (1964). *The Physique of the Olympic Athlete*. London: Unwin.

Tanner, J. M. (1965). Radiographic studies of body composition in children and adults. In *Human Body Composition. Approaches and Applications*, ed. J. Brozek, pp. 211–36. Oxford: Pergamon Press.

Tanner, J. M. (1966). Growth and physique in different populations of mankind. In *The Biology of Human Adaptability*, ed. P. T. Baker & J. S. Weiner, pp. 45–66. Oxford: Clarendon Press.

Tanner, J. M. & Israelsohn, W. (1963). Parent–child correlations for body measurements of children between the ages of one month and seven years. *Ann. Hum. Genet.*, **26**, 245–59.

Tanner, J. M. & Weiner, J. S. (1949). Reliability of the photogrammetric method of anthropometry with a description of a miniature camera technique. *Am. J. Phys. Anthrop.*, **7**, 145–86.

Tanner, J. M. & Whitehouse, R. H. (1967). The effect of human growth hormone on subcutaneous fat thickness in hyposomatotrophic and panhypopituitary dwarfs. *J. Endocrinol.*, **39**, 263–75.

Tanner, J. M. & Whitehouse, R. H. (1975). Revised standards for triceps and subscapular skinfolds in British children. *Arch. Dis. Childh.*, **50**, 142–5.

Tanner, J. M., Healy, M. J. R., Lockhart, R. D., MacKenzie, J. D. & Whitehouse, R. H. (1956). Aberdeen growth study. *Arch. Dis. Childh.*, **31**, 372–81.

Tanner, J. M., Healy, M. J. R. & Whitehouse, R. H. (1959). Fat, bone and muscle in the limbs of young men and women; their quantitative relationships studied radiographically. *J. Anat.*, **93**, 563 (Abstract).

Tanner, J. M., Hayashi, T., Preece, M. A. & Cameron, N. (1982). Increase in length of leg relative to trunk in Japanese children and adults from 1957 to 1977: comparison with British and with Japanese Americans. *Ann. Hum. Biol.*, **9**, 411–24.

Taylor, A., Aksoy, Y., Scopes, J. W., Mont, G. du & Taylor, B. A. (1985). Development of an air displacement method for whole body volume measurement of infants. *J. Biomed. Eng.*, **7**, 9–17.

Telkka, A., Pere, S. & Kunnas, M. (1951). Anthropometric studies of Finnish athletes and wrestlers. *Ann. Acad. Sci. Fennic.* Series A, V (28), 1. Cited by Maas (1974).

Templeton, G. H, Padalino, M., Manton, J., LeConey, T., Hagler, H. & Glasberg, M. (1984). The influence of rat suspension-hypokinesia on the gastrocnemius muscle. *J. Appl. Physiol.*, **56**, 278–86.

Tepperman, J. (1980). *Metabolic and Endocrine Physiology*. London: Year Book Medical Publishers.

Thomas, B. M. & Miller, A. T. (1958). Adaptation to forced exercise in the rat. *Am. J. Physiol.*, **193**, 350–4.

Thomas, L. W. (1962). The chemical composition of adipose tissue of man and mice. *Quart. J. Exp. Physiol.*, **47**, 179–88.

Thomas, R. D., Silverton, N. P., Burkinshaw, L. & Morgan, D. B. (1979). Potassium depletion and tissue loss in chronic heart disease. *Lancet*, (ii), 9–11.

Thomas, T. R. & Etheridge, G. L. (1980). Hydrostatic weighing at residual volume and functional residual capacity. *J. Appl. Physiol.*, **49**, 157–9.

Thomas, T. R., Etheridge, G. L., Londeree, B. R. & Shannon, W. (1979). Prolonged exercise and changes in percent fat determinations by hydrostatic weighing and scintillation counting. *Res. Quart.*, **50**, 709–14.

Thompson, D. D., Harper, A. B., Laughlin, W. S. & Jorgensen, J. B. (1981). Bone loss in Eskimos. In *Circumpolar Health* 81, ed. B. Harvald & J. P. Hart Hansen, pp. 327–30. Copenhagen: Nordic Council for Arctic Medical Research.

Thorland, W. G., Johnson, G. O., Tharp, G. D., Housh, T. J. & Cisar, C. J. (1984). Estimation of body density in adolescent athletes. *Hum. Biol.*, **56**, 339–48.

Thorstensson, A. (1976). Muscle strength, fibre types and enzyme activities in man. *Acta Physiol. Scand.* Suppl. 443, 1–45.

Thurlby, P. L. & Trayhurn, P. (1978). The development of obesity in preweanling ob/ob mice. *Br. J. Nutr.*, **39**, 397–402.

Timiras, P. S. (1972). *Developmental Physiology and Aging*. New York: Macmillan.

Timson, B. F. & Coffman, J. L. (1984). Body composition by hydrostatic weighing at total lung capacity and residual volume. *Med. Sci. Sports Exerc.*, **16**, 411–14.

Tipton, C. M. & Tcheng, T. K. (1970). Iowa wrestling study: weight loss in high school students. *J. Am. Med. Assoc.*, **214**, 1269–74.

Tittel, K. (1965). Zur Biotypologie und funktionellen Anatomie der Leistungs-sportlers. *Novo Acta Leopoldiana* NF 172, Bd. 30 (cited by Maas, 1974).

Tobias (1971). Cited by Ghesquière & D'Hulst (1988).

Todd, T. W. & Lindala, A. (1928). Thickness of the subcutaneous tissue in the living and the dead. *Am. J. Anat.*, **412**, 153–69.

Tokunaga, K., Matsuzawa, Y., Ishikawa, K. & Tarui, S. (1983). A novel technique for the determination of body fat by computed tomography. *Int. J. Obesity*, **7**, 437–45.

Tomanek, R. J. & Wood, Y. K. (1970). Compensatory hypertrophy of the plantaris muscle in relation to age. *J. Gerontol.*, **25**, 23–9.

Tomas, F. M., Ballard, F. J. & Pope, L. M. (1979). Age-dependent changes in the rate of myofibrillar protein degradation in humans as assessed by 3-methylhistidine and creatinine excretion. *Clin. Sci.*, **56**, 341–6.

Toriola, A. L., Adeniran, S. A. & Ogunremi, P. T. (1987). Body composition and anthropometric chatacteristics of elite male basketball and volleyball players. *J. Spts. Med. Phys. Fitness*, **27**, 235–9.

Tran, Z. V. & Weltman, A. (1988). Predicting body composition from girth measurements. *Hum. Biol.*, **60**, 167–75.

Treasure, J. L. & Russell, G. F. M. (1988). Intrauterine growth and neonatal weight gain in babies of women with anorexia nervosa. *Br. Med. J.*, **296**, 1038.

Treasure, J. L., Russell, G. M., Fogelmann, I. & Murby, B. (1987). Reversible bone loss in anorexia nervosa. *Br. Med. J.*, **295**, 474–5.

Trotter, M. & Gleser, G. C. (1952). Estimation of stature from long bones of American Whites and Negroes. *Am. J. Phys. Anthrop.*, **10**, 463–514.

Trotter, M. & Hixon, B. B. (1974). Sequential changes in weight, density and percentage ash from an early fetal period through old age. *Anat. Rec.*, **179**, 1–18.

Trotter, M. & Peterson, R. R. (1969). Weight of bone during the fetal period. *Growth*, **33**, 167–84.

Trotter, M. & Peterson, R. R. (1970). Weight of the skeleton during post-natal development. *Am. J. Phys. Anthrop.*, **33**, 313–24.

Trotter, M., Broman, G. E. & Peterson, R. R. (1959). Density of cervical vertebrae and comparison with the densities of other bones. *Am. J. Phys. Anthrop.*, **17**, 19–25.

Trotter, M., Broman, G. E. & Peterson, R. R. (1960). Density of bones of white and negro skeletons. *J. Bone Joint Surg.*, **42**A, 50–8.

Trowbridge, F. L., Graham, G. G., Wong, W. W., Mellits, E. D., Rabold, J. D., Lee, L. S., Cabrera, M. P. & Klein, P. D. (1982). Body water measurements in premature and older infants using $H_2^{18}O$ isotopic determinations. *Pediatr. Res.*, **18**, 524–7.

Tulloa, H. S. & King, J. W. (1972). Lesions of the pitching arm in adolescents. *J. Am. Med. Assoc.*, **220**, 264–71.

Turner, W. J. & Cohn, S. (1975). Total body potassium and 24-hour creatinine excretion in healthy males. *Clin. Pharm. Therap.*, **18**, 405–12.

Tzankoff, S. P. & Norris, A. H. (1977). Effect of muscle mass decrease on age-related BMR changes. *J. Appl. Physiol.*, **43**, 1001–6.

Tzankoff, S. P. & Norris, A. H. (1978). Longitudinal changes in basal metabolism in man. *J. Appl. Physiol.*, **45**, 536–9.

Uezu, N., Yamamoto, S., Rikimaru, T., Kishi, K. & Inoue, G. (1983). Contributions of individual body tissues to nitrogen excretion in adult rats fed protein-deficient diets. *J. Nutr.*, **113**, 105–114.

Ushakov, A. S., Smirnova, T. A., Pitts, G. C., Pace, N., Smith, A. H. & Rahlmann, D. F. (1980). Body composition of rats flown aboard Cosmos 1129. *Physiologist*, **23** (6), S41–4.

Usher, R., Shephard, M. & Lind, J. (1963). The blood volume of the newborn infant and placental transfusion. *Acta Paediatr. Scand.*, **52**, 497–512.

Vague, J. (1956). Degree of masculine differentiation of obesities: a factor determining predisposition to diabetes. Atherosclerosis, gout and uric calculus disease. *Am. J. Clin. Nutr.*, **4**, 20–34.

Vague, J., Vague, P. H., Boyer, J. & Cloix, C. (1971). Anthropometry of obesity, diabetes and beta cell functions. Proc. 7th Congress of Intern. Diab. Fed. *Excerpta Medica International Congress Series*, **231**, 517–25.

Vajda, A. S., Hebbelinck, M. & Ross, W. D. (1977). Dimensional relationships of height and weight in Belgium. Cross-sectional studies since 1840. In *Frontiers of Activity and Child Health*, ed. H. Lavallée & R. J. Shephard, p. 269. Quebec City: Editions du Pélican.

Van der Walt, T. S. P., Blaauw, J. H., Desipre, M., Daehne, H. O. & Van Rensburg, J. P. (1986). Proportionality of muscle volume calculations of male and female participants in different sports. In *Perspectives in Kinanthropometry*, ed. J. A. P. Day, pp. 229–38. Champaign, Ill.: Human Kinetics Publishers.

Van Erp-Baart, M-A., Fredrix, L. W. H. M., Binkhorst, R. M., Lavaleye, T. C. L., Vergouwen, P. C. J. & Saris, W. H. M. (1985). Energy intake and energy expenditure in top female gymnasts. In *Children and Exercise* XI, ed. R. A. Binkhorst, H. C. G. Kemper & W. H. M. Saris, pp. 218–23. Champaign, Ill.: Human Kinetics Publishers.

Van Es, A. J. H., Vogt, J. E., Niessen, C. H., Veth, J., Rodenburg, L., Teeuwse, V. & Dhuyvetter, J. (1984). Human energy metabolism below, near and above energy equilibrium. *Br. J. Nutr.*, **52**, 429–42.

Van Itallie & Yang, M. U. (1984). Cardiac dysfunction in obese dieters: a potentially lethal complication of rapid, massive weight loss. *Am. J. Clin. Nutr.*, **38**, 695–702.

Van Itallie, T. B., Segal, K. R., Yang, M-U. & Funk, R. C. (1985). Clinical assessments of body fat content in adults: potential role of electrical impedance methods. In *Body Composition Assessment in Youth and Adults. Report of the Sixth Ross Conference on Medical Research*, ed. A. F. Roche, pp. 5–8, Columbus, Ohio: Ross Laboratories.

Van Loan, M. & Mayclin, P. (1987). Bioelectrical impedance analysis: is it a reliable estimator of lean body mass and total body water. *Hum. Biol.*, **59**, 299–309.

Van Uytvanck, P. & Vrijens, J. (1966). Investigations about some body circumference measurements for the appreciation of physical fitness in adolescence. *J. Spts. Med. Phys. Fitness*, **6**, 176–82.

Van Wieringen, J. C. (1972). *Secular Changes of Growth*. Leiden: Institute of Preventive Medicine.

Vandenberg, S. G. (1962). How stable are heritability estimates? A comparison of heritability estimates from six anthropometric studies. *Am. J. Phys. Anthrop.*, **20**, 331–8.

Vandervoort, A., McComas, A. J., Toi, A. & Viviani, K. (1983). Measurement of triceps surae muscle cross-sectional area in young and very old adults using ultrasonic imaging. *Can. J. Appl. Spt. Sci.*, **8**, 211 (Abstract).

Vartsky, D., Ellis, K. J. & Cohn, S. H. (1979). In vivo quantification of body nitrogen by neutron capture prompt gamma analysis. *J. Nucl. Med.*, **20**, 1158–65.

Vaswani, A. N., Vartsky, D., Ellis, K. J., Yasumura, S. & Cohn, S. H. (1983). Effects of caloric restriction on body composition and total body nitrogen as measured by neutron activation. *Metabolism*, **32**, 185–8.

Veicsteinas, A. (1987). Changes of muscle and fat insulation during exercise in water. In *Adaptive Physiology to Stressful Environments*, ed. S. Samueloff & M. K. Yousef, pp. 51–8. Boca Raton: Chemical Rubber Company Press.

Venkatchalam, P. S., Shankar, K. & Galapan, G. (1960). Changes in body weight and body composition during pregnancy. *Ind. J. Med. Res.*, **48**, 511–17.

Vestergaard, P. & Leverett, R. (1968). Constancy of urinary creatinine output. *J. Lab. Clin. Med.*, **51**, 211–18.

Vickery, S. R., Cureton, K. J. & Collins, M. A. (1988). Prediction of body density from skinfolds in black and white young men. *Hum. Biol.*, **60**, 135–49.

Vico, L., Chappard, D., Bakulin, A. K., Nivokov, V. E. & Alexandre, C. (1987). Effects of 7-day space flight on weight-bearing and non weight-bearing bones in rats (Cosmos 1667). *Physiologist*, **30** (1), S45–6.

Vierordt, H. (1906). *Anatomische, Physiologische und Physikalische Daten und Tabellen*. Jena: G. Fischer.

Virtama, P. & Helelä, T. (1969). Radiographic measurements of cortical bone. Variations in a normal population between 1 and 90 years of age. *Acta Radiol.*, Suppl. 293.

Viteri, F. E. (1971). Considerations on the effect of nutrition on the composition and physical working capacity of young Guatemalan adults. In *Amino Acid Fortification of Protein Foods*, ed. N. Scrimshaw & A. M. Altshull, pp. 350–75. Cambridge, Mass.: M.I.T. Press.

Viteri, F. E. & Alvarado, J. (1970). The creatinine height index: its use in the estimation of the degree of protein depletion and repletion in protein calorie malnourished children. *Pediatrics*, **46**, 696–706.

Volz, P. A. & Ostrove, S. M. (1984). Evaluation of a portable ultrasonoscope in assessing the body composition of college-age women. *Med. Sci. Sports Exerc.*, **16**, 96–102.

Von Döbeln, W. (1956). Human standard and maximal metabolic rate in relation to fat-free body mass. *Acta Physiol. Scand.*, Suppl. 126, 1–79.

Von Döbeln, W. (1966). Kroppstorlek, energieomsattning och kondition. In *Handbok i Ergonomi*, ed. G. Luthman, U. Aberg & N. Lundgren. Stockholm: Almqvist & Wiksell.

Von Döbeln, W. & Eriksson, B. O. (1972). Physical training, maximal oxygen uptake and dimensions of the oxygen transporting and metabolizing organs in boys 11–13 years of age. *Acta Paediatr. Scand.*, 61, 653–60.

Von Hevesy, G. & Hofer, E. (1934). Die Verweilzeit des Wassers im menschlichen Korper, Untersicht mit Hilfe von 'schwerem' Wasser als Indikator. *Klin. Wschr.*, 13, 1524–6.

Von Kriegel, W. & Airsherl, W. (1964). Zur Wirkung von Aldosteron und Corticosteron auf den Elektrolyt – und Wassergehalt bindegewebiger Organe der Ratte. *Acta Endocrinol.*, 46, 47–64.

Wagner, W. W., Mays, C. W., Lloyd, R. D., Pendleton, R. C. & Zundel, W. S. (1966). Potassium studies in humans. In *Research in Radiobiology*, pp. 107–36. Report C00–119–235, University of Utah, cited by Lloyd & Mays (1987).

Wakat, D. K., Johnson, R. E., Krzywicki, H. J. & Gerber, L. I. (1971). Correlation between body volume and body mass in men. *Am. J. Clin. Nutr.*, 24, 1308–12.

Walberg, J. (1986). Weight control and the athlete. In *Sport, Health and Nutrition*, ed. F. I. Katch, pp. 11–20. Champaign, Ill.: Human Kinetics Publishers.

Walker, J., Roberts, S., Halmi, K. & Goldberg, S. (1979). Caloric requirements for weight gain in anorexia nervosa. *Am. J. Clin. Nutr.*, 32, 1396–400.

Waller, R. E. & Brooks, A. G. F. (1972). Heights and weights of men visiting a public health examination. *Br. J. Prev. Soc. Med.*, 26, 180–5.

Wang, J. & Pierson, R. N. (1976). Disparate hydration of adipose and lean tissue require a new model for body water distribution in man. *J. Nutr.*, 106, 1687–93.

Warner, J. G., Yeater, R., Sherwood, L. & Weber, K. (1986). A hydrostatic weighing method using total lung capacity and a small tank. *Br. J. Spts Med.*, 20, 17–21.

Wartenweiler, J., Hess, A. & Wuest, B. (1974). Anthropologic measurements and performance. In *Fitness, Health and Work Capacity: International Standards for Assessment*, ed. L. A. Larson, pp. 211–40. New York: MacMillan.

Wassner, S. J. & Li, J. B. (1982). N-methyl histidine release: contributions of rat skeletal muscle, GI tract and skin. *Am. J. Physiol.*, 243, E293–7.

Waterlow, J. C. (1986). Global nutritional status. *Bull. World Health Org.*, 64, 929–41.

Watson, J. D. & Dako, D. Y. (1977). Anthropometric studies on African athletes who participated in the 1st African University Games. *Br. J. Nutr.*, 38, 353–60.

Watson, R. C. (1973). Bone growth and physical activity. In *International Conference on Bone Mineral Measurements*, ed. R. B. Mazess, pp. 75–683. Washington, DC: DHEW Publications NIH.

Waxman, M. & Stunkard, A. J. (1980). Caloric intake and expenditure of obese boys. *J. Pediatr.*, **96**, 187–93.

Webster, J. D., Hesp, J. R. & Garrow, J. S. (1984). The composition of excess weight in obese women estimated by body density, total body water, and total body potassium. *Hum. Nutr. Clin. Nutr.*, **37C**, 117–31.

Weech, A. A. (1954). Signposts on the highway of growth. *Am. J. Dis. Childr.*, **88**, 452–7.

Weil, W. B. & Wallace, W. M. (1960). The effect of alterations in extracellular fluid on the composition of connective tissue. *Pediatrics*, **26**, 915–24.

Weil, W. B. & Wallace, W. M. (1963). The effect of variable food intakes on growth and body composition. *Ann. N.Y. Acad. Sci.*, **110**, 358–73.

Weiner, J. S. & Lourie, J. A. (1981). *Practical Human Biology*. New York: Academic Press.

Welch, B. E. & Crisp, C. E. (1958). Effect of the level of expiration on body density measurement. *J. Appl. Physiol.*, **12**, 210–13.

Weldon, V. V., Kowarski, A. & Migeon, C. J. (1967). Aldosterone secretion rates in normal subjects from infancy to adulthood. *Pediatrics*, **39**, 713–23.

Welham, W. C. & Behnke, A. R. (1942). The specific gravity of healthy men. Body weight + volume and other physical characteristics of exceptional athletes and Naval personnel. *J. Am. Med. Assoc.*, **118**, 498–501.

Weltman, A. & Katch, V. (1975). Preferential use of casing girths for estimating body volume and density. *J. Appl. Physiol.*, **38**, 560–3.

Weltman, A. & Katch, V. L. (1978). A non-population specific method for predicting total body volume and per cent fat. *Hum. Biol.*, **50**, 151–8.

Weltman, A. & Katch, V. (1981). Comparison of hydrostatic weighing at residual volume and total lung capacity. *Med. Sci. Sports Exerc.*, **13**, 210–13.

Weltman, A., Janney, C., Huber, R., Rians, C. B. & Katch, F. (1987*a*). Comparison of hydrostatic weighing at residual volume and total lung capacity in pre-pubertal males. *Hum. Biol.*, **59**, 51–7.

Weltman, A., Seip, R. L. & Tran, Z. V. (1987*b*). Practical assessment of body composition in adult obese males. *Hum. Biol.*, **59**, 523–35.

Weredin, E. J. & Kyle, L. H. (1960). Estimation of the constancy of the fat-free body. *J. Clin. Invest*, **39**, 626–9.

Werner, J. (1987). Thermoregulatory mechanisms: adaptations to thermal loads. In: *Adaptive Physiology to Stressful Environments*, ed. S. Samueloff & M. K. Yousef, pp. 18–25. Boca Raton: Chemical Rubber Company Press.

West, K. M., Bailey, C. P., Coniglione, T. C., Smith, W. O., Scroggins, L. R., Ellison, G., Maxwell, E. L., Sanders, M. L. & Whayne, T. F. (1974). Diabetes in nineteen Indian Oklahoma tribes. *Diabetes*, **23** (Suppl. 1), 385 (Abstract).

West, R. R. (1973). The estimation of total skeletal mass from bone densitometry measurements using 60 keV photons. *Br. J. Radiol.*, **46**, 599–603.

Whedon, G. D., Lutwak, L., Rambaut, P. C., Whittle, M. W., Reid, J., Smith, M. C., Leach, C. Stadler, C. R. & Sanford, D. D. (1976). Mineral and nitrogen balance study observations: the second manned Skylab mission. *Aviation Space Env. Med.*, **47**, 391–6.

Whitelaw, A. G. L. (1979). Subcutaneous fat measurement as an indication of nutrition of the fetus and newborn. In *Nutrition and Metabolism of the Fetus and Infant*, ed. H. K. A. Visser, pp. 131–43. The Hague: Martinus Nijhoff.

Whyte, R. K., Bayley, H. S. & Schwarz, N. P. (1985). The measurement of whole body water by $H_2^{18}O$ dilution in newborn pigs. *Am. J. Clin. Nutr.*, **41**, 801–9.

Widdowson, E. M. (1965). Chemical analysis of the body. In *Human Body Composition. Approaches and Applications*, ed. J. Brozek, pp. 31–47. Oxford: Pergamon Press.

Widdowson, E. M. & Dickerson, J. W. T. (1964). Chemical composition of the body. In *Mineral Metabolism*, Vol. 2, Part A, ed. C. L. Comar & F. Bronner, pp. 1–247. New York: Academic Press.

Widdowson, E. M., McCance, R. A. & Spray, C. M. (1951). The chemical composition of the human body. *Clin. Sci.*, **10**, 113–25.

Willi, J. & Grossmann, S. (1983). Epidemiology of anorexia nervosa in a defined region of Switzerland. *Am. J. Psychiatr.*, **140**, 564–7.

Williams, D., Anderson, T. & Currier, D. (1984). Underwater weighing using the Hubbard tank vs. the standard tank. *Phys. Therap.*, **64**, 658–64.

Williams, E. D., Boddy, K., Brown, J. J., Cumming, A. M. M., Davies, D. L., Harvey, I. R., Haywood, J. K., Lever, A. F. & Robertson, J. I. S. (1982). Whole body elemental composition in patients with essential hypertension. *Eur. J. Clin. Invest.*, **12**, 321–5.

Williams, P. T., Fortmann, S. P., Terry, R. B., Garay, S. C., Vranizan, K. M., Ellsworth, N. & Wood, P. D. (1987). Associations of dietary fat, regional adiposity, and blood pressure in men. *J. Am. Med. Assoc.*, **257**, 3251–6.

Wilmore, D. W., Moylan, J. A., Bristow, B. F., Mason, A. D. & Pruitt, B. A. (1974). Anabolic effect of human growth hormone and high caloric feedings following thermal injury. *Surg. Gyn. Obstetr.*, **138**, 875–84.

Wilmore, J. H. (1969). The use of actual, predicted and constant residual volumes in the assessment of body composition by underwater weighing. *Med. Sci. Sports*, **1**, 87–90.

Wilmore, J. (1974). Alterations in strength, body composition and anthropometric measurements consequent to a 10-week weight training programme. *Med. Sci. Sports Exerc.*, **6**, 133–8.

Wilmore, J. (1983). Body composition in sport and exercise. Directions for future research. *Med. Sci. Sports Exerc.*, **15**, 21–31.

Wilmore, J. (1986). Testing the elite masters athlete. In *Sports Medicine for the Mature Athlete*, ed. J. R. Sutton & R. M. Brock, pp. 91–107. Indianapolis: Benchmark Press.

Wilmore, J. H., Vodak, P. A., Parr, R. B., Girandola, R. N. & Billing, J. E. (1980). Further simplification for a method of determination of residual lung volume. *Med. Sci. Sports. Exerc.*, **12**, 216–18.

Wilson, P. D. & Franks, L. M. (1975). The effect of age on mitochondria ultrastructure and enzymes. *Adv. Exp. Med. Biol.*, **53**, 171–83.

Wilson, R. S. (1986). Twins: genetic influences on growth. In *Sport and Human Genetics*, ed. R. M. Malina & C. Bouchard, pp. 1–21. Champaign, Ill.: Human Kinetics Publishers.

Winiarski, A. M., Roy, R. R., Alford, E. K., Chiang, P. C. & Edgerton, V. R. (1987). Mechanical properties of rat skeletal muscle after hind limb suspension. *Exp. Neurol.*, **96**, 650–60.

Wing, R. R., Epstein, L. H., Ossip, D. J. & LaPorte, R. E. (1979). Reliability and validity of self-report and observer's estimates of relative weight. *Addict. Behav.*, **4**, 133–40.

Winick, M. & Noble, A. (1967). Cellular response with increased feeding in neonatal rats. *J. Nutr.*, **91**, 179–82.

Withers, R. F. J. (1964). Problems in the genetics of human obesity. *Eugen. Rev.*, **56**, 81.

Witt, R. M. & Mazess, R. B. (1978). Photon absorptiometry of soft-tissue and fluid content: the method and its precision and accuracy. *Phys. Med. Biol.*, **23**, 620–9.

Wolanski, N. (1969a). An approach to the inheritance of systolic and diastolic blood pressure. *Gen. Pol.*, **10**, 263–8.

Wolanski, N. (1969b). Arm circumference standards. Cited in Jelliffe, D. B. & Jelliffe, E. F. P. Arm circumference as a public health index of PCM in childhood. *J. Trop. Pediatr.*, **51**, 253–60.

Wolanski, N. (1970). Review article: genetic and ecological factors in human growth. *Hum. Biol.*, **42**, 349–68.

Wolanski, N. (1977). Genetic and ecological control of human growth. In *Growth and Development Physique*, ed. O. G. Eiben, pp. 19–33. Budapest: Akademia Kiado.

Wolff, J. (1892). *Das Gesetz der Transformation der Knochen*. Berlin: Hirschwald.

Womersley, J. & Durnin, J. V. G. A. (1973). An experimental study of variability of measurements of skinfold thicknesses in young adults. *Hum. Biol.*, **45**, 281–92.

Womersley, J. & Durnin, J. V. G. A. (1977). A comparison of the skinfold method with the extent of 'overnight' and various weight–height relationships in the assessment of obesity. *Br. J. Nutr.*, **38**, 271–84.

Womersley, J., Durnin, J. V. G. A., Armstrong, W. L. & Friskey, M. (1973). An experimental study on variability of measurements of skinfold thickness on young adults. *Hum. Biol.*, **45**, 281–92.

Wood, P. D., Haskell, W. L., Blair, S. N., Williams, P. T., Krauss, R. M., Lindgren, F. T., Albers, J. J., Ho., P. H. & Farquhar, J. W. (1983). Increased exercise level and plasma lipoprotein concentrations: a one year, randomized controlled study in sedentary middle-aged men. *Metabolism*, **32**, 31–9.

World Health Organization (1983). *Measuring Change in Nutritional Status. Guidelines for Assessing the Nutritional Impact of Supplementary Feeding Programmes for Vulnerable Groups*. Geneva: World Health Organization.

World Health Organization (1985). *Energy and Protein Requirements*. Tech. Rept. 724. Geneva: World Health Organization.

Wright, H. F., Dotson, C. O. & Davis, P. O. (1981). A simple technique for measurement of per cent body fat in man. *US Navy Med. Newsl.*, **72**, 23–7.

Wronski, T. J., Morey-Holton, E. & Jee, W. S. S. (1980). Bone resorption and calcium absorption in rats. *Physiologist*, **23** (6), S83–6.

Wronski, T. J., Morey-Holton, E. R., Doty, S. B., Maese, A. C. & Walsh, C. C. (1987). Histomorphometric analysis of rat skeleton following spaceflight. *Am. J. Physiol.*, **252**, R252–5.

Wurst, F. (1964). *Umwelteinflusse auf Wachstum und Entwicklung*. Munich: J. A. Barth.

Wyndham, C. H. (1966). Southern African ethnic adaptation to temperature and exercise. In *The Biology of Human Adaptability,* ed. P. T. Baker & J. S. Weiner, pp. 201–44. Oxford: Clarendon Press.

Wyndham, C. H. & Morrison, J. F. (1958). Adjustment to cold of bushmen in the Kalahari desert. *J. Appl. Physiol.,* **13,** 219–25.

Wyndham, C. H., Strydom, N. B., Morrison, J. F., Williams, C. G., Bredell, G. A. G. & Heyns, H. (1966). The capacity for endurance effort of Bantu males from different tribes. *S. Afr. J. Sci.,* **62,** 259–63.

Wyndham, C. H., Williams, C. G. & Loots, H. (1968). Reactions to cold. *J. Appl. Physiol.,* **24,** 282–7.

Wynn, V., Abraham, R. R. & Densem, J. W. (1985). Method for estimating rate of fat loss during treatment of obesity by calorie restriction. *Lancet,* (i), 482–6.

Wyshak, G. & Frisch, R. E. (1982). Evidence for a secular trend in age of menarche. *New Engl. J. Med.,* **306,** 1033–5.

Yamaoka, S. (1975). Physical constitution and fitness of obese children. In *Human Adaptability* Vol. 3. *Physiological Adaptability and Nutritional Status of the Japanese. B. Growth, Work Capacity and Nutrition of Japanese,* ed. K. Asahina & R. Shigiya, pp. 31–40. Tokyo: Japanese Committee for the International Biological Programme, Science Council of Japan.

Yang, M. U., Wang, J., Pierson, R. M. & Van Itallie, T. B. (1977). Estimation of composition of weight loss in man: a comparison of methods. *J. Appl. Physiol. Resp. Environ. Ex. Physiol.,* **43,** 331–8.

Yasumura, S., Cohn, S. H. & Ellis, K. J. (1983). Measurement of extracellular space by total body neutron activation. *Am. J. Physiol.,* **244,** R36–40.

Yokoyama, T. & Iwasaki, S. (1975). Ecology of the Japanese Ama. In *Physiological Adaptability and Nutritional Status of the Japanese. A. Thermal Adaptability of the Japanese and Physiology of the Ama,* ed. H. Yoshimura & S. Kobayashi, pp. 199–210. Tokyo: Japanese Committee for the International Biological Program, Science Council of Japan.

York, D. A., Bray, G. A. & Yukimura, Y. (1978). An enzymatic defect in the obese (ob/ob) mouse. Loss of thyroid-induced sodium and potassium dependent adenosine triphosphate. *Proc. Nat. Acad. Sci.,* USA, **75,** 477–81

Yoshikawa, M., Okano, K., Nakai, R., Tomori, T. & Takenaka, M. (1978). Aging and nutrition. *Asian Med. J.,* **21,** 359–78.

Youmans, J. B. (1936). Nutritional oedema. *Internat. Clin.,* **4,** 120–45.

Young, A. & Stokes, M. (1986). Non-invasive measurement of muscle in the rehabilitation of masters athletes. In *Sports Medicine for the Mature Athlete,* ed. J. R. Sutton & R. M. Brock, pp. 45–55. Indianapolis: Benchmark Press.

Young, A., Hughes, I., Russell, P., Parker, M. J. & Nichols, P. J. R. (1980). Measurements of quadriceps muscle wasting by ultrasonography. *Rheumatol. Rehabil.,* **19,** 141–8.

Young, C. M., Blondin, J., Tensuan, R. & Fryer, J. H. (1963). Body composition studies of 'older' women, thirty to seventy years of age. *Ann. N.Y. Acad. Sci.,* **110,** 589–607.

Young, C. M., Bogan, A. D., Roe, D. A. & Lutwak, L. (1968). Body composition of preadolescent and adolescent girls. IV. Body water and creatinine. *J. Am. Diet. Assoc.*, **53**, 579–87.

Young, H. B. (1965). Body composition, culture and sex: two components. In *Human Body Composition. Approaches and Applications*, ed. J. Brozek, pp. 139–60. Oxford: Pergamon Press.

Young, K., McDonagh, M. J. & Davies, C. T. M. (1985). The effect of two forms of isometric training on the mechanical properties of the triceps surae in man. *Pflüg. Archiv.*, **405**, 384–8.

Young, V. R. & Munro, H. N. (1978). N-methylhistidine (3-methyl histidine) and muscle protein turnover. An overview. *Fed. Proc.*, **37**, 2291–300.

Yousef, M. K. (1985). *Stress Physiology in Livestock.* Boca Raton: Chemical Rubber Company Press.

Yousef, M. K. & Samueloff, S. (1987). Adaptive physiology: introductory remarks and glossary. In *Adaptive Physiology to Stressful Environments*, ed. S. Samueloff & M. K. Yousef, pp. 3–9. Boca Raton: Chemical Rubber Company Press.

Zak, R. (1974). Development and proliferative capacity of cardiac muscle cells. *Circ. Res.*, **34–35** (Suppl.), 17–26.

Zambraski, E. J., Tipton, C. M., Jordan, H. R., Palmer, W. K. & Tcheng, T. K. (1974). Iowa wrestling study: Urinary profiles of state finalists prior to competition. *Med. Sci. Spts.*, **6**, 129–32.

Ziegler, E. E., O'Donnell, A. M., Nelson, S. E. & Fomon, S. J. (1976). Body composition of the reference fetus. *Growth*, **40**, 329–41.

Zuti, W. B. & Golding, L. A. (1973). Equations for estimating per cent fat and body density of active adult males. *Med. Sci. Sports*, **5**, 262–6.

Index

Page numbers in italic type refer to figures, those in bold to tables.